Advances in
COMPUTERS
VOLUME 69

Advances in COMPUTERS

Architectural Issues

EDITED BY

MARVIN V. ZELKOWITZ

Department of Computer Science
and Institute for Advanced Computer Studies
University of Maryland
College Park, Maryland

VOLUME 69

ELSEVIER

AMSTERDAM • BOSTON • HEIDELBERG • LONDON • NEW YORK • OXFORD
PARIS • SAN DIEGO • SAN FRANCISCO • SINGAPORE • SYDNEY • TOKYO
Academic Press is an imprint of Elsevier

ACADEMIC PRESS

Academic Press is an imprint of Elsevier
84 Theobald's Road, London WC1X 8RR, UK
Radarweg 29, PO Box 211, 1000 AE Amsterdam, The Netherlands
The Boulevard, Langford Lane, Kidlington, Oxford OX5 1GB, UK
30 Corporate Drive, Suite 400, Burlington, MA 01803, USA
525 B Street, Suite 1900, San Diego, CA 92101-4495, USA

First edition 2007

Copyright © 2007 Elsevier Inc. All rights reserved

No part of this publication may be reproduced, stored in a retrieval system or transmitted in any form or by any means electronic, mechanical, photocopying, recording or otherwise without the prior written permission of the publisher

Permissions may be sought directly from Elsevier's Science & Technology Rights Department in Oxford, UK: phone (+44) (0) 1865 843830; fax (+44) (0) 1865 853333; email: permissions@elsevier.com. Alternatively you can submit your request online by visiting the Elsevier web site at http://elsevier.com/locate/permissions, and selecting *Obtaining permission to use Elsevier material*

Notice
No responsibility is assumed by the publisher for any injury and/or damage to persons or property as a matter of products liability, negligence or otherwise, or from any use or operation of any methods, products, instructions or ideas contained in the material herein. Because of rapid advances in the medical sciences, in particular, independent verification of diagnoses and drug dosages should be made

ISBN-13: 978-0-12-373745-8

ISSN: 0065-2458

For information on all Academic Press publications
visit our website at books.elsevier.com

Printed and bound in USA

07 08 09 10 11   10 9 8 7 6 5 4 3 2 1

Contents

CONTRIBUTORS ix
PREFACE xiii

The Architecture of Efficient Multi-Core Processors: A Holistic Approach

Rakesh Kumar and Dean M. Tullsen

1. Introduction 3
2. The Move to Multi-Core Processors 5
3. Holistic Design for Adaptability: Heterogeneous Architectures 9
4. Amortizing Overprovisioning through Conjoined Core Architectures . . . 45
5. Holistic Design of the Multi-Core Interconnect 62
6. Summary and Conclusions 81
   Acknowledgements 82
   References 82

Designing Computational Clusters for Performance and Power

Kirk W. Cameron, Rong Ge, and Xizhou Feng

1. Introduction 90
2. Background 91
3. Single Processor System Profiling 96
4. Computational Cluster Power Profiling 98
5. Low Power Computational Clusters 112
6. Power-Aware Computational Clusters 125
7. Conclusions 149
   References 149

Compiler-Assisted Leakage Energy Reduction for Cache Memories

Wei Zhang

1. Introduction 156
2. Related Work 159
3. Static Next Sub-Bank Prediction for Drowsy Instruction Caches 161
4. Compiler-Assisted Loop-Based Data Cache Leakage Reduction 170
5. Evaluation Methodology 172
6. Conclusion 186
   References 187

Mobile Games: Challenges and Opportunities

Paul Coulton, Will Bamford, Fadi Chehimi, Reuben Edwards, Paul Gilbertson, and Omer Rashid

1. Introduction 192
2. Challenges 193
3. Opportunities 217
4. Conclusions 239
   Acknowledgements 239
   References 240

Free/Open Source Software Development: Recent Research Results and Methods

Walt Scacchi

1. Introduction 244
2. Individual Participation in FOSSD Projects 248
3. Resources and Capabilities Supporting FOSSD 253
4. Cooperation, Coordination, and Control in FOSS Projects 260
5. Alliance Formation, Inter-project Social Networking and Community Development 265
6. FOSS as a Multi-project Software Ecosystem 270
7. FOSS as a Social Movement 274
8. Research Methods for Studying FOSS 277
9. Discussion 284

10. Conclusions	286
Acknowledgements	287
References	287
AUTHOR INDEX	297
SUBJECT INDEX	307
CONTENTS OF VOLUMES IN THIS SERIES	319

Contributors

Will Bamford graduated from Lancaster University in 2005 with a BSc in IT and Media Communications. Currently in his second year of a PhD at Lancaster University performing research into novel mobile applications, with a focus on harnessing new mobile technologies such as GPS, RFID, Bluetooth and other mobile sensors such as 3D accelerometers and combining these with emerging web technologies and services to create new social/gaming experiences for mobile devices.

Kirk W. Cameron is an associate professor in the Department of Computer Science and director of the Scalable Performance (SCAPE) Laboratory at Virginia Polytechnic Institute and State University. His research interests include high-performance and grid computing, parallel and distributed systems, computer architecture, power-aware systems, and performance evaluation and prediction. Cameron received a PhD in computer science from Louisiana State University. He is a member of the IEEE and the IEEE Computer Society. Contact him at cameron@vt.edu.

Fadi Chehimi is an experienced Symbian programmer working for Mobica Ltd and studying part-time for a PhD at Lancaster University. His research interests are based around novel tools and methodologies for 3-D graphics, imaging and interactive mobile advertising on mobile phones.

Paul Coulton is a Senior Lecturer in Mobile Applications and was one of fifty mobile developers worldwide, selected from a community of two million, to be a Forum Nokia Champion. He has been a pioneer of mobile games particularly those incorporating novel interfaces and location based information and continues to drive innovation in the sector. Paul is a regular speaker at international conferences, including the Mobile GDC, and is highly respected both in the industrial and academic communities.

Reuben Edwards is Lecturer in games and m-commerce and has been at the forefront of innovative multimedia application development for over ten years. He is renowned for his extensive knowledge of development environments across all platforms and has also pioneered the use of many platforms in technology education and produced many cutting edge courses.

Xizhou Feng is a Research Associate in the Department of Computer Science at Virginia Polytechnic Institute and State University. His research interests include bioinformatics, computational biology, distributed system, high performance computing, and parallel algorithms. He received a PhD in Computer Science and Engineering from the University of South Carolina, and MS degree in Engineering Thermophysics from Tsinghua University, China. He is a member of the IEEE and the IEEE Computer Society. Contact him at fengx@cs.vt.edu.

Rong Ge is a PhD candidate in the Department of Computer Science and a researcher at the SCAPE Laboratory at Virginia Tech. Her research interests include performance modeling and analysis, parallel and distributed systems, power-aware systems, high-performance computing, and computational science. Ge received the BS degree and MS degree in Fluid Mechanics from Tsinghua University, China, and the MS degree in computer science from the University of South Carolina. She is a member of the IEEE, ACM and Upsilon Pi Epsilon. Contact her at ge@cs.vt.edu.

Paul Gilbertson is currently studying for his PhD at the Department of Communication Systems at Lancaster University in the UK. He has worked with Windows systems of all sizes, from Active Directory networks, to Windows Mobile phones and PDAs both as a programmer and whilst at APT in London. Paul is currently developing location aware software using .NET, Java, and Symbian OS technologies as part of his PhD.

Rakesh Kumar is an Assistant Professor in the Department of Electrical and Computer Engineering at the University of Illinois, Urbana-Champaign. His research interests include multicore and multithreaded architectures, low-power architectures, and on-chip interconnects. Kumar received a PhD in computer engineering from the University of California at San Diego, and a BS in computer science and engineering from the Indian Institute of Technology, Kharagpur. He is a member of the ACM. Contact him at rakeshk@uiuc.edu.

Omer Rashid is a PhD student at Lancaster University with over 3 years experience in research and mobile application development with Java. His research mainly focuses on novel networked mobile entertainment applications and his work has gained accreditation from Nokia and recently ACM. He is also a member of Mobile Radicals research initiative and the IEEE.

Walt Scacchi received a PhD in Information and Computer Science at the University of California, Irvine in 1981. From 1981 until 1998, he was on the faculty at the University of Southern California. In 1999, he joined the Institute for Software Research at UC Irvine, and in 2002 became associate director for research at the Computer Game Culture and Technology Laboratory (http://www.ucgamelab.net). His research interests include open source software development, networked and grid-based computer games, knowledge-based systems for modeling and simulating complex engineering and business processes, developing decentralized heterogeneous information systems, and electronic commerce/business. Dr. Scacchi is a member of ACM, IEEE, AAAI, and the Software Process Association (SPA). He is an active researcher with more than 150 research publications and has directed 45 externally funded research projects. He also has had numerous consulting and visiting scientist positions with a variety of firms and research laboratories. In 2007, he serves as General Chair of the 3rd IFIP International Conference on Open Source Systems, Limerick, IE.

Dean Tullsen is a Professor in the Computer Science and Engineering Department at the University of California, San Diego. He received his PhD from the University of Washington in 1996, where his dissertation was *Simultaneous Multithreading*. He received his BS and MS degrees in Computer Engineering from UCLA. He was a computer architect for AT&T Bell Labs, and taught Computer Science for a year at a University in China prior to returning for his PhD. His research interests include the architecture of multithreading processors of all types (including simultaneous multi-threading, multi-core, or the gray area in between), compiling for such architectures, and high performance architectures in general.

Wei Zhang received the BS degree in computer science from the Peking University in China in 1997, the MS from the Institute of Software, Chinese Academy of Sciences in 2000, and the PhD degree in computer science and engineering from the Pennsylvania State University in 2003. He is an assistant professor in the Electrical and Computer Engineering Department at Southern Illinois University Carbondale. His current research interests are in embedded systems, low-power computing, computer architecture and compiler. His research has been supported by NSF, Altera and SIUC. He is a member of the IEEE and ACM. He has served as a member of the technical program committees for several IEEE/ACM conferences and workshops.

Preface

In volume 69 of the **Advances in Computers** we present five chapters that discuss significant changes to both the hardware and software of present day computers. These chapters in the ever-changing landscape of information technology address how computers are evolving to address our differing needs for information technology as the computer becomes more ubiquitous in our everyday life. This series began back in 1960 and annually we present three volumes that offer approximately 18 chapters that describe the latest technology in the use of computers today.

The first chapter, "The Architecture of Efficient Multi-Core Processors: A Holistic Approach" by Rakesh Kumar and Dean M. Tullsen discusses the recent development of multicore processors. Moore's Law, essentially the doubling of computer power every 18 months, while not a natural law of the universe, has been amazingly true for over 30 years. However, as processor speeds have been increasing to clock rates of over 3 GHz (10^9 cycles per second), it is not clear how much faster current technology can push processors. One way to increase this power is to put multiple processors on the same chip, thus using each clock cycle to run an instruction in each separate "core," and hence double processing power. Kumar and Tullsen discuss how efficient multicore processors can be developed.

Chapter 2, "Designing Computational Clusters for Performance and Power" by Kirk W. Cameron, Rong Ge and Xizhou Feng, continues with the development of high-performance processors discussed in Chapter 1. An important attribute, not often cited when discussing newer high-speed computers, is the generation of heat. As machines get faster and their clock rates increase, they generate more heat, thus requiring more powerful cooling devices like fans to get rid of this heat. This also means it takes more energy (and hence cost) to run these computers. In this chapter the authors discuss an architectural process that considers both maximal performance as well as minimizing power requirements in order to improve on the development of high speed processors.

In Chapter 3,"Compiler-Assisted Leakage Energy Reduction for Cache Memories," Wei Zhang considers the same heat problem discussed in Chapter 2, but addresses the issue of cache memories. Since processor speeds (on the order of 10^9 cycles per second) is much faster than the speeds of accessing memory (on the order of 10^8 cycles per second), one way to increase processor performance is to keep a copy of part of main memory in the processor as a high-speed *cache* memory. This cache memory has the same heat generation problem discussed in the previous chapter. Dr. Wei discusses mechanisms to minimize energy (i.e., heat) losses in these memories by using compilation techniques to control cache accesses.

In "Mobile Games: Challenges and Opportunities" (Chapter 4) by Paul Coulton, Will Bamford, Fadi Chehimi, Reuben Edwards, Paul Gilbertson, and Omer Rashid, the authors discuss a new computer technology that is probably the fastest growing market—that of mobile phones. Phones have grown far beyond the simple concept of making a telephone call when not tethered by a wire to a land-based telephone network. Phones are powerful miniature computers with tiny console screens and only a few keys (the digits plus a few others) rather than a full desktop computer keyboard. Users often have the desire to play a game while waiting for other events to happen. So the design problem for mobile games is quite different from the established video game industry. You need a game that works on a small screen, only has a few keys for input, and can be played in short bursts of time. This chapter discusses these design constraints and future directions for this growing industry.

In the final chapter of this volume, "Free/Open Source Software Development: Recent Research Results and Methods" (Chapter 5), Walt Scacchi discusses open source development. Rather than have software developed and sold by a company, a group of individuals organize to develop and give away the software for free. The basic question is "Why?". In this chapter, the author discusses the sociology of how such groups form, why they do it, and how companies still make a profit on this form of development. *Open Source* is probably the most significant change in building some complex systems over the past 10 years, and it promises to have a profound effect on new software development in the future.

I hope that you find these chapters to be of value to you. I am always looking for new topics to explore. If we have not covered a relevant topic for several years, or if you wish to contribute a topic you believe you are an expert on, please let me know. I can be reached at mvz@cs.umd.edu.

Marvin Zelkowitz
University of Maryland and Fraunhofer Center, Maryland
College Park, Maryland

The Architecture of Efficient Multi-Core Processors: A Holistic Approach

RAKESH KUMAR

Coordinated Science Laboratory
University of Illinois at Urbana-Champaign
Urbana, IL 61801
USA
rakeshk@uiuc.edu

DEAN M. TULLSEN

Department of Computer Science and Engineering
University of California, San Diego
La Jolla, CA 92093-0404
USA
tullsen@cs.ucsd.edu

Abstract

The most straightforward methodology for designing a multi-core architecture is to replicate an off-the-shelf core design multiple times, and then connect the cores together using an interconnect mechanism. However, this methodology is "multi-core oblivious" as subsystems are designed/optimized unaware of the overall chip-multiprocessing system they would become parts of. The chapter demonstrates that this methodology is inefficient in terms of area/power. It recommends a holistic approach where the subsystems are designed from the ground up to be effective components of a complete system.

The inefficiency in "multi-core oblivious" designs comes from many sources. Having multiple replicated cores results in an inability to adapt to the demands of execution workloads, and results in either underutilization or overutilization of processor resources. *Single-ISA (instruction-set architecture) heterogeneous multi-core architectures* host cores of varying power/performance characteristics on the die, but all cores are capable of running the same ISA. Such a processor can result in significant power savings and performance improvements if the applications are mapped to cores judiciously. The paper also presents holistic design methodologies for such architectures. Another source of inefficiency is blind replication of over-provisioned hardware structures. *Conjoined-core chip multiprocessing* allows adjacent cores of a multi-core architecture to share some resources. This can result in significant area savings with little performance degradation. Yet another source of inefficiency is the interconnect. The interconnection overheads can be very significant for a "multi-core oblivious" multi-core design—especially as the number of cores increases and the pipelines get deeper. The paper demonstrates the need to co-design the cores, the memory and the interconnection to address the inefficiency problem, and also makes several suggestions regarding co-design.

1. Introduction 3
   1.1. A Naive Design Methodology for Multi-Core Processors 3
   1.2. A Holistic Approach to Multi-Core Design 4
2. The Move to Multi-Core Processors 5
   2.1. Early Multi-Core Efforts 7
3. Holistic Design for Adaptability: Heterogeneous Architectures 9
   3.1. Workload Diversity 10
   3.2. Single-ISA Heterogeneous Multi-Core Architectures 10
   3.3. Power Advantages of Architectural Heterogeneity 19
   3.4. Overview of Other Related Proposals 31
   3.5. Designing Multi-Cores from the Ground Up 32
4. Amortizing Overprovisioning through Conjoined Core Architectures 45
   4.1. Baseline Architecture 46
   4.2. Conjoined-Core Architectures 47
   4.3. Simple Sharing 53
   4.4. Intelligent Sharing of Resources 58
   4.5. A Unified Conjoined-Core Architecture 61
5. Holistic Design of the Multi-Core Interconnect 62
   5.1. Interconnection Mechanisms 62
   5.2. Shared Bus Fabric 62
   5.3. P2P Links 65
   5.4. Crossbar Interconnection System 66
   5.5. Modeling Interconnect Area, Power, and Latency 67
   5.6. Modeling the Cores 69
   5.7. Shared Bus Fabric: Overheads and Design Issues 71
   5.8. Shared Caches and the Crossbar 76
   5.9. An Example Holistic Approach to Interconnection 79
6. Summary and Conclusions 81
   Acknowledgements 82
   References 82

1. Introduction

The microprocessor industry has seen a tremendous performance growth since its inception. The performance of processors has increased by over 5000 times since the time Intel introduced the first general-purpose microprocessor [8]. This increase in processor performance has been fueled by several technology shifts at various levels of the processor design flow—architecture, tools and techniques, circuits, processes, and materials. Each of these technology shifts not only provide a jump in performance, but also require us to rethink the basic assumptions and processes that govern how we architect systems.

Specifically, at the architectural level, we have moved from scalar processing, where a processor executes one instruction every clock cycle, to superscalar processing, to out-of-order instruction execution, to on-chip hardware multithreading (e.g., simultaneous multithreading). We are now at the cusp of another major technology shift at the architectural level. This technology shift is the introduction of multi-core architectures—i.e., architectures with multiple processing nodes on the same die. Such processors, also called chip multiprocessors (CMPs), not only support multiple streams of program execution at the same time, but provide productivity advantages over monolithic processors due to the relative simplicity of the cores and the shorter design cycles for the processor core. Such processors can also make better use of the hardware resources, as the marginal utility of transistors is higher for a smaller processing node with a smaller number of transistors.

But like many other technological shifts, we must also change the way we design these architectures to take full advantage of the new technology. How, then, do we approach the design of such an architecture? What traditional architectural assumptions no longer apply? How do we design the components of such a system so that the whole processor meets our design goals?

This chapter seeks to address these questions.

1.1 A Naive Design Methodology for Multi-Core Processors

One suggested (as well as practiced) methodology for multi-core design is to take an off-the-shelf core design, perhaps optimize it for power and/or performance, replicate it multiple times, and then connect the cores together using a good interconnection mechanism in a way that maximizes performance for a given area and/or power budget. This is a clean, relatively easy way to design a multi-core processor because one design team can work on the core, another can work on the caches, the third can work on the interconnect, and then there can be a team of chip integrators who will put them all together to create a multi-core processor. Such a methodology encourages modularity as well as reuse, and serves to keep the design costs manageable.

However, this methodology is "multi-core oblivious." This is because each subsystem that constitutes the final chip multiprocessing system is designed and optimized without any cognizance of the overall system it would become a part of. For example, a methodology like the above forces each subsystem to target the entire universe of applications (i.e., a set of all possible applications that a processor is expected to run).

This chapter shows that "multi-core oblivious" designs result in inefficient processors in terms of area and power. This is because the above constraint results in either overutilization or underutilization of processor resources. For example, Pentium Extreme is a dual-core Intel processor that is constructed by replicating two identical off-the-shelf cores. While the area and power cost of duplicating cores is 2X (in fact, even more considering the glue logic required), the performance benefits are significantly lower [66]. That paper shows that while the costs are superlinear with the number of cores for all "multi-core oblivious" designs, the benefits tend to be highly sublinear.

1.2 A Holistic Approach to Multi-Core Design

In the era of uniprocessor microprocessors, each CPU core was expected to provide high general-purpose performance (that is, provide high performance for a wide range of application characteristics), high energy efficiency, reliability, and high availability. In a multi-core processor, the user will have the same expectations, but they will be applied to the entire processor. In this architecture, then, no single component (e.g., processor cores, caches, interconnect components) need meet any of those constraints, as long as the whole system (the processor) meets those constraints. This presents a degree of freedom not previously available to architects of mainstream high-performance processors.

The multi-core oblivious approach, for example, would build a system out of reliable processor cores. The holistic approach designs each core to be part of a reliable system. The difference between these two approaches is significant. The implications are actually relatively well understood in the realm of reliability. We can build a reliable and highly available multi-core system, even if the individual cores are neither. However, the implications of a holistic approach to delivering high general-purpose performance and energy efficiency have not been explored, and that is the focus of this chapter. As an example of our holistic approach to processor design, we show that a processor composed of general-purpose cores does not deliver the same average general-purpose performance as a processor composed of specialized cores, none optimized to run every program well.

Section 3 discusses how "multi-core oblivious" designs fail to adapt to workload diversity. It presents *single-ISA heterogeneous multi-core architectures* as a holistic solution to adapting to diversity. These architectures can provide significantly higher throughput for a given area or power budget. They can also be used to reduce processor power dissipation. That section also discusses methodologies for holistic, ground-up design of multi-core architecture and demonstrates their benefits over processors designed using off-the-shelf components.

Section 4 introduces overprovisioning as another source of inefficiency in multi-core architectures that are designed by blindly replicating cores. Such architectures unnecessarily multiply the cost of overprovisioning by the number of compute nodes. *Conjoined-core chip-multiprocessing* is a holistic approach to addressing overprovisioning. Conjoined-core multi-core processors have adjacent cores sharing large, overprovisioned structures. Intelligently scheduling accesses to the shared resources enables conjoined-core architectures to achieve significantly higher efficiency (*throughput/area*) than their "multi-core oblivious" counterparts.

Section 5 details the overheads that conventional interconnection mechanisms entail, especially as the number of cores increases and as transistors get faster. It shows that overheads become unmanageable very soon and require a holistic approach to designing multi-cores where the interconnect is co-designed with the cores and the caches. A high-bandwidth interconnect, for example, can actually decrease performance if it takes resources needed for cores and caches. Several examples are presented for the need to co-design.

2. The Move to Multi-Core Processors

This section provides background information on multi-core architectures and explains why we are seeing the entire industry now move in this architectural direction. We also provide an overview of some groundbreaking multi-core efforts.

The processor industry has made giant strides in terms of speed and performance. The first microprocessor, the Intel 4004 [8], ran at 784 KHz while the microprocessors of today run easily in the GHz range due to significantly smaller and faster transistors. The increase in performance has been historically consistent with Moore's law [70] that states that the number of transistors on the processor die doubles every eighteen months due to shrinking of the transistors every successive process technology.

However, the price that one pays for an increment in performance has been going up rapidly as well. For example, as Horowitz et al. [47] show, the power cost for squeezing a given amount of performance has been going up linearly with the performance of the processor. This represents, then, a super-exponential increase in power over time (performance has increased exponentially, power/performance increases linearly, and total power is the product of those terms). Similarly, the area cost for squeezing a given amount of performance has been going up as well.

An alternative way to describe the same phenomenon is that the marginal utility of each transistor we add to a processor core is decreasing. While area and power are roughly linear with the number of transistors, performance is highly sublinear with the number of transistors. Empirically, it has been close to the square root of the number of transistors [46,45]. The main reason why we are on the wrong side of the square law is that we have already extracted the easy ILP (instruction-level parallelism) through techniques like pipelining, superscalar processing, out-of-order processing, etc. The ILP that is left is difficult to extract. However, technology keeps making transistors available to us at the rate predicted by Moore's Law (though it has slowed down, of late). We have reached a point where we have more transistors available than we know how to make effective use of in a conventional monolithic processor environment.

Multi-core computing, however, allows us to cheat the square law. Instead of using all the transistors to construct a monolithic processor targeting high single-thread performance, we can use the transistors to construct multiple simpler cores where each core can execute a program (or a thread of execution). Such cores can collectively provide higher many-thread performance (or throughput) than the baseline monolithic processor at the expense of single-thread performance.

Consider, for example, the Alpha 21164 and Alpha 21264 cores. Alpha 21164 (also called, and henceforth referred to as, EV5) is an in-order processor that was originally implemented in 0.5 micron technology [17]. Alpha 21264 (also called, and henceforth referred to as, EV6) is an out-of-order processor that was originally implemented in 0.35 micron technology [18]. If we assume both the processors to be mapped to the same technology, an EV6 core would be roughly five times bigger than an EV5 core. If one were to take a monolithic processor like EV6, and replace it with EV5 cores, one could construct a multi-core that can support five streams of execution for the same area budget (ignoring the cost of interconnection and glue logic). However, for the same technology, the single-thread performance of an EV6 core is only roughly 2.0–2.2 times that of an EV5 core (assuming performance is proportional to the square root of the number of transistors). Thus, if we replaced an EV6 monolithic processor by a processor with five EV5 cores, the aggregate throughput would be more than a factor of two higher than the monolithic design for the same area budget. Similar throughput gains can be shown even for a fixed power budget. This potential to get significantly higher aggregate performance for the same budget is the main motivation for multi-core architectures.

Another advantage of multi-core architectures over monolithic designs is improved design productivity. The more complex the core, the higher the design and verification costs in terms of time, opportunity, and money. Several recent monolithic designs have taken several thousand man years worth of work. A multi-core approach enables deployment of pre-existing cores, thereby bringing down the design and verification costs. Even when the cores are designed from scratch, the simplicity of the cores can keep the costs low. With increasing market competition and declining hardware profit margins, the time-to-market of processors is more important than before, and multi-cores should improve that metric.

2.1 Early Multi-Core Efforts

This section provides an overview of some of the visible general-purpose multi-core projects that have been undertaken in academia and industry. It does not describe all multi-core designs, but is biased towards the first few general-purpose multi-core processors that broke ground for mainstream multi-core computing.

The first multi-core description was the Hydra [43,44], first described in 1994. Hydra was a 4-way chip multiprocessor that integrated four 250 MHz MIPS cores on the same die. The cores each had 8 KB private instruction and data caches and shared a 128 KB level-2 cache. Hydra was focused not only on providing hardware parallelism for throughput-oriented applications, but also on providing high single-thread performance for applications that can be parallelized into threads by a compiler. A significant amount of support was also provided in the hardware to enable thread-level speculation efforts. Hydra was never implemented, but an implementation of Hydra (0.25 micron technology) was estimated to take up 88 mm^2 of area.

One of the earliest commercial multi-core proposals, Piranha [23] (description published in 2000) was an 8-way chip multiprocessor designed at DEC/Compaq WRL. It was targeted at commercial, throughput-oriented workloads whose performance is not limited by instruction-level parallelism. It integrated eight simple, in-order processor cores on the die. Each core contained private 64 KB instruction and data caches and shared a 1 MB L2 cache. The processor also integrated on the chip functionality required to support scalability of the processor to large multiprocessing systems. Piranha also was never implemented.

Around the same time as Piranha, Sun started the design of the MAJC 5200 [86]. It was a two-way chip multiprocessor where each core was a four-issue VLIW (very large instruction word) processor. The processor was targeted at multimedia and Java applications. Each core had a private 16 KB L1 instruction cache. The cores shared a 16 KB dual-ported data cache. Small L1 caches and the lack of an on-chip L2 cache made the processor unsuitable for many commercial workloads. One implementation of MAJC 5200 (0.22 micron technology) took 15 W of power and 220 mm^2 of area. Sun also later came out with other multi-core products, like UltraSparc-IV [83] and Niagara [55].

IBM's multi-core efforts started with Power4 [49]. Power4, a contemporary of MAJC 5200 and Piranha, was a dual-core processor running at 1 GHz where each core was a five-issue out-of-order superscalar processor. Each core consisted of a private direct-mapped 32 KB instruction cache and a private 2-way 32 KB data cache. The cores were connected to a shared triply-banked 8-way set-associative L2 cache. The connection was through a high-bandwidth crossbar switch (called crossbar-interface unit). Four Power4 chips could be connected together within a multi-chip module and made to logically share the L2. One implementation of Power4 (in 0.13 micron technology) consisted of 184 million transistors and took up 267 mm^2 in die-area.

IBM also later came up with successors to Power4, like Power5 [50] and Power6. However, IBM's most ambitious multi-core offering arguably has been Cell [52]. Cell is a multi-ISA heterogeneous chip multiprocessor that consists of one two-way simultaneous multithreading (SMT) [90,89] dual-issue Power core and eight dual-issue SIMD (single instruction, multiple data) style Synergistic Processing Element (SPE) cores on the same die. While the Power core executes the PowerPC instruction set (while supporting the vector SIMD instruction set at the same time), the SPEs execute a variable width SIMD instruction set architecture (ISA). The Power core has a multi-level storage hierarchy—32 KB instruction and data caches, and a 512 KB L2. Unlike the Power core, the SPEs operate only on their local memory (local store, or LS). Code and data must be transferred into the associated LS for an SPE to operate on. LS addresses have an alias in the Power core address map, and transfers between an individual LS and the global memory are done through DMAs (direct memory accesses). An implementation of Cell in 90 nm operates at 3.2 GHz, consists of 234 million transistors and takes 229 mm^2 of die area.

The first x86 multi-core processors were introduced by AMD in 2005. At the time of writing this chapter, AMD offers Opteron [2], Athlon [3], and Turion [1] dual-core processors serving different market segments. Intel's dual-core offerings include Pentium D [10], Pentium Extreme [11], Xeon [12], and Core Duo [9] processors. The approach Intel has taken so far is to take last-generation, aggressive designs and put two on a die when space allows, rather than to target less aggressive cores. The Pentium Processor Extreme includes designs that are both dual-core and hardware multithreaded on each core.

One constraint for all the above multi-core designs was that they had to be capable of running legacy code in their respective ISAs. This restricted the degree of freedom in architecture and design of these processors. Three academic multi-core projects that were not bound by such constraints were RAW, TRIPS, and WaveScalar.

The RAW [91] processor consists of sixteen identical tiles spread across the die in a regular two-dimensional pattern. Each tile consists of communication routers, one scalar core, an FPU (floating-point unit), a 32 KB DCache, and a software-managed

96 KB ICache. Tiles are sized such that the latency of communication between two adjacent tiles is always one cycle. Tiles are connected using on-chip networks that interface with the tiles through the routers. Hardware resources on a RAW processor (tiles, pins, and the interconnect) are exposed to the ISA. This enables the compiler to generate aggressive code that maps more efficiently to the underlying computation substrate.

The TRIPS [76] processor consists of four large cores that are constructed out of small decentralized, polymorphous structures. The cores can be partitioned into small execution nodes when a program with high data-level parallelism needs to be executed. These execution nodes can also be logically chained when executing streaming programs. Like RAW, the microarchitecture is exposed to the ISA. Unlike RAW and other multi-cores discussed above, the unit of execution is a hyperblock. A hyperblock commits only when all instructions belonging to a hyperblock finish execution.

WaveScalar [84] attempts to move away from Von-Neumann processing in order to get the full advantage of parallel hardware. It has a dataflow instruction set architecture that allows for traditional memory ordering semantics. Each instruction executes on an ALU (arithmetic logic unit) that sits inside a cache, and explicitly forwards the result to the dependent instructions. The $ALU + cache$ is arranged as regular tiles, thereby allowing the communication overheads to be exposed to hardware. Like RAW and TRIPS, wavescalar also supports all the traditional imperative languages.

There have been multi-core offerings in non-mainstream computing markets as well. A few examples are Broadcom SiByte (SB1250, SB1255, SB1455) [5], PA-RISC (PA-8800) [7], Raza Microelectronics' XLR processor [13] that has eight MIPS cores, Cavium Networks' Octeon [6] processor that has 16 MIPS cores, Arm's MPCore processor [4], and Microsoft's Xbox 360 game console [14] that uses a triple-core PowerPC microprocessor.

3. Holistic Design for Adaptability: Heterogeneous Architectures

The rest of this chapter discusses various holistic approaches to multi-core design. Each of these approaches results in processors that are significantly more efficient than their "multi-core oblivious" counterparts.

First, we discuss the implication of workload diversity on multi-core design and present a new class of holistically-designed architectures that deliver significant gains in efficiency. Section 3.1 discusses the diversity present in computer workloads and how naively-designed multi-core architectures do not adapt well. Section 3.2

introduces *single-ISA heterogeneous multi-core architectures* that can adapt to workload diversity. That section also discusses the scheduling policies and mechanisms that are required to effect adaptability. Section 3.3 discusses the power advantages of the architecture and the corresponding scheduling policies. Section 3.4 provides an overview of other work evaluating advantages of heterogeneous multi-core architectures. Section 3.5 discusses a holistic design methodology for such architectures.

3.1 Workload Diversity

The amount of diversity among applications that a typical computer is expected to run can be considerable. For example, there can often be more than a factor of ten difference in the average performance of SPEC2000 applications *gcc* and *mcf* [79]. Even in the server domain there can be diversity among threads. Here the diversity can be because of batch processing, different threads processing different inputs, or threads having different priorities. Even in such a homogeneous workload, if each thread experiences phased behavior, different phases will typically be active in different threads at the same time.

A multi-core oblivious design (i.e., a homogeneous design that consists of identical cores) can target only a single point efficiently in the diversity spectrum. For example, a decision might need to be made beforehand if the core should be designed to target *gcc* or *mcf*. In either case, an application whose resource demands are different from those provided by the core will suffer—the resource mismatch will either result in underutilization of the resulting processor or it will result in low program performance.

Similarly, there is diversity within applications [79], as different phases of the same application often have different execution resource needs. Note that this is the same problem that a general-purpose uniprocessor faces as well, as the same design is expected to perform well for the entire universe of applications. There is a new form of diversity, as well, that uniprocessor architectures did not need to account for. A general-purpose multi-core should also provide good performance whether there is one thread (program) running, eight threads running, or one hundred.

3.2 Single-ISA Heterogeneous Multi-Core Architectures

To address the poor adaptability of homogeneous multi-core architectures, we present single-ISA heterogeneous multi-core architectures. That is, architectures with multiple core types on the same die. The cores are all capable of executing the same ISA (instruction-set architecture), but represent different points in the power/performance continuum—for example, a low-power, low-performance core and a high-power, high-performance core on the same die.

The advantages of single-ISA heterogeneous architectures stem from two sources. The first advantage results from a more *efficient adaptation to application diversity*. Given a set of diverse applications and heterogeneous cores, we can assign applications (or phases of applications) to cores such that those that benefit the most from complex cores are assigned to them, while those that benefit little from complex cores are assigned to smaller, simpler cores. This allows us to approach the performance of an architecture with a larger number of complex cores.

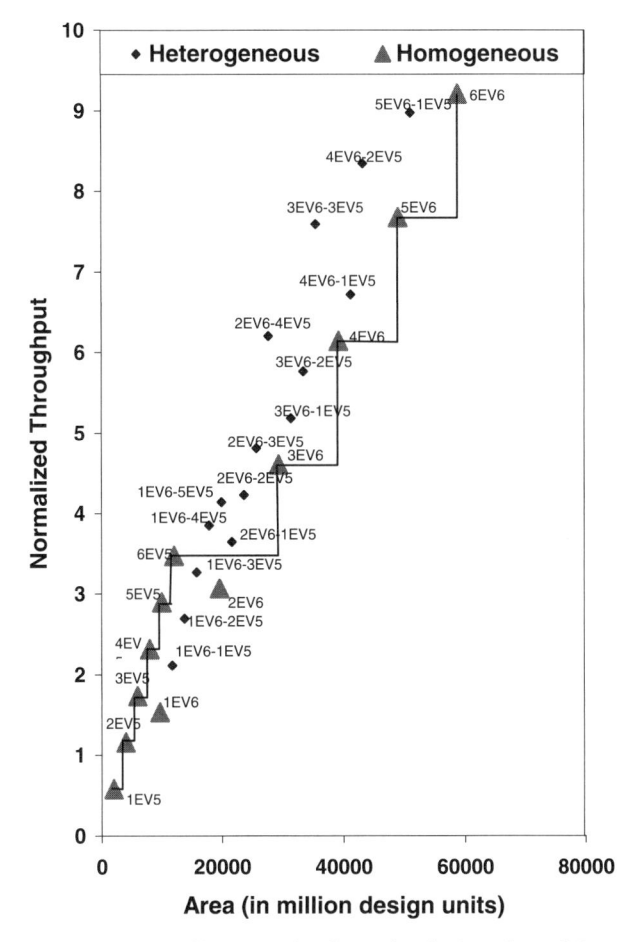

FIG. 1. Exploring the potential of heterogeneity: Comparing the throughput of six-core homogeneous and heterogeneous architectures for different area budgets.

The second advantage from heterogeneity results from a more *efficient use of die area for a given thread-level parallelism*. Successive generations of microprocessors have been providing diminishing performance returns per chip area [47] as the performance improvement of many microprocessor structures (e.g., cache memories) is less than linear with their size. Therefore, in an environment with large amounts of thread-level parallelism (TLP, the number of programs or threads currently available to run on the system), higher throughputs could be obtained by building a large number of small processors, rather than a small number of large processors. However, in practice the amount of thread level parallelism in most systems will vary with time. This implies that building chip-level multiprocessors with a mix of cores—some large cores with high single-thread performance and some small cores with high throughput per die area—is a potentially attractive approach.

To explore the potential from heterogeneity, we model a number of chip multiprocessing configurations that can be derived from combinations of two existing off-the-shelf processors from the Alpha architecture family—the EV5 (21164) and the EV6 (21264) processors. Figure 1 compares the various combinations in terms of their performance and their total chip area. In this figure, performance is that obtained from the best static mapping of applications constituting multiprogrammed SPEC2000 workloads to the processor cores. The solid staircase line represents the maximum throughput obtainable using a homogeneous configuration for a given area.

We see from this graph that over a large portion of the graph the highest performance architecture for a given area limit, often by a significant margin, is a heterogeneous configuration. The increased throughput is due to the increased number of contexts as well as improved processor utilization.

3.2.1 Methodology to Evaluate Multi-Core Architectures

This section discusses the methodological details for evaluating the benefits of heterogeneous multi-core architectures over their homogeneous counterparts. It also details the hardware assumptions made for the evaluation and provides the methodology for constructing multiprogrammed workloads for evaluation. We also discuss the simulation details and the evaluation metrics.

3.2.1.1 Supporting Multi-Programming.

The primary issue when using heterogeneous cores for greater throughput is with the scheduling, or assignment, of jobs to particular cores. We assume a scheduler at the operating system level that has the ability to observe coarse-grain program behavior over particular intervals, and move jobs between cores. Since the phase lengths of applications are typically large [78], this enables the cost of core switching to be piggybacked with the operating system context-switch overhead. Core switching overheads are modeled in detail for the evaluation of dynamic scheduling policies presented in this chapter—ones where jobs can move throughout execution.

3.2.1.2 Hardware Assumptions.
Table I summarizes the configurations used for the EV5 and the EV6 cores that we use for our throughput-related evaluations. The cores are assumed to be implemented in 0.10 micron technology and are clocked at 2.1 GHz.

In addition to the individual L1 caches, all the cores share an on-chip 4 MB, 4-way set-associative, 16-way L2 cache. The cache line size is 128 bytes. Each bank of the L2 cache has a memory controller and an associated RDRAM channel. The memory bus is assumed to be clocked at 533 MHz, with data being transferred on both edges of the clock for an effective frequency of 1 GHz and an effective bandwidth of 2 GB/s per bank. A fully-connected matrix crossbar interconnect is assumed between the cores and the L2 banks. All L2 banks can be accessed simultaneously, and bank conflicts are modeled. The access time is assumed to be 10 cycles. Memory latency was set to be 150 ns. We assume a snoopy bus-based MESI coherence protocol and model the writeback of dirty cache lines for every core switch.

Table I also presents the area occupied by the core. These were computed using a methodology outlined in Section 3.3. As can be seen from the table, a single EV6 core occupies as much area as 5 EV5 cores.

To evaluate the performance of heterogeneous architectures, we perform comparisons against homogeneous architectures occupying equivalent area. We assume that the total area available for cores is around 100 mm^2. This area can accommodate a maximum of 4 EV6 cores or 20 EV5 cores. We expect that while a 4-EV6 homogeneous configuration would be suitable for low-TLP (thread-level parallelism) environments, the 20-EV5 configuration would be a better match for the cases where TLP is high. For studying heterogeneous architectures, we choose a configuration with 3 EV6 cores and 5 EV5 cores with the expectation that it would perform well

TABLE I
CONFIGURATION AND AREA OF THE EV5 AND EV6 CORES

Processor	EV5	EV6
Issue-width	4	6 (OOO)
I-Cache	8 KB, DM	64 KB, 2-way
D-Cache	8 KB, DM	64 KB, 2-way
Branch Pred.	2K-gshare	hybrid 2-level
Number of MSHRs	4	8
Number of threads	1	1
Area (in mm^2)	5.06	24.5

over a wide range of available thread-level parallelism. It would also occupy roughly the same area.

3.2.1.3 Workload Construction.
All our evaluations are done for various thread counts ranging from one through a maximum number of available processor contexts. Instead of choosing a large number of benchmarks and then evaluating each number of threads using workloads with completely unique composition, we instead choose a relatively small number of SPEC2000 [82] benchmarks (8) and then construct workloads using these benchmarks. These benchmarks are evenly distributed between integer benchmarks (*crafty, mcf, eon, bzip2*) and floating-point benchmarks (*applu, wupwise, art, ammp*). Also, half of them (*applu, bzip2, mcf, wupwise*) have a large memory footprint (over 175 MB), while the other half (*ammp, art, crafty, eon*) have memory footprints of less than 30 MB.

All the data points are generated by evaluating 8 workloads for each case and then averaging the results. The 8 workloads at each level of threading are generated by constructing subsets of the workload consistent with the methodology in [90,81].

3.2.1.4 Simulation Approach.
Benchmarks are simulated using SMTSIM [87], a cycle-accurate execution-driven simulator that simulates an out-of-order, simultaneous multithreading processor [90]. SMTSIM executes unmodified, statically linked Alpha binaries. The simulator was modified to simulate the various multi-core architectures.

The Simpoint tool [78] was used to find good representative fast-forward distances for each benchmark (how far to advance the program before beginning measured simulation). Unless otherwise stated, all simulations involving n threads were done for $500 \times n$ million instructions. All the benchmarks are simulated using the SPEC *ref* inputs.

3.2.1.5 Evaluation Metrics.
In a study like this, IPC (number of total instructions committed per cycle) is not a reliable metric as it would inordinately bias all the heuristics (and policies) against inherently slow-running threads. Any policy that favors high-IPC threads boosts the reported IPC by increasing the contribution from the favored threads. But this does not necessarily represent an improvement. While the IPC over a particular measurement interval might be higher, in a real system the machine would eventually have to run a workload inordinately heavy in low-IPC threads, and the artificially-generated gains would disappear. Therefore, we use weighted speedup [81,88] for our evaluations. In this section, weighted speedup measures the arithmetic sum of the individual IPCs of the threads constituting a workload divided by their IPC on a baseline configuration when running alone. This

metric makes it difficult to produce artificial speedups by simply favoring high-IPC threads.

3.2.2 Scheduling for Throughput: Analysis and Results

In this section, we demonstrate the performance advantage of the heterogeneous multi-core architectures for multithreaded workloads and demonstrate job-to-core assignment mechanisms that allow the architecture to deliver on its promise. The first two subsections focus on the former, and the rest of the section demonstrates the further gains available from a good job assignment mechanism.

3.2.2.1 Static Scheduling for Inter-thread Diversity.
A heterogeneous architecture can exploit two dimensions of diversity in an application mix. The first is diversity between applications. The second is diversity over time within a single application. Prior work [78,92] has shown that both these dimensions of diversity occur in common workloads. In this section, we attempt to separate these two effects by first looking at the performance of a static assignment of applications to cores. Note that the static assignment approach may not eliminate the need for core switching in several cases, because the best assignment of jobs to cores will change as jobs enter and exit the system.

Figure 2 shows the results comparing one heterogeneous architecture against two homogeneous architectures all requiring approximately the same area. The heterogeneous architecture that we evaluate includes 3 EV6 cores and 5 EV5 cores, while the two homogeneous architectures that we study have 4 EV6 cores or 20 EV5 cores,

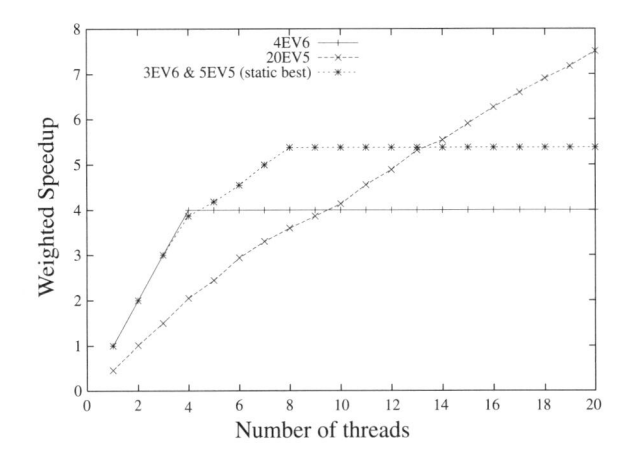

FIG. 2. Benefits from heterogeneity—static scheduling for inter-thread diversity.

respectively. For each architecture, the graph shows the average weighted speedup for varying number of threads.

For the homogeneous CMP configuration, we assume a straightforward scheduling policy, where as long as a core is available, any workload can be assigned to any core. For the heterogeneous case, we use an assignment that seeks to match the optimal static configuration as closely as possible. The optimal configuration would factor in both the effect of the performance difference between executing on a different core and the potential shared L2 cache interactions. However, determining this configuration is only possible by running *all possible* combinations. Instead, as a simplifying assumption, our scheduling policy assumes no knowledge of L2 interactions (only for determining core assignments—the interactions are still simulated) when determining the static assignment of workloads to cores. This simplification allows us to find the best configuration (defined as the one which maximizes weighted speedup) by simply running each job alone on each of our unique cores and using that to guide our core assignment. This results in consistently good, if not optimal, assignments. For a few cases, we compared this approach to an exhaustive exploration of all combinations; our results indicated that this results in performance close to the optimal assignment.

The use of weighted speedup as the metric ensures that those jobs assigned to the EV5 are those that are least affected (in relative IPC) by the difference between EV6 and EV5. In both the homogeneous and heterogeneous cases, once all the contexts of a processor get used, we just assume that the weighted speedup will level out as shown in Fig. 2. The effect when the number of jobs exceeds the number of cores in the system (e.g., additional context switching) is modeled more exactly in the following section.

As can be seen from Fig. 2, even with a simple static approach, the results show a strong advantage for heterogeneity over the homogeneous designs, for most levels of threading. The heterogeneous architecture attempts to combine the strengths of both the homogeneous configurations—CMPs of a few powerful processors (EV6 CMP) and CMPs of many less powerful processors (EV5 CMP). While for low threading levels, the applications can run on powerful EV6 cores resulting in high performance for each of the few threads, for higher threading levels, more applications can run on the added EV5 contexts enabled by heterogeneity, resulting in higher overall throughput.

The results in Fig. 2 show that the heterogeneous configuration achieves performance identical to the homogeneous EV6 CMP from 1 to 3 threads. At 4 threads, the optimum point for the EV6 CMP, that configuration shows a slight advantage over the heterogeneous case. However, this advantage is very small because with 4 threads, the heterogeneous configuration is nearly always able to find one thread that is impacted little by having to run on an EV5 instead of EV6. As soon as we have more than 4 threads, however, the heterogeneous processor shows clear advantage.

The superior performance of the heterogeneous architecture is directly attributable to the diversity of the workload mix. For example, *mcf* underutilizes the EV6 pipeline due to its poor memory behavior. On the other hand, benchmarks like *crafty* and *applu* have much higher EV6 utilization. Static scheduling on heterogeneous architectures enables the mapping of these benchmarks to the cores in such a way that overall processor utilization (average of individual core utilization values) is high.

The heterogeneous design remains superior to the EV5 CMP out to 13 threads, well beyond the point where the heterogeneous architecture runs out of processors and is forced to queue jobs. Beyond that, the raw throughput of the homogeneous design with 20 EV5 cores wins out. This is primarily because of the particular heterogeneous designs we chose. However, more extensive exploration of the design space than we show here confirms that we can always come up with a different configuration that is competitive with more threads (e.g., fewer EV6s, more EV5s), if that is the desired design point.

Compared to a homogeneous processor with 4 EV6 cores, the heterogeneous processor performs up to 37% better with an average 26% improvement over the configurations considering 1–20 threads. Relative to 20 EV5 cores, it performs up to 2.3 times better, and averages 23% better over that same range.

These results demonstrate that over a range of threading levels, a heterogeneous architecture can outperform comparable homogeneous architectures. Although the results are shown here only for a particular area and two core types, our experiments with other configurations (at different processor areas and core types) indicate that these results are representative of other heterogeneous configurations as well.

3.2.2.2 *Dynamic Scheduling for Intra-thread Diversity.*

The previous section demonstrated the performance advantages of the heterogeneous architecture when exploiting core diversity for inter-workload variation. However, that analysis has two weaknesses—it used unimplementable assignment policies in some cases (e.g., the static assignment oracle) and ignored variations in the resource demands of individual applications. This section solves each of these problems, and demonstrates the importance of good dynamic job assignment policies.

Prior work has shown that an application's demand for processor resources varies across phases of the application. Thus, the best match of applications to cores will change as those applications transition between phases. In this section, we examine implementable heuristics that dynamically adjust the mapping to improve performance.

These heuristics are sampling-based. During the execution of a workload, every so often, a trigger is generated that initiates a *sampling phase*. In the *sampling* phase, the scheduler permutes the assignment of applications to cores. During this phase, the dynamic execution profiles of the applications being run are gathered by referencing hardware performance counters. These profiles are then used to create a new assignment, which is then employed during a much longer phase of execution, the *steady* phase. The steady phase continues until the next trigger. Note that applications continue to make forward progress during the sampling phase, albeit perhaps non-optimally.

In terms of the core sampling strategies, there are a large number of application-to-core assignment permutations possible. We prune the number of permutations significantly by assuming that we would never run an application on a less powerful core when doing so would leave a more powerful core idle (for either the sampling phase or the steady phase). Thus, with four threads on our 3 EV6/5 EV5 configuration, four possible assignments are possible based on which thread gets allocated to the EV5. With more threads, the number of permutations increase, up to 56 potential choices with eight threads. Rather than evaluating all these possible alternatives, our heuristics only sample a subset of possible assignments. Each of these assignments are run for 2 million cycles. At the end of the sampling phase, we use the collected data to make assignments.

We experimented with several dynamic heuristics (more details on those heuristics can be found in [63]). Our best dynamic scheduling policy was what we call *bounded-global-event*. With this policy, we sum the absolute values of the percent changes in IPC for each application constituting a workload, and trigger a sampling phase when this value exceeds 100% (average 25% change for each of the four threads). The policy also initiates a sampling phase if more than 300 million cycles has elapsed since the last sampling phase, and avoids sampling if the global event trigger occurs within 50 million cycles since the last sampling phase. This strategy guards against oversampling and undersampling.

Figure 3 shows the results for the *bounded-global-event* dynamic scheduling heuristic comparing it against random scheduling as well as the best static mapping. The results presented in Fig. 3 indicate that dynamic heuristics which intelligently adapt the assignment of applications to cores can better leverage the diversity advantages of a heterogeneous architecture. Compared to the base homogeneous architecture, the best dynamic heuristic achieves close to a 63% improvement in throughput in the best case (for 8 threads) and an average improvement in throughput of 17% over configurations running 1–8 threads. Even more interesting, the best dynamic heuristic achieves a weighted speedup of 6.5 for eight threads, which is close to 80% of the optimal speedup (8) achievable for this configuration (despite the fact that over half of our cores have roughly half the raw computation power of the baseline

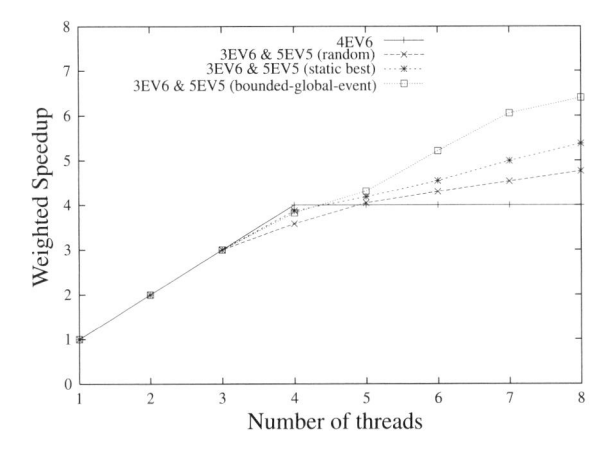

FIG. 3. Benefits of dynamic scheduling.

core!). In contrast, the homogeneous configuration achieves only 50% of the optimal speedup. The heterogeneous configuration, in fact, beats the homogeneous configuration with 20 EV5 cores till the number of threads exceeds thirteen. We have also demonstrated the importance of the intelligent dynamic assignment, which achieves up to 31% improvement over a baseline random scheduler.

3.3 Power Advantages of Architectural Heterogeneity

The ability of *single-ISA heterogeneous multi-core architectures* to adapt to workload diversity can also be used for improving the power and energy efficiency of processors. The approach again relies on a chip-level multiprocessor with multiple, diverse processor cores. These cores all execute the same instruction set, but include significantly different resources and achieve different performance and energy efficiency on the same application. During an application's execution, the operating system software tries to match the application to the different cores, attempting to meet a defined objective function. For example, it may be trying to meet a particular performance requirement or goal, but doing so with maximum energy efficiency.

3.3.1 Discussion of Core Switching

There are many reasons, some discussed in previous sections, why the best core for execution may vary over time. The demands of executing code vary widely between applications; thus, the best core for one application will often not be the best for the next, given a particular objective function (assumed to be some combination of energy and performance). In addition, the demands of a single application can also vary across phases of the program.

Even the objective function can change over time, as the processor changes power conditions (e.g., plugged vs. unplugged, full battery vs. low battery, thermal emergencies), as applications switch (e.g., low priority vs. high priority job), or even within an application (e.g., a real-time application is behind or ahead of schedule).

The experiments in this section explore only a subset of these possible changing conditions. Specifically, we examine adaptation to phase changes in single applications. However, by simulating multiple applications and several objective functions, it also indirectly examines the potential to adapt to changing applications and objective functions. We believe a real system would see far greater opportunities to switch cores to adapt to changing execution and environmental conditions than the narrow set of experiments exhibited here.

3.3.2 Choice of Cores

In this study, we consider a design that takes a series of previously implemented processor cores with slight changes to their interface—this choice reflects one of the key advantages of the CMP architecture, namely the effective amortization of design and verification effort. We include four Alpha cores—EV4 (Alpha 21064), EV5 (Alpha 21164), EV6 (Alpha 21264) and a single-threaded version of the EV8 (Alpha 21464), referred to as EV8-. These cores demonstrate strict gradation in terms of complexity and are capable of sharing a single executable. We assume the four

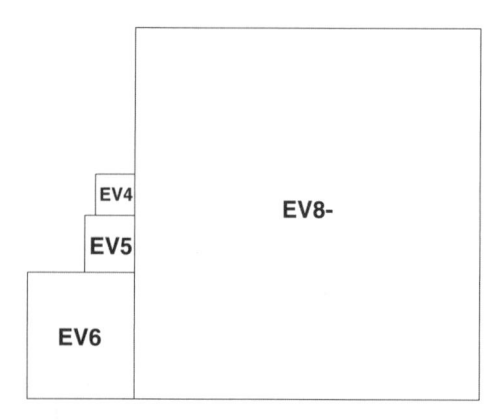

FIG. 4. Relative sizes of the Alpha cores when implemented in 0.10 micron technology.

cores have private L1 data and instruction caches and share a common L2 cache, phase-lock loop circuitry, and pins.

We chose the cores of these off-the-shelf processors due to the availability of real power and area data for these processors, except for the EV8 where we use projected numbers [30,35,54,69]. All these processors have 64-bit architectures. Note that technology mapping across a few generations has been shown to be feasible [57].

Figure 4 shows the relative sizes of the cores used in the study, assuming they are all implemented in a 0.10 micron technology (the methodology to obtain this figure is described in the next section). It can be seen that the resulting core is only modestly (within 15%) larger than the EV8- core by itself.

3.3.3 Switching Applications between Cores

There is a cost to switching cores, so we must restrict the granularity of switching. One method for doing this would switch only at operating system timeslice intervals, when execution is in the operating system, with user state already saved to memory. If the OS decides a switch is in order, it powers up the new core, triggers a cache flush to save all dirty cache data to the shared L2, and signals the new core to start at a predefined OS entry point. The new core would then power down the old core and return from the timer interrupt handler. The user state saved by the old core would be loaded from memory into the new core at that time, as a normal consequence of returning from the operating system. Alternatively, we could switch to different cores at the granularity of the entire application, possibly chosen statically. In this study, we consider both these options.

In this work, we assume that unused cores are completely powered down, rather than left idle. Thus, unused cores suffer no static leakage or dynamic switching power. This does, however, introduce a latency for powering a new core up. We estimate that a given processor core can be powered up in approximately one thousand cycles of the 2.1 GHz clock. This assumption is based on the observation that when we power down a processor core we do not power down the phase-lock loop that generates the clock for the core. Rather, in our multi-core architecture, the same phase-lock loop generates the clocks for all cores. Consequently, the power-up time of a core is determined by the time required for the power buses to charge and stabilize. In addition, to avoid injecting excessive noise on the power bus bars of the multi-core processor, we assume a staged power up would be used.

In addition, our experiments confirm that switching cores at operating-system timer intervals ensures that the switching overhead has almost no impact on performance, even with the most pessimistic assumptions about power-up time, software overhead, and cache cold start effects. However, these overheads are still modeled in our experiments.

3.3.4 Modeling Power, Area, and Performance

This section discusses the various methodological challenges of this study, including modeling the power, the real estate, and the performance of the heterogeneous multi-core architecture.

3.3.4.1 Modeling of CPU Cores.

The cores we simulate are roughly modeled after cores of EV4 (Alpha 21064), EV5 (Alpha 21164), EV6 (Alpha 21264) and EV8-. EV8- is a hypothetical single-threaded version of EV8 (Alpha 21464). The data on the resources for EV8 was based on predictions made by Joel Emer [35] and Artur Klauser [54], conversations with people from the Alpha design team, and other reported data [30,69]. The data on the resources of the other cores are based on published literature on these processors [16–18].

The multi-core processor is assumed to be implemented in a 0.10 micron technology. The cores have private first-level caches, and share an on-chip 3.5 MB 7-way set-associative L2 cache. At 0.10 micron, this cache will occupy an area just under half the die size of the Pentium 4. All the cores are assumed to run at 2.1 GHz. This is the frequency at which an EV6 core would run if its 600 MHz, 0.35 micron implementation was scaled to a 0.10 micron technology. In the Alpha design, the amount of work per pipe stage was relatively constant across processor generations [24,31, 35,40]; therefore, it is reasonable to assume they can all be clocked at the same rate when implemented in the same technology (if not as designed, processors with similar characteristics certainly could). The input voltage for all the cores is assumed to be 1.2 V.

Table II summarizes the configurations that were modeled for various cores. All architectures are modeled as accurately as possible, given the parameters in Table II, on a highly detailed instruction-level simulator. However, we did not faithfully model every detail of each architecture; we were most concerned with modeling the approximate spaces each core covers in our complexity/performance continuum.

The various miss penalties and L2 cache access latencies for the simulated cores were determined using CACTI [80]. Memory latency was set to be 150 ns.

TABLE II
CONFIGURATION OF THE CORES USED FOR POWER EVALUATION OF HETEROGENEOUS MULTI-CORES

Processor	EV4	EV5	EV6	EV8-
Issue-width	2	4	6 (OOO)	8 (OOO)
I-Cache	8 KB, DM	8 KB, DM	64 KB, 2-way	64 KB, 4-way
D-Cache	8 KB, DM	8 KB, DM	64 KB, 2-way	64 KB, 4-way
Branch Pred.	2 KB,1-bit	2 K-gshare	hybrid 2-level	hybrid 2-level (2X EV6 size)
Number of MSHRs	2	4	8	16

TABLE III
POWER AND AREA STATISTICS OF THE ALPHA CORES

Core	Peak-power (W)	Core-area (mm^2)	Typical-power (W)	Range (%)
EV4	4.97	2.87	3.73	92–107
EV5	9.83	5.06	6.88	89–109
EV6	17.80	24.5	10.68	86–113
EV8-	92.88	2.36	46.44	82–128

3.3.4.2 Modeling Power. Modeling power for this type of study is a challenge. We need to consider cores designed over the time span of more than a decade. Power depends not only on the configuration of a processor, but also on the circuit design style and process parameters. Also, actual power dissipation varies with activity, though the degree of variability again depends on the technology parameters as well as the gating style used.

No existing architecture-level power modeling framework accounts for all of these factors. Current power models like Wattch [25] are primarily meant for activity-based architectural level power analysis and optimizations within a single processor generation, not as a tool to compare the absolute power consumption of widely varied architectures. Therefore we use a hybrid power model that uses estimates from Wattch, along with additional scaling and offset factors to calibrate for technology factors. This model not only accounts for activity-based dissipation, but also accounts for the design style and process parameter differences by relying on measured datapoints from the manufacturers. Further details of the power model can be found in [60].

Table III shows our power and area estimates for the cores. As can be seen from the table, the EV8- core consumes almost 20 times the peak power and more than 80 times the real estate of the EV4 core. The table also gives the derived typical power for each of our cores. Also shown, for each core, is the range in power demand for the actual applications run, expressed as a percentage of typical power.

3.3.4.3 Estimating Chip Area. Table III also summarizes the area occupied by the cores at 0.10 micron (also shown in Fig. 4). The area of the cores (except EV8-) is derived from published photos of the dies after subtracting the area occupied by I/O pads, interconnection wires, the bus-interface unit, L2 cache, and control logic. Area of the L2 cache of the multi-core processor is estimated using CACTI.

The die size of EV8 was predicted to be 400 mm^2 [75]. To determine the core size of EV8-, we subtract out the estimated area of the L2 cache (using CACTI). We also account for reduction in the size of register files, instruction queues, the reorder buffer, and renaming tables to account for the single-threaded EV8-. For this, we use detailed models of the register bit equivalents (rbe) [73] for register files, the reorder buffer, and renaming tables at the original and reduced sizes. The sizes of the original and reduced instruction queue sizes were estimated from examination of MIPS R10000 and HP PA-8000 data [27,59], assuming that the area grows more than linear with respect to the number of entries ($num\_entries^{1.5}$). The area data is then scaled for the 0.10 micron process.

3.3.4.4 Modeling Performance.
In this study, we simulate the execution of 14 benchmarks from the SPEC2000 benchmark suite, including 7 from SPECint (*bzip2, crafty, eon, gzip, mcf, twolf, vortex*) and 7 from SPECfp (*ammp, applu, apsi, art, equake, fma3d, wupwise*).

Benchmarks are again simulated using SMTSIM [87], modified to simulate a multi-core processor comprising four heterogeneous cores sharing an on-chip L2 cache and the memory subsystem. The simpoint tool [79] is used to determine the number of committed instructions which need to be fast-forwarded so as to capture the representative program behavior during simulation. After fast-forwarding, we simulate 1 billion instructions. All benchmarks are simulated using *ref* inputs.

3.3.5 Scheduling for Power: Analysis and Results

This section examines the effectiveness of single-ISA heterogeneous multi-core designs in reducing the power dissipation of processors. We first examine the relative energy efficiency across cores, and how it varies by application and phase. Later sections use this variance, demonstrating both oracle and realistic core switching heuristics to maximize particular objective functions.

3.3.5.1 Variation in Core Performance and Power.
As discussed in Section 3.3.1, this work assumes that the performance ratios between our processor cores is not constant, but varies across benchmarks, as well as over time on a single benchmark. This section verifies that premise.

Figure 5(a) shows the performance measured in million instructions committed per second (IPS) of one representative benchmark, *applu*. In the figure, a separate curve is shown for each of the five cores, with each data point representing the IPS over the preceding 1 million committed instructions.

With *applu*, there are very clear and distinct phases of performance on each core, and the relative performance of the cores varies significantly between these phases. Nearly all programs show clear phased behavior, although the frequency and variety of phases varies significantly.

If relative performance of the cores varies over time, it follows that energy efficiency will also vary. Figure 6 shows one metric of energy efficiency (defined in this

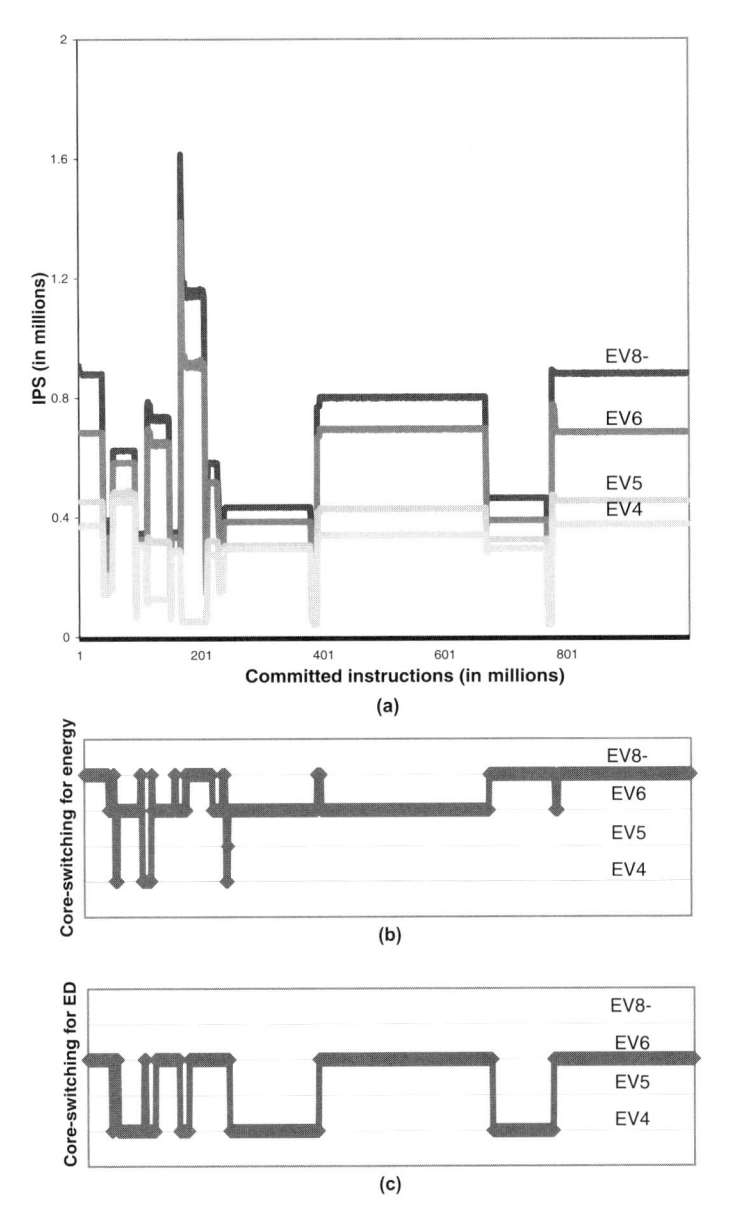

FIG. 5. (a) Performance of *applu* on the four cores; (b) oracle switching for energy; (c) oracle switching for energy-delay product.

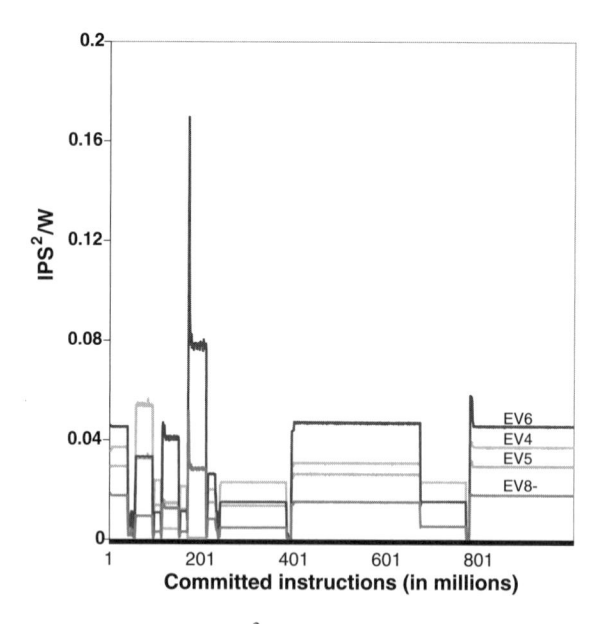

FIG. 6. *applu* energy efficiency. IPS^2/W varies inversely with energy-delay product.

case as IPS^2/W) of the various cores for the same benchmark. IPS^2/W is merely the inverse of Energy-Delay product. As can be seen, the relative value of the energy-delay product among cores, and even the ordering of the cores, varies from phase to phase.

3.3.5.2 Oracle Heuristics for Dynamic Core Selection.

This section examines the limits of power and efficiency improvements possible with a heterogeneous multi-core architecture. The ideal core-selection algorithm depends heavily on the particular goals of the architecture or application. This section demonstrates oracle algorithms that maximize two sample objective functions. The first optimizes for energy efficiency with a tight performance threshold. The second optimizes for energy-delay product with a looser performance constraint.

These algorithms assume perfect knowledge of the performance and power characteristics at the granularity of intervals of one million instructions (corresponding roughly to an OS time-slice interval). It should be noted that choosing the core that minimizes energy or the energy-delay product over each interval subject to performance constraints does not give an optimal solution for the global energy or energy-delay product; however, the algorithms do produce good results.

The first oracle that we study seeks to minimize the energy per committed instruction (and thus, the energy used by the entire program). For each interval, the oracle chooses the core that has the lowest energy consumption, given the constraint that performance has always to be maintained within 10% of the EV8- core for each interval. This constraint assumes that we are willing to give up performance to save energy but only up to a point. Figure 5(b) shows the core selected in each interval for *applu*.

For *applu*, we observe that the oracle chooses to switch to EV6 in several phases even though EV8- performs better. This is because EV6 is the less power-consuming core and still performs within the threshold. The oracle even switches to EV4 and EV5 in a small number of phases. Table IV shows the results for all benchmarks. In this, and all following results, performance degradation and energy savings are always given relative to EV8- performance. As can be seen, this heuristic achieves an average energy reduction of 38% (see column 8) with less than 4% average performance degradation (column 9). Five benchmarks (*ammp, fma3d, mcf, twolf, crafty*) achieve no gain because switching was denied by the performance constraint. Excluding these benchmarks, the heuristic achieves an average energy reduction of 60% with about 5% performance degradation.

Our second oracle utilizes the *energy-delay product* metric. The energy-delay product seeks to characterize the importance of both energy and response time in a

TABLE IV
SUMMARY FOR DYNAMIC *oracle* SWITCHING FOR ENERGY ON HETEROGENEOUS MULTI-CORES

Benchmark	Total switches	% of instructions per core				Energy savings (%)	ED savings (%)	ED^2 savings (%)	Perf. loss (%)
		EV4	EV5	EV6	EV8-				
ammp	0	0	0	0	100	0	0	0	0
applu	27	2.2	0.1	54.5	43.2	42.7	38.6	33.6	7.1
apsi	2	0	0	62.2	37.8	27.6	25.3	22.9	3.1
art	0	0	0	100	0	74.4	73.5	72.6	3.3
equake	20	0	0	97.9	2.1	72.4	71.3	70.1	3.9
fma3d	0	0	0	0	100	0	0	0	0
wupwise	16	0	0	99	1	72.6	69.9	66.2	10.0
bzip	13	0	0.1	84.0	15.9	40.1	38.7	37.2	2.3
crafty	0	0	0	0	100	0	0	0	0
eon	0	0	0	100	0	77.3	76.3	75.3	4.2
gzip	82	0	0	95.9	4.1	74.0	73.0	71.8	3.9
mcf	0	0	0	0	100	0	0	0	0
twolf	0	0	0	0	100	0	0	0	0
vortex	364	0	0	73.8	26.2	56.2	51.9	46.2	9.8
Average	1 (median)	0.2%	0%	54.8%	45.0%	38.5%	37.0%	35.4%	3.4%

TABLE V
SUMMARY FOR DYNAMIC *oracle* SWITCHING FOR ENERGY-DELAY ON HETEROGENEOUS MULTI-CORES

Benchmark	Total switches	% of instructions per core				ED savings (%)	Energy savings (%)	ED^2 savings (%)	Perf. loss (%)
		EV4	EV5	EV6	EV8-				
ammp	0	0	0	100	0	63.7	70.3	55.7	18.1
applu	12	32.3	0	67.7	0	69.8	77.1	59.9	24.4
apsi	0	0	0	100	0	60.1	69.1	48.7	22.4
art	619	65.4	0	34.5	0	78.0	84.0	69.6	27.4
equake	73	55.8	0	44.2	0	72.3	81.0	59.2	31.7
fma3d	0	0	0	100	0	63.2	73.6	48.9	28.1
wupwise	0	0	0	100	0	68.8	73.2	66.9	10.0
bzip	18	0	1.2	98.8	0	60.5	70.3	47.5	24.8
crafty	0	0	0	100	0	55.4	69.9	33.9	32.5
eon	0	0	0	100	0	76.2	77.3	75.3	4.2
gzip	0	0	0	100	0	74.6	75.7	73.5	4.2
mcf	0	0	0	100	0	46.9	62.8	37.2	24.3
twolf	0	0	0	100	0	26.4	59.7	−34.2	45.2
vortex	0	0	0	100	0	68.7	73.0	66.7	9.9
Average	0 (median)	11.0%	0.1%	88.9%	0%	63.2%	72.6%	50.6%	22.0%

single metric, under the assumption that they have equal importance. Our oracle minimizes energy-delay product by always selecting the core that maximizes IPS^2/W over an interval. We again impose a performance threshold, but relax it due to the fact that energy-delay product already accounts for performance degradation. In this case, we require that each interval maintains performance within 50% of EV8-.

Figure 5(c) shows the cores chosen for *applu*. Table V shows the results for all benchmarks. As can be seen, the average reduction in energy-delay is about 63%; the average energy reductions are 73% and the average performance degradation is 22%. That is nearly a *three-fold* increase in energy efficiency, powered by a four-fold reduction in actual expended energy with a relatively small performance loss. All but one of the fourteen benchmarks have fairly significant (47 to 78%) reductions in energy-delay savings. The corresponding reductions in performance ranges from 4 to 45%. As before, switching activity and the usage of the cores varies. This time, EV8 never gets used. EV6 emerges as the dominant core. Given our relaxed performance constraint, there is a greater usage of the lower-power cores compared to the previous experiment.

Both Tables IV and V also show results for Energy-Delay^2 [95] improvements. Improvements are 35–50% on average. This is instructive because chip-wide voltage/frequency scaling can do no better than break even on this metric, demonstrating that this approach has the potential to go well beyond the capabilities of that technique. In other experiments specifically targeting the ED^2 metric (again with the 50% performance threshold), we saw 53.3% reduction in energy-delay^2 with 14.8% degradation in performance.

3.3.5.3 Realistic Dynamic Switching Heuristics.
This section examines the extent to which the energy benefits in the earlier sections can be achieved with a real system implementation that does not depend on oracular future knowledge. We do, however, assume an ability to track both the accumulated performance and energy over a past interval. This functionality either already exists or is easy to implement. This section is intended to be an existence proof of effective core selection algorithms, rather than a complete evaluation of the scheduling design space. We only demonstrate a few simple heuristics for selecting the core to run on. The heuristics seek to minimize overall energy-delay product during program execution.

Our previous oracle results were idealized not only with respect to switching algorithms, but also ignored the cost of switching (power-up time, flushing dirty pages to the L2 cache and experiencing cold-start misses in the new L1 cache and TLB) both in performance and power. The simulations in this section account for both, although our switching intervals are long enough and switchings infrequent enough that the impact of both effects is under 1%.

In this section, we measure the effectiveness of several heuristics for selecting a core. The common elements of each of the heuristics are these: every 100 time intervals (one time interval consists of 1 million instructions in these experiments), one or more cores are sampled for five intervals each (with the results during the first interval ignored to avoid cold start effects). Based on measurements done during sampling, the heuristic selects one core. For the case when one other core is sampled, the switching overhead is incurred once if the new core is selected, or twice if the old core is chosen. The switching overhead is greater if more cores are sampled. The dynamic heuristics studied here are:

- *Neighbor.* One of the two neighboring cores in the performance continuum is randomly selected for sampling. A switch is made if that core has lower energy-delay product over the sample interval than the current core over the last run interval.
- *Neighbor-global.* Similar to neighbor, except that the selected core is the one that would be expected to produce the lowest accumulated energy-delay product to this point in the application's execution. In some cases this is different than the core that minimizes the energy-delay product for this interval.
- *Random.* One other randomly-chosen core is sampled, and a switch is made if that core has lower energy-delay over the sample interval.
- *All.* All other cores are sampled.

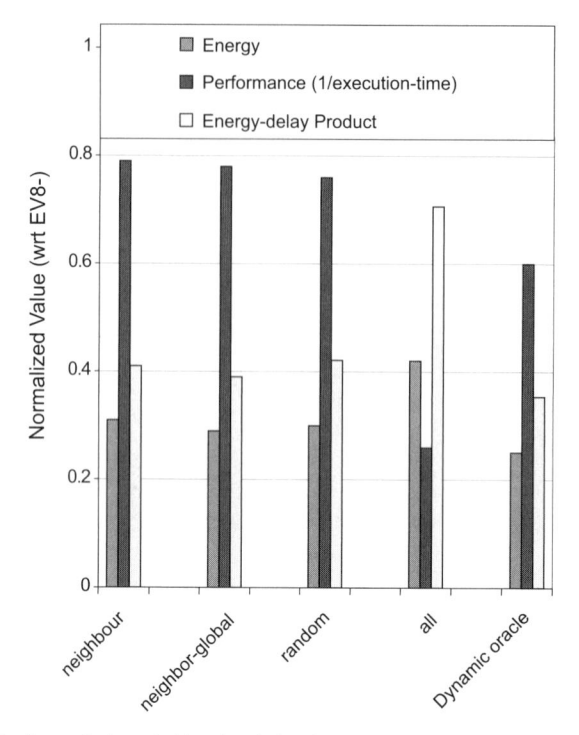

FIG. 7. Results for realistic switching heuristics for heterogeneous multi-cores—the last one is a constraint-less dynamic oracle.

The results are shown in Fig. 7. The results are all normalized to EV8- values. This figure also includes oracle results for dynamic switching based on the energy-delay metric when core selection is not hampered with performance constraints. Lower bars for energy and energy-delay, and higher bars for performance are desirable.

Our heuristics achieve up to 93% of the energy-delay gains achieved by the oracle-based switcher, despite modeling the switching overhead, sampling overhead, and non-oracle selection. The performance degradation on applying our dynamic heuristics is, on average, *less* than the degradation found by the oracle-based scheme. Also, although not shown in the figure, there is a greater variety in core-usage between applications.

It should be noted that switching for this particular objective function is not heavy; thus, heuristics that find the best core quickly, and minimize sampling overhead after that, tend to work best. The best heuristic for a different objective function, or a dynamically varying objective function may be different. These results do show, however, that for a given objective function, very effective realtime and predictive core switching heuristics can be found.

3.3.5.4 Practical Heterogeneous Architectures.
Although our use of existing cores limits design and verification overheads, these overheads do scale with the number of distinct cores supported. Some of our results indicate that in specific instances, two cores can introduce sufficient heterogeneity to produce significant gains. For example the (minimize energy, maintain performance within 10%) objective function relied heavily on the EV8- and the EV6 cores. The (energy-delay, performance within 50%) objective function favored the EV6 and EV4. However, if the objective function is allowed to vary over time, or if the workload is more diverse than what we model, wider heterogeneity than 2 cores will be useful. Presumably, other objective functions than those we model may also use more than 2 cores.

3.4 Overview of Other Related Proposals

There have been other proposals studying the advantages of on-chip heterogeneity. This section provides an overview of other work directly dealing with heterogeneous multi-core architectures. We also discuss previous work with similar goals—i.e., adapting to workload diversity to improve processor efficiency.

Morad et al. [71,72] explore the theoretical advantages of placing asymmetric core clusters in multiprocessor chips. They show that asymmetric core clusters are expected to achieve higher performance per area and higher performance for a given power envelope. Annavaram et al. [20] evaluate the benefits of heterogeneous multiprocessing in minimizing the execution times of multi-threaded programs containing non-trivial parallel and sequential phases, while keeping the CMP's total power consumption within a fixed budget. They report significant speedups. Balakrishanan et al. [22] seek to understand the impact of such an architecture on software. They show, using a hardware prototype, that asymmetry can have significant impact on the performance, stability, and scalability of a wide range of commercial applications. They also demonstrate that in addition to heterogeneity-aware kernels, several commercial applications may themselves need to be aware of heterogeneity at the hardware level.

The energy benefits of heterogeneous multi-core architectures is also explored by Ghiasi and Grunwald [38]. They consider single-ISA, heterogeneous cores of different frequencies belonging to the x86 family for controlling the thermal characteristics of a system. Applications run simultaneously on multiple cores and the operating system monitors and directs applications to the appropriate job queues. They report significantly better thermal and power characteristics for heterogeneous processors.

Grochowsky et al. [41] compare voltage/frequency scaling, asymmetric (heterogeneous) cores, variable-sized cores, and speculation as means to reduce the energy per instruction (EPI) during the execution of a program. They find that the EPI range for asymmetric chip-multiprocessors using x86 cores was 4–6X, significantly more than the next best technique (which was voltage/frequency scaling).

There have also been proposals for multi-ISA heterogeneous multi-core architectures. The proposed Tarantula processor [36] is one such example of integrated heterogeneity. It consists of a large vector unit sharing the die with an EV8 core. The Alpha ISA is extended to include vector instructions that operate on the new architectural state. The unit is targeted towards applications with high data-level parallelism. IBM Cell [52] (see Section 2.1) is another example of a heterogeneous chip multiprocessor with cores belonging to different ISAs.

3.5 Designing Multi-Cores from the Ground Up

While the previous sections demonstrate the benefits of heterogeneity, they gave no insight into what constitutes, or how to arrive at, a good heterogeneous design. Previous work assumed a given heterogeneous architecture. More specifically, those architectures were composed of existing architectures, either different generations of the same processor family [63,60,38,41], or voltage and frequency scaled editions of a single processor [20,22,39,56]. While these architectures surpassed similar homogeneous designs, they failed to reach the full potential of heterogeneity, for three reasons. First, the use of pre-existing designs presents low flexibility in choice of cores. Second, core choices maintain a monotonic relationship, both in design and performance—for example, the most powerful core is bigger or more complex in every dimension and the performance-ordering of the cores is the same for every application. Third, all cores considered perform well for a wide variety of applications—we show in this section that the best heterogeneous designs are composed of specialized core architectures.

Section 3.5.1 describes the approach followed to navigate the design space and arrive at the best designs for a given set of workloads. Section 3.5.2 discusses the benefits of customization. Section 3.5.3 discusses the methodology followed for our evaluations. Section 3.5.4 presents the results of our experiments.

3.5.1 From Workloads to Multi-Core Design

The goal of this research is to identify the characteristics of cores that combine to form the best heterogeneous architectures, and also demonstrate principles for designing such an architecture. Because this methodology requires that we accurately reflect the wide diversity of applications (their parallelism, their memory behavior), running on widely varying architectural parameters, there is no real shortcut to using simulation to characterize these combinations.

The design space for even a single processor is large, given the flexibility to change various architectural parameters; however, the design space explodes when considering the combined performance of multiple different cores on arbitrary permutations of the applications. Hence, we make some simplifying assumptions that make this problem tractable so that we navigate through the search space faster; however, we show that the resulting methodology still results in the discovery of very effective multi-core design points.

First, we assume that the performance of individual cores is separable—that is, that the performance of a four-core design, running four applications, is the sum (or the sum divided by a constant factor) of the individual cores running those applications in isolation. This is an accurate assumption if the cores do not share L2 caches or memory controllers (which we assume in this study). This assumption dramatically accelerates the search because now the single-thread performance of each core (found using simulation) can be used to estimate the performance of the processor as a whole without the need to simulate all 4-thread permutations.

Since we are interested in the highest performance that a processor can offer, we assume good static scheduling of threads to cores. Thus, the performance of four particular threads on four particular cores is the performance of the best static mapping. However, this actually represents a *lower bound* on performance. We have already shown that the ability to migrate threads dynamically during execution only increases the benefits of heterogeneity as it exploits intra-thread diversity—this continues to hold true for the best heterogeneous designs that we come up with under the static scheduling assumption.

To further accelerate the search, we consider only major blocks to be configurable, and only consider discrete points. For example, we consider 2 instruction queue sizes (rather than all the intermediate values) and 4 cache configurations (per cache). But we consider only a single branch predictor, because the area/performance tradeoffs of different sizes had little effect in our experiments. Values that are expected to be correlated (e.g., size of re-order buffer and number of physical registers) are scaled together instead of separately. This methodology might appear to be crude for an important commercial design, but we believe that even in that environment this methodology would find a design very much in the neighborhood of the best design. Then, a more careful analysis could be done of the immediate neighborhood, considering structure sizes at a finer granularity and considering particular choices for smaller blocks we did not vary.

We only consider and compare processors with a fixed number (4) of cores. It would be interesting to also relax that constraint in our designs, but we did not do so for the following reasons. Accurate comparisons would be more difficult, because the interconnect and cache costs would vary. Second, it is shown both in this work (Section 3.5.4) and in previous work [63] that heterogeneous designs are much more tolerant than homogeneous when running a different number of threads than the processor is optimized for (that is, it is less important in a heterogeneous design to get the number of cores right). However, the methodology shown here need only be applied multiple times (once for each possible core count) to fully explore the larger design space, assuming that an accurate model of the off-core resources was available.

The above assumptions allow us to model performance for various combinations of cores for various permutations of our benchmarks, and thereby evaluate the expected performance of all the possible homogeneous and heterogeneous processors for various area and power budgets.

3.5.2 Customizing Cores to Workloads

One of the biggest advantages of creating a heterogeneous processor as a custom design is that the cores can be chosen in an unconstrained manner as long as the processor budgetary constraints are satisfied. We define *monotonicity* to be a property of a multi-core architecture where there is a total ordering among the cores in terms of performance and this ordering remains the same for all applications. For example, a multiprocessor consisting of EV5 and EV6 cores is a monotonic multiprocessor. This is because EV6 is strictly superior to EV5 in terms of hardware resources and virtually always performs better than EV5 for a given application given the same cycle time and latencies. Similarly, for a multi-core architecture with identical cores, if the voltage/frequency of a core is set lower than the voltage/frequency of some other core, it will always provide less performance, regardless of application. Fully customized monotonic designs represent the upper bound (albeit a high one) on the benefits possible through previously proposed heterogeneous architectures.

As we show in this section, monotonic multiprocessors may not provide the "best fit" for various workloads and hence result in inefficient mapping of applications to cores. For example, in the results shown in Section 3.2.2, *mcf*, despite having very low ILP, consistently gets mapped to the EV6 or EV8- core for various energy-related objective functions, because of the larger caches on these cores. Yet it fails to take advantage of the complex execution capabilities of these cores, and thus still wastes energy unnecessarily.

Doing a custom design of a heterogeneous multi-core architecture allows us to relax the monotonicity constraint and finally take full advantage of our holistic design approach. That is, it is possible for a particular core of the multiprocessor to be the highest performing core for some application but not for others. For example, if one core is in-order, scalar, with 32 KB caches, and another core is out-of-order, dual-issue, with larger caches, applications will always run best on the latter. However, if the scalar core had larger L1 caches, then it might perform better for applications with low ILP and large working sets, while the other would likely be best for jobs with high ILP and smaller working sets.

The advantage of non-monotonicity is that now different cores on the same die can be customized to different classes of applications, which was not the case with previously studied designs. The holistic approach gives us the freedom to use cores not well suited for all applications, as long as the processor as a whole can meet the needs of all applications.

3.5.3 Modeling the Custom Design Process

This section discusses the various methodological challenges of this research, including modeling power, real estate, and performance of the heterogeneous multi-core architectures.

3.5.3.1 Modeling of CPU Cores.

For all our studies in this paper, we model 4-core multiprocessors assumed to be implemented in 0.10 micron, 1.2 V technology. Each core on a multiprocessor, either homogeneous or heterogeneous, has a private L2 cache and each L2 bank has a corresponding memory controller.

We consider both in-order cores and out-of-order cores for this study. We base our OOO processor microarchitecture model on the MIPS R10000, and our in-order cores on the Alpha EV5 (21164). We evaluate 480 cores as possible building blocks for constructing the multiprocessors. This represents all possible distinct cores that can be constructed by changing the parameters listed in Table VI. The various values that were considered are listed in the table as well. We assumed a gshare branch predictor with 8k entries for all the cores. Out of these 480 cores, there are 96 distinct in-order cores and 384 distinct out-of-order cores. The number of distinct 4-core multiprocessors that can be constructed out of 480 distinct cores is over 2.2 billion.

TABLE VI
VARIOUS PARAMETERS AND THEIR POSSIBLE VALUES FOR CONFIGURATION OF THE CORES

Issue-width	1, 2, 4
I-Cache	8 KB-DM, 16 KB 2-way, 32 KB 4-way, 64 KB 4-way
D-Cache	8 KB-DM, 16 KB 2-way, 32 KB 4-way, 64 KB 4-way dual ported
FP-IntMul-ALU units	1-1-2, 2-2-4
IntQ-fpQ (OOO)	32-16, 64-32
Int-FP PhysReg-ROB (OOO)	64-64-32, 128-128-64
L2 Cache	1 MB/core, 4-way, 12 cycle access
Memory Channel	533 MHz, doubly-pumped, RDRAM
ITLB-DTLB	64, 28 entries
Ld/St Queue	32 entries

Other parameters that are kept fixed for all the cores are also listed in Table VI. The various miss penalties and L2 cache access latencies for the simulated cores were determined using CACTI [80].

All evaluations are done for multiprocessors satisfying a given aggregate area and power budget for the 4 cores. We do not concern ourselves with the area and power consumption of anything other than the cores for this study.

3.5.3.2 Modeling Power and Area.

The area budget refers to the sum of the area of the 4 cores of a processor (the L1 cache being part of the core), and the power budget refers to the sum of the worst case power of the cores of a processor. Specifically, we consider peak activity power, as this is a critical constraint in the architecture and design phase of a processor. Static power is not considered explicitly in this paper (though it is typically proportional to area, which we do consider).

We model the peak activity power and area consumption of each of the key structures in a processor core using a variety of techniques. Table VII lists the methodology and assumptions used for estimating area and power overheads for various structures.

To get total area and power estimates, we assume that the area and power of a core can be approximated as the sum of its major pieces. In reality, we expect that the unaccounted-for overheads will scale our estimates by constant factors. In that case, all our results will still be valid.

Figure 8 shows the area and power of the 480 cores used for this study. As can be seen, the cores represent a significant range in terms of power (4.1–16.3 W) as well as area (3.3–22 mm^2). For this study, we consider 4-core multiprocessors with different area and peak power budgets. There is a significant range in the area and

TABLE VII
AREA AND POWER ESTIMATION METHODOLOGY AND RELEVANT ASSUMPTIONS FOR VARIOUS HARDWARE STRUCTURES

Structure	Methodology	Assumptions
L1 caches	[80]	Parallel data/tag access
TLBs	[80], [42]	
RegFiles	[80], [73]	$2 \times IW$ RP, IW WP
Execution Units	[42]	
RenameTables	[80], [73]	$3 \times IW$ RP, IW WP
ROBs	[80]	IW RP, IW WP, 20b-entry, 6b-tag
IQs(CAM arrays)	[80]	IW RP, IW WP, 40b-entry, 8b-tag
Ld/St Queues	[80]	64b-addressing, 40b-data

Notes. Renaming for OOO cores is assumed to be done using RAM tables. *IW* refers to issue-width, WP to a write-port, and RP to a read-port.

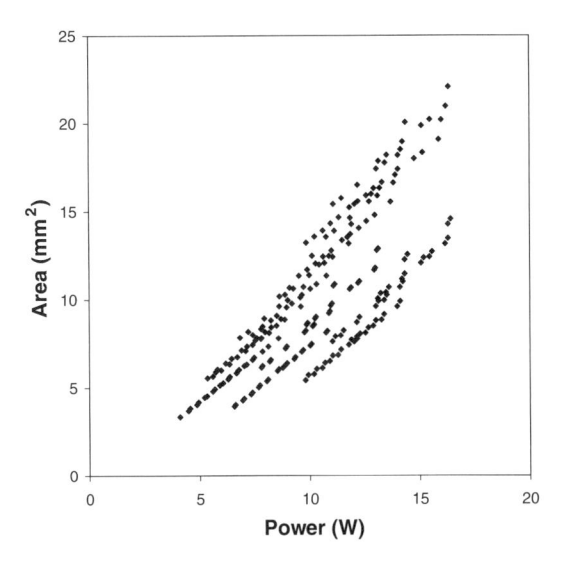

FIG. 8. Area and power of the cores.

power budget of the 4-core multiprocessors that can be constructed out of these cores. Area can range from 13.2 to 88 mm^2. Power can range from 16.4 to 65.2 W.

3.5.3.3 Modeling Performance.
All our evaluations are done for multiprogrammed workloads. Workloads are construction from a set of ten benchmarks (*ammp, crafty, eon, mcf, twolf, mgrid*, and *mesa* from SPEC2000, and *groff, deltablue*, and *adpcmc*). Every multiprocessor is evaluated on two classes of workloads. The *all different* class consists of all possible 4-threaded combinations that can be constructed such that each of the 4 threads running at a time is different. The *all same* consists of all possible 4-threaded combinations that can be constructed such that all the 4 threads running at a time are the same. For example, *a,b,c,d* is an *all different* workload while *a,a,a,a* is an *all same* workload. This effectively brackets the expected diversity in any workload—including server, parallel, and multithreaded workloads. Hence, we expect our results to be generalizable across a wide range of applications.

As discussed before, there are over 2.2 billion distinct 4-core multiprocessors that can be constructed using our 480 distinct cores. We assume that the performance of a multiprocessor is the sum of the performance of each core of the multiprocessor, as described in Section 3.5.1. Note that this is a reasonable assumption because each core is assumed to have a private L2 cache as well as a memory channel.

We find the single thread performance of each application on each core by simulating for 250 million cycles, after fast-forwarding an appropriate number of instructions [77]. This represents 4800 simulations. Simulations use a modified version of SMTSIM [87]. Scripts are used to calculate the performance of the multiprocessors using these single-thread performance numbers.

All results are presented for the best (oracular) static mapping of applications to cores. Note that realistic dynamic mapping can do better, as demonstrated earlier; in fact, dynamic mapping continues being useful for the best heterogeneous designs that our methodology produces—this result, and other more detailed results of this research can be found in [62]. However, evaluating 2.2 billion multiprocessors becomes intractable if dynamic mapping is assumed.

3.5.4 Analysis and Results

This section presents the results of our heterogeneous multi-core architecture design space search. We present these results for a variety of different area and power constraints, allowing us to observe how the benefits of heterogeneity vary across area and power domains. We also examine the effect of different levels of thread level parallelism and the impact of dynamic thread switching mechanisms. Last, we quantify the gains observed due to allowing non-monotonic cores on the processor.

3.5.4.1 Fixed Area Budget.

This section presents results for fixed area budgets. For every fixed area limit, a complete design space exploration is done to find the highest performing 4-core multiprocessor. In fact, for each area budget, we find the best architectures across a range of power constraints—the best architecture overall for a given area limit, regardless of power, will always be the highest line on the graph.

Figure 9 shows the weighted speedup for the highest performing 4-core multiprocessors within an area budget of 40 mm^2. The three lines correspond to different power budgets for the cores. The top line represents the highest performing 4-core multiprocessors with total power due to cores *not exceeding* 50 W. The middle line corresponds to 4-core multiprocessors with total power due to cores not exceeding 40 W. The line at the bottom corresponds to 4-core multiprocessors with total power due to cores not exceeding 30 W.

Figure 10 shows the results for different area budgets.

The performance of these 4-core multiprocessors is shown for different amounts of on-chip diversity. *One core type*, for example, implies that all cores on the die are of the same type, and refers to a homogeneous multiprocessor. Points that represent two core types could either have two of each core type or three of one type and one of another core type. *3 core types* refers to multiprocessors with three types of

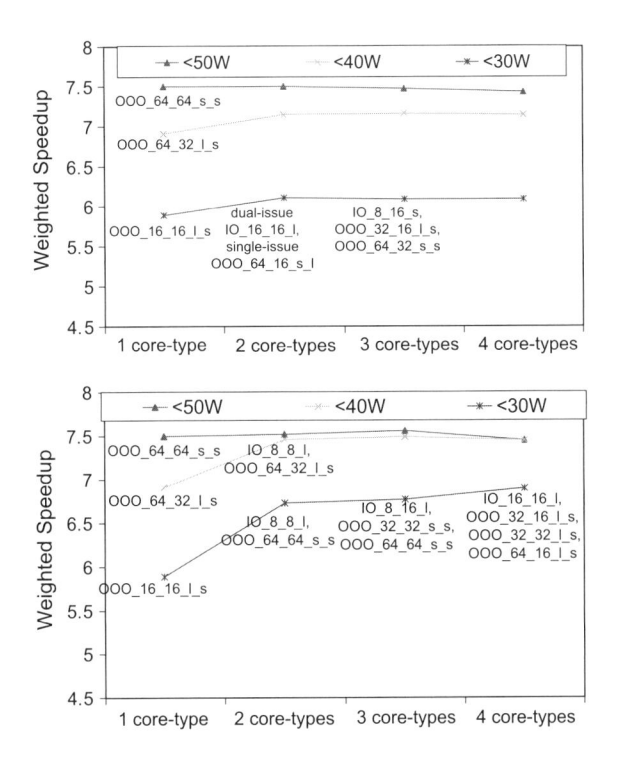

FIG. 9. Throughput for *all-same* (top) and *all-different* (bottom) workloads, area budget = 40 mm².

cores on the die and *4 core types* refers to multiprocessors with all different cores. Note that the *4 core types* result, for example, only considers processors with four unique cores. Thus, if the best heterogeneous configuration has two unique cores, the three-core and four-core results will show as lower.

For each area budget, the results shown assume that all contexts are busy (TLP = 4).

The results lead to several interesting observations. First, we notice that the advantages of diversity are lower with *all same* than the *all different* workload, but they do exist. This is non-intuitive for our artificially-homogeneous workload; however, we find that even these workloads achieve their best performance when at least one of the cores is well suited for the application—a heterogeneous design ensures that whatever application is being used for the homogeneous runs, such a core likely exists. For example, the best homogeneous CMP for *all same* workloads for an area budget of 40 mm² and a power budget of 30 W consists of 4 single-issue OOO cores with 16 KB L1 caches and double the functional units than the simplest core. This

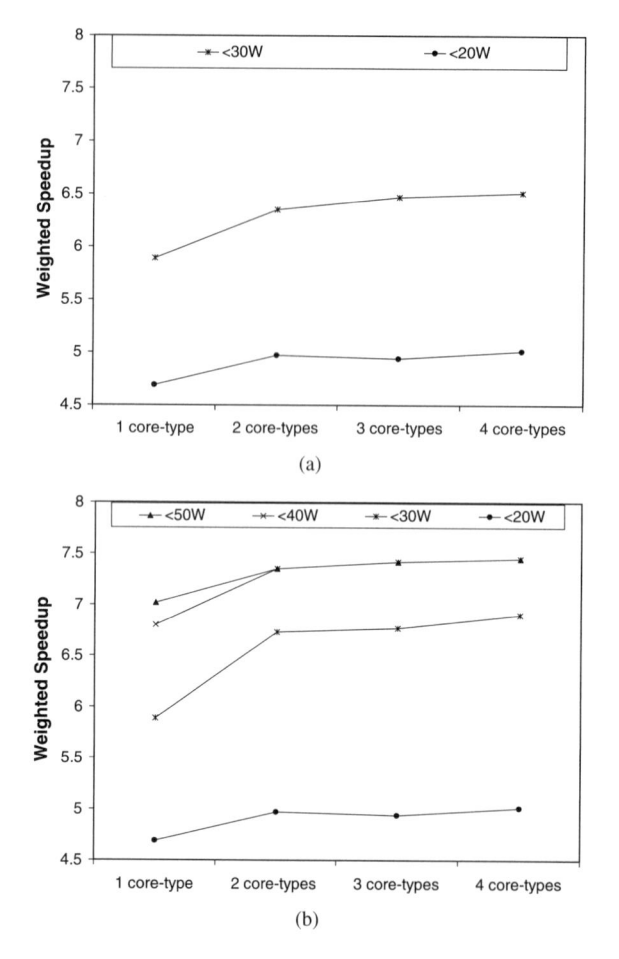

FIG. 10. Throughput for *all-different* workloads for an area budget of (a) 20 mm^2, (b) 30 mm^2, (c) 50 mm^2, and (d) 60 mm^2.

multiprocessor does not perform well when running applications with high cache requirements. On the other hand, the best heterogeneous multiprocessor with 3 core types for *all same* workloads for identical budgets consists of two single-issue in-order cores with 8 KB ICache and 16 KB DCache, one scalar OOO core with 32 KB ICache, 16 KB DCache and double the functional units, and one scalar OOO core with 64 KB ICache and 32 KB DCache. The three core types cover the spectrum of application requirements better and result in outperforming the best homogeneous

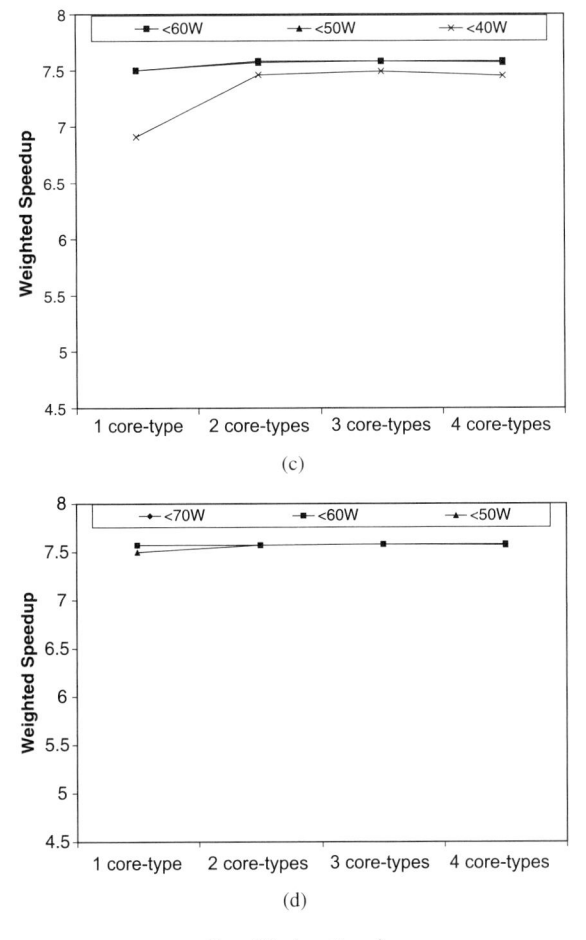

FIG. 10. (*continued*)

CMP by 3.6%. We do not show all of the *all same* results, but for other area budgets the results were similar—if there is benefit to be had from heterogeneity (as shown in the *all-different* results), it typically also exists in the *all same* case, but to a lesser degree.

Second, we observe that the advantages due to heterogeneity for a fixed area budget depend largely on the power budget available—as shown by the shape of the lines corresponding to different power budgets. In this case (Figure 9), heterogeneity buys little additional performance with a generous power budget (50 W), but is increasingly important as the budget becomes more tightly constrained. We see this pattern throughout our results, whenever either power or area is constrained. What we find is that without constraints, the homogeneous architecture can create "envelope" cores—cores that are over-provisioned for any single application, but able to run most applications with high performance. For example, for an area budget of 40 mm^2, if the power budget is set high (50 W), the "best" homogeneous architectures consists of 4 OOO cores with 64 KB ICache, 32 KB DCache and double the number of functional units than the simplest core. This architecture is able to run both the memory-bound as well as processor-bound applications well. When the design is more constrained, we can only meet the needs of each application through heterogeneous designs that are customized to subsets of the applications. It is likely that in the space where homogeneous designs are most effective, a heterogeneous design that contained more cores would be even better; however, we did not explore this axis of the design space.

We see these same trends in Fig. 10, which shows results for four other area budgets. There is significant benefit to a diversity of cores as long as either area or power are reasonably constrained.

The power and area budgets also determine the amount of diversity needed for a multi-core architecture. In general, the more constrained the budget, the more benefits are accrued due to increased diversity. For example, considering the *all different* results in Fig. 9, while having 4 core types results in the best performance when the power limit is 30 W, two core types (or one) are sufficient to get all the potential benefits for higher power limits. In some of the regions where moderate diversity is sufficient, two unique cores not only matches configurations with higher diversity, but even beats it. In cases where higher diversity is optimal, the gains must still be compared against the design and test costs of more unique cores. For example, in the example above, the marginal performance of 4 core types over the best 2-type result is 2.5%, and probably does not justify the extra effort in this particular example.

These results underscore the increasing importance of single-ISA heterogeneous multi-core architectures for current and future processor designs. As designs become more aggressive, we will want to place more cores on the die (placing area pressure on the design), and power budgets per core will likely tighten even more severely. Our results show that while having two core types is sufficient for getting most of the potential out of moderately power-limited designs, increased diversity results in significantly better performance for highly power-limited designs.

Another way to interpret these results is that heterogeneous designs dampen the effects of constrained power budgets significantly. For example, in the 40 mm^2 results, both homogeneous and heterogeneous solutions provide good performance with a 50 W budget. However, the homogeneous design loses 9% performance with a 40 W budget and 23% with a 30 W budget. Conversely, with a heterogeneous design, we can drop to 40 W with only a 2% penalty and to 30 W with a 9% loss.

Perhaps more illuminating than the raw performance of the best designs is what architectures actually provide the best designs for a given area and power budget. We observe that there can be a significant difference between the cores of the best heterogeneous multiprocessor and the cores constituting the best homogeneous CMP. That is, the best heterogeneous multiprocessors cannot be constructed only by making slight modifications to the best homogeneous CMP design. Rather, they need to be designed from the ground up. Consider, for example, the best multiprocessors for an area budget of 40 mm^2 and a power budget of 30 W. The best homogeneous CMP consists of single-issue OOO cores with 16 KB L1 caches, few functional units (1-1-2) and a large number of registers (128). On the other hand, the best heterogeneous CMP with two types of cores, for *all different* workloads, consists of two single-issue in-order cores with 8 KB L1 caches and two single-issue OOO cores with 64 KB ICache, 32 KB DCache and double the number of functional units. Clearly, these cores are significantly different from each other.

Another interesting observation is the reliance on non-monotonicity. Prior work on heterogeneous multi-core architectures, including the work described in previous sections and other research [39,67,41], assumed configurations where every core was either a subset or superset of every other core (in terms of processor parameters). However, in several of our best heterogeneous configurations, we see that no core is a subset of any other core. For example, in the same example as above, the best heterogeneous CMP for two core types for *all same* workloads, consists of superscalar in-order cores (issue-width = 2) and scalar out-of-order cores (issue-width = 1). Even when all the cores are different, the "best" multiprocessor for *all different* workloads consists of one single-issue in-order core with 16 KB L1 caches, one single-issue OOO core with 32 KB ICache and 16 KB DCache, one single-issue in-order core with 32 KB L1 caches and one single-issue OOO core with 64 KB ICache and 16 KB DCache. Thus, the real power of heterogeneity is not in combining "big" and "little" cores, but rather in providing cores each well tuned for a class of applications. This was a common result, and we will explore the importance of non-monotonic designs further in Section 3.5.4.2.

3.5.4.2 Impact of Non-monotonic Design.
As discussed above, we observed non-monotonicity in several of the highest performing multiprocessors for various area and power budgets. In this section, we analyze this phenomenon further and also try to quantify the advantages due to non-monotonic design.

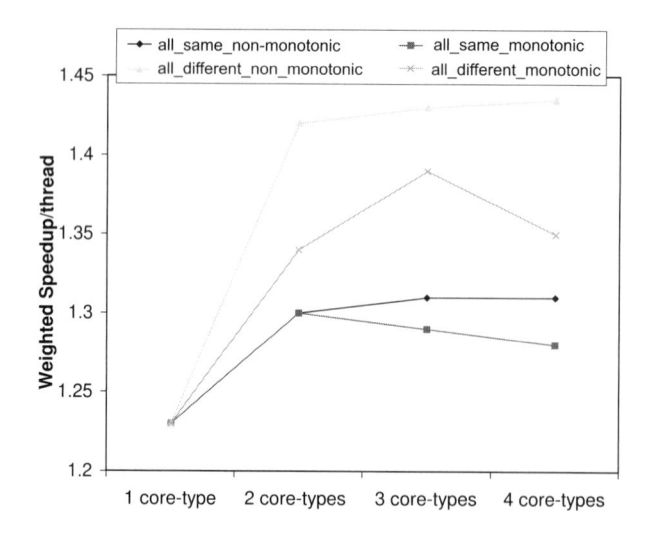

FIG. 11. Benefits due to non-monotonicity of cores; area budget = 40 mm^2, power budget = 30 W.

The reason this feature is particularly interesting is that any design that starts with pre-existing cores from a given architectural family is likely to be monotonic [63, 60]. Additionally, a heterogeneous design that is achieved with multiple copies of a single core, but each with separate frequencies, is also by definition monotonic. Thus, the monotonic results we show in this section serve as a generous upper bound (given the much greater number of configurations we consider) to what can be achieved with an architecture constructed from existing same-ISA cores that are monotonic.

Our results in Fig. 11 show the results for a single set of area and power budgets. In this case, we see that for the *all-same* workload, the benefits from non-monotonic configurations is small, but with the heterogeneous workload, the non-monotonic designs outperform the monotonic much more significantly. More generally (results not shown here), we find that the cost of monotonicity in terms of performance is greater when budgets are constrained. In fact, diversity beyond two core types has benefits only for non-monotonic designs for very constrained power and area budgets.

With custom design of heterogeneous cores, we have the ability to take full advantage of the holistic architecture approach. By creating non-monotonic cores, we are able to specialize each core to a class of applications. In many cases, our best processor has no truly general-purpose core, yet overall general-purpose performance surpasses that available with general-purpose cores.

4. Amortizing Overprovisioning through Conjoined Core Architectures

Most modern processors are highly overprovisioned. Designers usually provision the CPU for a few important applications that stress a particular resource. For example, the vector processing unit (VMX) of a processor can often take up more than 10% of the die-area, gets used by only a few applications, but the functionality still needs to be there.

"Multi-core-oblivious" designs exacerbate the overprovisioning problem because the blind replication of cores results in multiplying the cost of overprovisioning by the number of cores. What is really needed is the same level of overprovisioning for any single thread without multiplying the cost by the number of cores. We can achieve some of this with a heterogeneous architecture, where some cores might be overloaded with a particular resource, and others not. However, in this section, we describe an alternate, more aggressive, solution to the problem.

This section presents a holistic approach to designing chip-multiprocessors where the adjacent cores of a multi-core processor share large, over-provisioned resources [61]. There are several benefits to sharing hardware between more than one processor or thread. Time-sharing a lightly-utilized resource saves area, increases efficiency, and reduces leakage. Dynamically sharing a large resource can also yield better performance than having distributed small private resources, statically partitioned [90,58,26,29,32].

Topology is a significant factor in determining what resources are feasible to share and what the area, complexity, and performance costs of sharing are. Take, for example, the case of sharing entire floating-point units (FPUs). Since processor floorplans often have the FPU on one side and the integer datapath on the other side, we can mirror adjacent processors and FPU sharing would present minimal disruption to the floorplan. For the design of a resource-sharing core, the floorplan must be co-designed with the architecture, otherwise the architecture may specify sharings that are not physically possible or have high communication costs. In general, resources to be shared should be large enough that the additional wiring needed to share them does not outweigh the area benefits obtained by sharing.

With these factors in mind we have investigated the possible sharing of FPUs, crossbar ports, first-level instruction caches, and first-level data caches between adjacent pairs of processors. Resources could potentially be shared among more than two processors, but this creates more topological problems. Because we primarily investigate sharing between pairs of processors, we call our approach *conjoined-core chip multiprocessors*.

4.1 Baseline Architecture

Conjoined-core chip multiprocessing deviates from a conventional chip multiprocessor (multi-core) design by sharing selected hardware structures between adjacent cores to improve processor efficiency. The choice of the structures to be shared depends not only on the area occupied by the structures but also whether it is topologically feasible without significant disruption to the floorplan or wiring overheads. In this section, we discuss the baseline chip multiprocessor architecture and derive a reasonable floorplan for the processor, estimating area for the various on-chip structures.

For our evaluations, we assume a processor similar to Piranha [23], with eight cores sharing a 4 MB, 8-banked, 4-way set-associative, 128 byte line L2 cache. The cores are modeled after Alpha 21164 (EV5). EV5 is a 4-issue in-order processor. The various parameters of the processor are given in Table VIII. The processor was assumed to be implemented in 0.07 micron technology and clocked at 3.0 GHz.

Each core has a private FPU. Floating point divide and square root are non-pipelined. All other floating point operations are fully pipelined. The latency for all operations is modeled after EV5 latencies.

Cores are connected to the L2 cache using a point-to-point fully-connected blocking matrix crossbar such that each core can issue a request to any of the L2 cache banks every cycle. However, one bank can entertain a request from only one of the cores any given cycle. Crossbar link latency is assumed to be 3 cycles, and the data transfer time is 4 cycles. Links are assumed to be implemented in 4X plane [51] and are allowed to run over L2 banks.

Each bank of the L2 cache has a memory controller and an associated RDRAM channel. The memory bus is assumed to be clocked at 750 MHz, with data being

TABLE VIII
SIMULATED BASELINE PROCESSOR FOR STUDYING CONJOINING

2K-gshare branch predictor
Issues 4 integer instrs per cycle, including up to 2 Load/Store
Issues 2 FP instructions per cycle
4 MSHRs
64 B linesize for L1 caches, 128 B linesize for L2 cache
64k 2-way 3 cycle L1 Instruction cache (1 access/cycle)
64k 2-way 3 cycle L1 Data cache (2 access/cycle)
4 MB 4-way set-associative, 8-bank 10 cycle L2 cache (3 cycle/access)
4 cycle L1-L2 data transfer time plus 3 cycle transfer latency
450 cycle memory access time
64 entry DTLB, fully associative, 256 entry L2 DTLB
48 entry ITLB, fully associative
8 KB pages

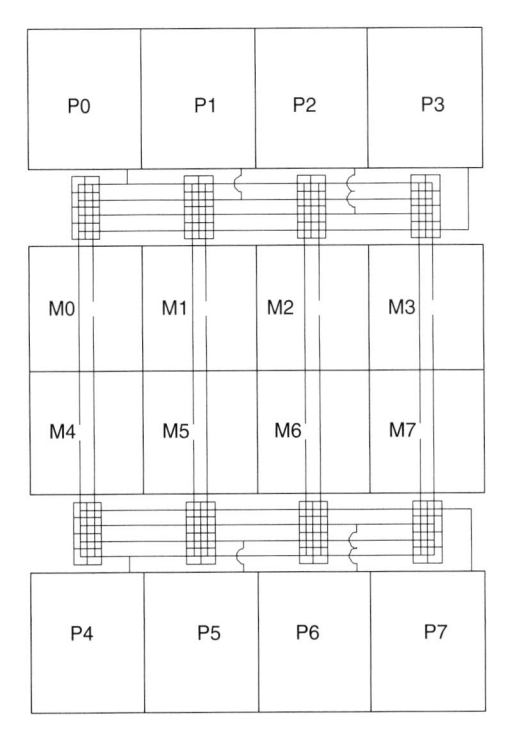

THE ARCHITECTURE OF EFFICIENT MULTI-CORE PROCESSORS 47

FIG. 12. Baseline die floorplan for studying conjoining, with L2 cache banks in the middle of the cluster, and processor cores (including L1 caches) distributed around the outside.

transferred on both edges of the clock for an effective frequency of 1.5 GHz and an effective bandwidth of 3 GB/s per bank (considering that each RDRAM memory channel supports 30 pins and 2 data bytes). Memory latency is set to 150 ns.

Figure 12 shows the die floorplan.

4.2 Conjoined-Core Architectures

For the conjoined-core chip multiprocessor, we consider four optimizations—instruction cache sharing, data cache sharing, FPU sharing, and crossbar sharing. For each kind of sharing, two adjacent cores share the hardware structure. In this section, we investigate the mechanism for each kind of sharing and discuss the area benefits that they accrue. We talk about the performance impact of sharing in Section 4.3. The usage of the shared resource can be based on a policy decided either statically, such that it can be accessed only during fixed cycles by a certain core, or the accesses can be determined based on certain dynamic conditions visible to both cores (given adequate propagation time). The initial mechanisms discussed in this section all assume the simplest and most naive static scheduling, where one of the cores has access to the shared resource during odd cycles while the other core gets access during even cycles. More intelligent sharing techniques/policies are discussed in Section 4.4. All of our sharing policies, however, maintain the assumption that communication distances between cores are too great to allow any kind of dynamic cycle-level arbitration for shared resources.

4.2.1 ICache sharing

We implement instruction cache (ICache) sharing between two cores by providing a shared fetch path from the ICache to both the pipelines. Figure 13 shows a floorplan

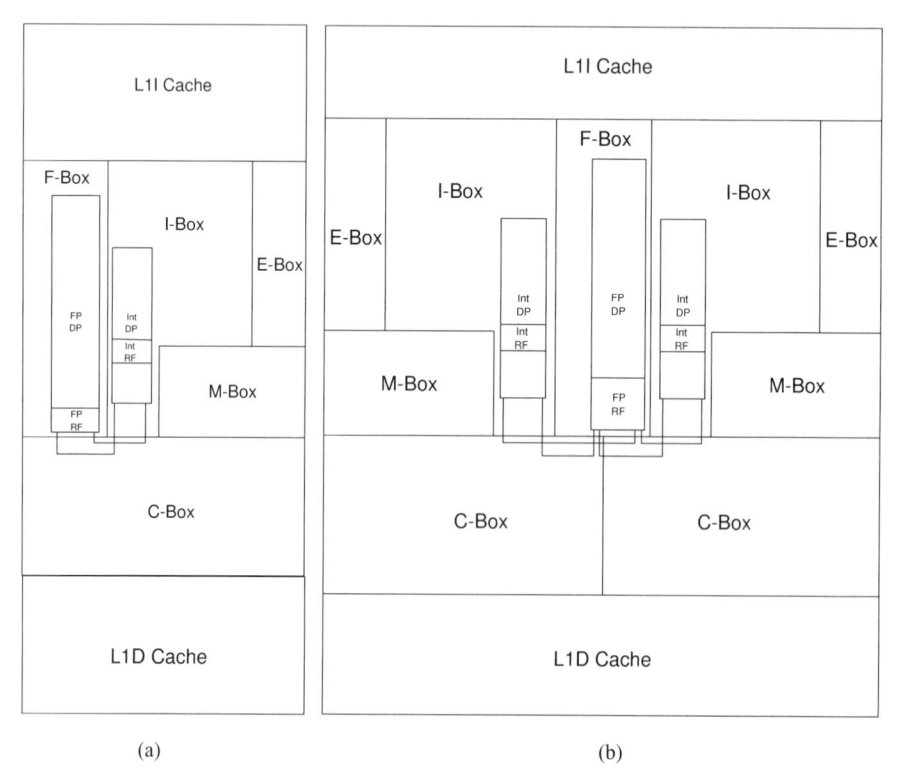

FIG. 13. (a) Floorplan of the original core; (b) layout of a conjoined-core pair, both showing FPU routing. Routing and register files are schematic and not drawn to scale.

of two adjacent cores sharing a 64 KB, 2-way associative ICache. Because the layout of memories is a function of the number of rows and columns, we have increased the number of columns but reduced the number of rows in the shared memory. This gives a wider aspect ratio that can span two cores.

As mentioned, the ICache is time-shared every other cycle. We investigate two ICache fetch widths. In the *double fetch width* case, the fetch width is changed to 8 instructions every other cycle (compared to 4 instructions every cycle in the unshared case). The time-averaged effective fetch bandwidth (ignoring branch effects) remains unchanged in this case. In the *original structure fetch width* case, we leave the fetch width to be the same. In this case the effective per-core fetch bandwidth is halved. Finally, we also investigate a banked architecture, where cores can fetch 4 instructions every cycle, but only if their desired bank is allocated to them that cycle.

In the *double fetch width* case, sharing results in a wider instruction fetch path, wider multiplexors and extra instruction buffers before decode for the instruction front end. We have modeled this area increase and we also assume that sharing may increase the access latency by 1 cycle. The double fetch width solution would also result in higher power consumption per fetch. Furthermore, since longer fetch blocks are more likely to include taken branches out of the block, the fetch efficiency is somewhat reduced. We evaluate two latency scenarios—one with the access time extended by a cycle and another where it remains unchanged.

Based on modeling with CACTI, in the baseline case each ICache takes up 1.15 mm^2. In the double fetch width case, the ICache has double the bandwidth (BITOUT = 256), and requires 1.16 mm^2. However, instead of 8 ICaches on the die, there are just four of them. This results in a core area savings of 9.8%. In the normal fetch width case (BITOUT = 128), sharing results in core area savings of 9.9%.

4.2.2 DCache sharing

Even though the data caches (DCaches) occupy a significant area, DCache sharing is not an obvious candidate for sharing because of its relatively high utilization. In our DCache sharing experiments, two adjacent cores share a 64 KB, 2-way set-associative L1 DCache. Each core can issue memory instructions only every other cycle.

Sharing entails lengthened wires that increase access latency slightly. This latency may or may not be able to be hidden in the pipeline. Thus, we again evaluate two cases—one where the access time is lengthened by one cycle and another where the access time remains unchanged.

Based on modeling with CACTI, each dual-ported DCache takes up 2.59 mm^2 in the baseline processor. In the shared case, it takes up the area of just one cache for every two cores, but with some additional wiring. This results in core area savings of 22.09%.

4.2.3 Crossbar sharing

As shown in Section 4.1, the crossbar occupies a significant fraction (13%) of the die area. The configuration and complexity of the crossbar is strongly tied to the number of cores, so we also study how crossbar sharing can be used to free up die area. We examine the area and performance costs of the crossbar and interconnect in much greater detail in Section 5.

Crossbar sharing involves two adjacent cores sharing an input port to the L2 cache's crossbar interconnect. This halves the number of rows (or columns) in the crossbar matrix resulting in linear area savings. Crossbar sharing means that only one of the two conjoined cores can issue a request to a particular L2 cache bank in a given cycle. Again, we assume a baseline implementation where one of the conjoined cores can issue requests to a bank every odd cycle, while the other conjoined core can issue requests only on even cycles. There would also be some overhead in routing signal and data to the shared input port. Hence, we assume the point-to-point

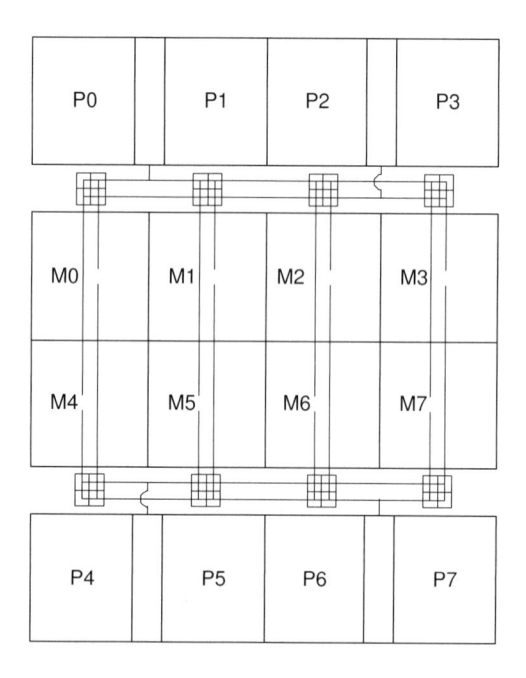

FIG. 14. A die floorplan with crossbar sharing.

communication latency will be lengthened by one cycle for the conjoined core case. Figure 14 shows conjoined core pairs sharing input ports to the crossbar.

Crossbar sharing results in halving the area occupied by the interconnect and results in 6.43% die area savings. This is equivalent to 1.38 times the size of a single core.

Note that this is not the only way to reduce the area occupied by the crossbar interconnect. One can alternatively halve the number of wires for a given point-to-point link to (approximately) halve the area occupied by that link. This would, though, double the transfer latency for each connection. In Section 4.3, we compare both these approaches and show that this performs worse than our port-sharing solution.

Finally, if the DCache and ICache are already shared between two cores, sharing the crossbar port between the same two cores is very straightforward since the cores and their accesses have already been joined together before reaching the crossbar.

4.2.4 FPU Sharing

Processor floorplans often have the FPU on one side and the integer datapath on the other side. So, FPU sharing can be enabled by simply mirroring adjacent processors without significant disruption to the floorplan. Wires connecting the FPU to the left core and the right core can be interdigitated, so no additional horizontal wiring tracks are required (see Fig. 13). This also does not significantly increase the length of wires in comparison the non-conjoined case. In our baseline FPU sharing model, each conjoined core can issue floating-point instructions to the fully-pipelined floating-point sub-units only every other cycle. Based on our design experience, we believe that there would be no operation latency increase when sharing pipelined FPU sub-units between the cores. This is because for arithmetic operations the FP registers remain local to the FPU. For transfers and load/store operations, the routing distances from the integer datapath and caches to the FPU remain largely unchanged (see Fig. 13). For the non-pipelined sub-units (e.g., divides and square root) we assume alternating three cycle scheduling windows for each core. If a non-pipelined unit is available at the start of its three-cycle window, the core may start using it, and has the remainder of the scheduling window to communicate this to the other core. Thus, when the non-pipelined units are idle, each core can only start a non-pipelined operation once every six cycles. However, since operations have a known long latency, there is no additional scheduling overhead needed at the end of non-pipelined operations. Thus, when a non-pipelined unit is in use, another core waiting for it can begin using the non-pipelined unit on the first cycle it becomes available.

The FPU area for EV5 is derived from published die photos, scaling the numbers to 0.07 micron technology and then subtracting the area occupied by the FP register file. The EV5 FPU takes up 1.05 mm^2 including the FP register file. We estimate the area taken up by a 5 read port, 4 write port, 32-entry FP register file using *register-bit equivalents* (rbe). The total area of the FPU (excluding the register file) is 0.72 mm^2. Sharing results in halving the number of units and results in area savings of 6.1%.

We also consider a case where each core has its own copy of the divide sub-unit, while the other FPU sub-units are shared. We estimated the area of the divide sub-unit to be 0.0524 mm^2. Total area savings in that case is 5.7%.

To sum up, ICache sharing results in core area savings of 9.9%, DCache sharing results in core area savings of 22%, FPU sharing saves 6.1% of the core area, and sharing the input ports to the crossbar can result in a savings equivalent to the area of 1.4 cores. Statically deciding to let each conjoined core access a shared hardware structure only every other cycle provides an upper-bound on the possible degradation. As our results in Section 4.3 indicate, even these conservative assumptions lead to relatively small performance degradation and thus reinforce the argument for conjoined-core chip multiprocessing.

4.2.5 Modeling Conjoined Cores

Benchmarks are simulated using SMTSIM [87]. The simulator was modified to simulate the various chip multiprocessor (conjoined as well as conventional) architectures.

Several of our evaluations are done for various numbers of threads ranging from one through a maximum number of available processor contexts. Each result corresponds to one of three sets of eight benchmarks, where each data point is the average of several permutations of those benchmarks.

For these experiments, the following SPEC CPU2000 benchmarks were used: *bzip2, crafty, eon, gzip, mcf, perl, twolf, vpr, applu, apsi, art, equake, facerec, fma3d, mesa, wupwise*. The methodology for workload construction is similar to that used to explore the heterogeneous multi-core design space in the previous sections. More details can be found in [61].

We also perform evaluations using the parallel benchmark *water* from the SPLASH benchmark suite and use the STREAM benchmark for crossbar evaluations. We change the problem size of STREAM to 16,384 elements. At this size, when running eight copies of STREAM, the working set fits into the L2-cache and hence it acts as a worst-case test of L1–L2 bandwidth (and hence crossbar interconnect). We also removed the timing statistics collection routines.

The Simpoint tool [78] was used to find good representative fast-forward distances for each SPEC benchmark. Early simpoints are used. For *water*, fast-forwarding is done just enough so that the parallel threads get forked. We do not fast forward for STREAM.

All simulations involving n threads are preceded by a warmup of $10 \times n$ million cycles. Simulation length was 800 million cycles. All the SPEC benchmarks are simulated using *ref* inputs. All the performance results are in terms of throughput.

4.3 Simple Sharing

This section examines the performance impact of conjoining cores assuming simple time-slicing of the shared resources on alternate cycles. More intelligent sharing techniques are discussed in the next section.

In this section, we show results for various threading levels. We schedule the workloads statically and randomly such that two threads are run together on a conjoined-core pair only if one of them cannot be placed elsewhere. Hence, for the given architecture, for 1 to 4 threads, there is no other thread that is competing for the shared resource. If we have 5 runnable threads, one of the threads needs to be put on a conjoined-core pair that is already running a thread. And so on. However, even if there is no other thread running on the other core belonging to a conjoined-core pair, we still assume, in this section, that accesses can be made to the shared resource by a core only every other cycle.

4.3.1 Sharing the ICache

Results are shown as performance degradation relative to the baseline conventional CMP architecture. Performance degradation experienced with ICache sharing comes from three sources: increased access latency, reduced effective fetch bandwidth, and inter-thread conflicts. Effective fetch bandwidth can be reduced even if the fetch width is doubled because of the decreased likelihood of filling an eight-wide fetch with useful instructions, relative to a four-wide fetch.

Figure 15 shows the performance impact of ICache sharing for varied threading levels for SPEC-based workloads. The results are shown for a fetch width of 8 instructions and assuming that there is an extra cycle latency for ICache access due to sharing. We assume the extra cycle is required since in the worst case the round-trip distance to read an ICache bit has gone up by two times the original core width, due to sharing. We observe a performance degradation of 5% for integer workloads, 1.2% for FP workloads and 2.2% for mixed workloads. The performance degradation does not change significantly when the number of threads is increased from 1 to 8. This indicates that inter-thread conflicts are not a problem for this workload and these caches. The SPEC benchmarks are known to have relatively small instruction working sets.

To identify the main cause for performance degradation on ICache sharing, we also show results assuming that there is no extra cycle increase in the latency. Figure 16 shows the 8-thread results for both integer and floating-point workloads.

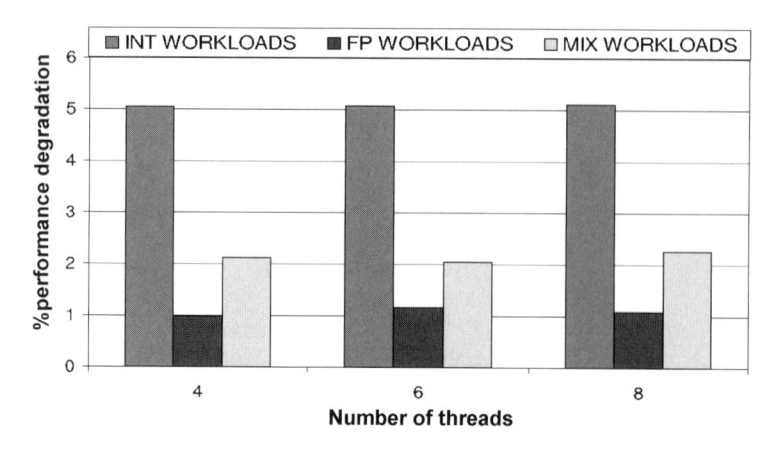

FIG. 15. Impact of ICache sharing for various threading levels.

Performance degradation becomes less than 0.25%. So the extra latency is the main reason for degradation on ICache sharing (note that the latency does not introduce a bubble in the pipeline—the performance degradation comes from the increased branch mispredict penalty due to the pipeline being extended by a cycle). The integer benchmarks are most affected by the extra cycle latency, being more sensitive to the branch mispredict penalty.

Increasing fetch width to 8 instructions ensures that the potential fetch bandwidth remains the same for the sharing case as the baseline case, but it increases the size of the ICache (relative to a single ICache in the base case) and results in increased power consumption per cache. This is because doubling the output width doubles both the number of sense amps and the data output lines being driven, and these structures account for much of the power in the original cache. Thus, we also investigate the case where fetch width is kept the same. Hence, only up to 4 instructions can be fetched every other cycle (effectively halving the per-core fetch bandwidth). Figure 16 shows the results for 8-thread workloads. As can be seen, degradation jumps up to 16% for integer workloads and 10.2% for floating-point workloads. This is because at effective fetch bandwidth of 2 instructions every cycle (per core), the execution starts becoming fetch limited.

We also investigate the impact of partitioning the ICache vertically into two equal sized banks. A core can alternate accesses between the two banks. It can fetch 4 instructions every cycle but only if their desired bank is available. A core has access to bank 0 one cycle, bank 1 the next, etc., with the other core having the opposite allocation. This allows both threads to access the cache in some cycles. It is also possible for both threads to be blocked in some cycles. However, bandwidth is guar-

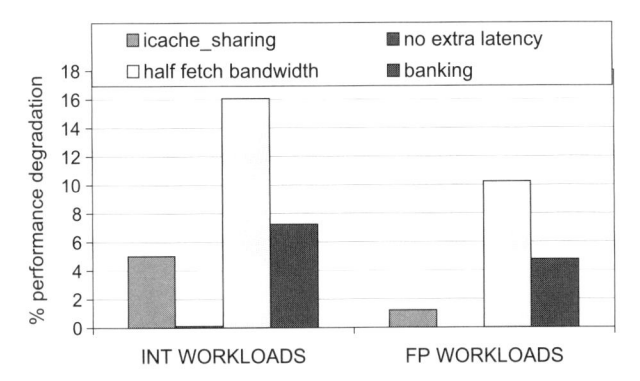

FIG. 16. ICache sharing when no extra latency overhead is assumed, cache structure bandwidth is not doubled, and cache is doubly banked.

anteed to exceed the previous case (ignoring cache miss effects) of one 4-instruction fetch every other cycle, because every cycle that both threads fail to get access will be immediately followed by a cycle in which they both can access the cache. Figure 16 shows the results. The degradation of conjoined sharing is reduced by 55% for integer workloads and 53% for FP workloads (relative the original fetch width solution), due to an overall improvement in fetch bandwidth.

4.3.2 DCache sharing

Similar to the ICache, performance degradation due to DCache sharing comes from: increased access latency, reduced cache bandwidth, and inter-thread conflicts. Unlike the ICache, the DCache latency has a direct effect on performance, as the latency of the load is effectively increased if it cannot issue on the first cycle it is ready.

Figure 17 shows the impact on performance due to DCache sharing for SPEC workloads. The results are shown for various threading levels. We observe a performance degradation of 4–10% for integer workloads, 1–9% for floating point workloads and 2–13% for mixed workloads. Degradation is higher for integer workloads than floating point workloads for small numbers of threads. This is because the typically higher instruction level parallelism of the FP workloads allows them to hide a small increase in latency more effectively. Also, inter-thread conflicts are higher, resulting in increased performance degradation for higher numbers of threads.

We also studied the case where the shared DCache has the same access latency as the unshared DCache. Figure 18 shows the results for the 8-thread case. Degradation lessens for both integer workloads as well as floating-point workloads, but less so

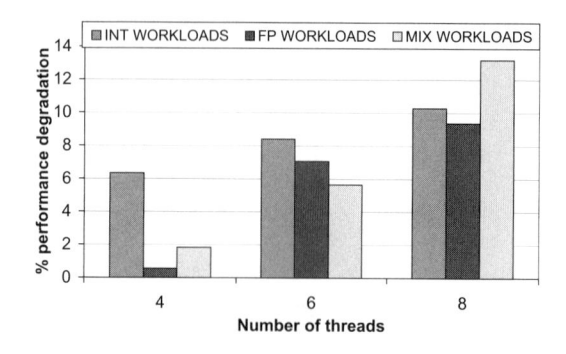

FIG. 17. Impact of DCache sharing for various threading levels.

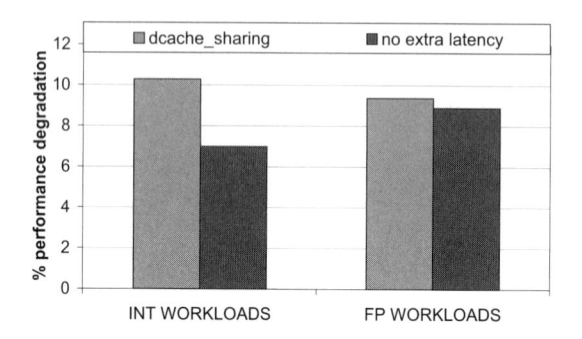

FIG. 18. DCache sharing when no extra latency overhead is assumed, with eight threads.

in the case of FP workloads as inter-thread conflict misses and cache bandwidth pressure remain.

4.3.3 FPU Sharing

Floating point units (FPUs) may be the most obvious candidates for sharing. For SPEC CINT2000 benchmarks only 0.1% of instructions are floating point, while even for CFP2000 benchmarks, only 32.3% of instructions are floating-point instructions [19]. Also, FPU bandwidth is a performance bottleneck only for specialized applications.

We evaluate FPU sharing for integer workloads, FP workloads, and mixed workloads, but only present the FP and mixed results (Fig. 19) here. The degradation is less than 0.5% for all levels of threading, even in these cases.

One reason for these results is that the competition for the non-pipelined units (divide and square root) is negligible in the SPEC benchmarks. To illustrate code

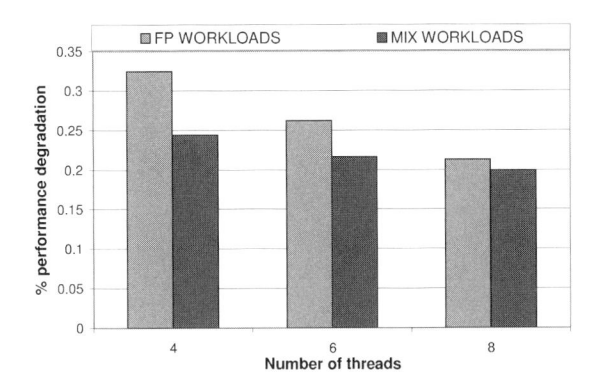

FIG. 19. Impact of FPU sharing for various threading levels.

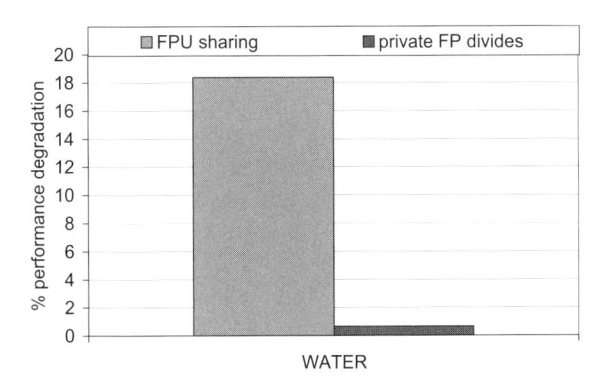

FIG. 20. Impact of private FP divide sub-units.

where non-pipelined units are more heavily used, Fig. 20 shows the performance of the SPLASH benchmark *water* (which has a non-trivial number of divides) running eight threads. It shows performance with a shared FP divide unit vs. unshared FP divide units. In this case, unless each core has its own copy of the FP divide unit, performance degradation can be significant.

4.3.4 *Crossbar Sharing*

We implement the L1–L2 interconnect as a blocking fully-connected matrix crossbar, based on the initial Piranha design. As the volume of traffic between L1 and L2 increases, the utilization of the crossbar goes up. Since there is a single path from a core to a bank, high utilization can result in contention and queuing delays.

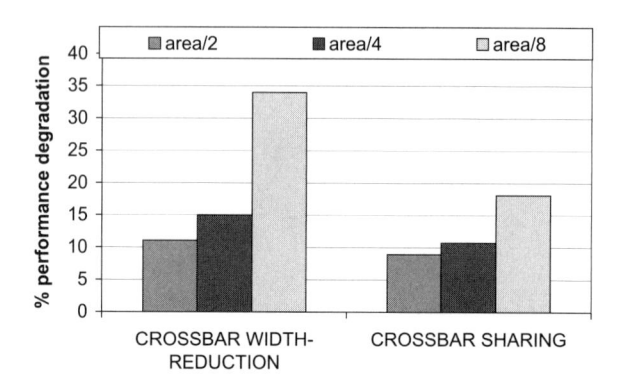

FIG. 21. Reducing crossbar area through width reduction and port sharing.

As discussed in Section 4.2, the area of the crossbar can be reduced by decreasing the width of the crossbar links or by sharing the ports of the crossbar, thereby reducing the number of links. We examine both techniques. Crossbar sharing involves the conjoined cores sharing an input port of the crossbar. Figure 21 shows the results for eight copies of the STREAM benchmark. It must be noted that this is a component benchmark we have tuned for worst-case utilization of the crossbar. The results are shown in terms of performance degradation caused for achieving certain area savings. For example, for achieving crossbar area savings of 75% (*area/4*), we assume that the latency of every crossbar link has been doubled for the *crossbar sharing* case while the latency has been quadrupled for the *crossbar width reduction* case.

We observe that crossbar sharing outperforms crossbar width reduction in all cases. Even though sharing results in increased contention at the input ports, it is the latency of the links that is primarily responsible for queuing of requests and hence overall performance degradation.

We also conducted crossbar exploration experiments using SPEC benchmarks. However, most of the benchmarks do not exercise L1–L2 bandwidth much, resulting in relatively low crossbar utilization rates. The performance degradation in the worst case was less than 5% for an area reduction factor of 2.

4.4 Intelligent Sharing of Resources

The previous section assumed a very basic sharing policy and thus gave an upper bound on the degradation for each kind of sharing. In this section, we discuss more advanced techniques for minimizing performance degradation.

4.4.1 ICache Sharing

This section focuses on that configuration that minimized area, but maximized slowdown—the four-wide fetch shared ICache, assuming an extra cycle of latency. In that case, both access latency and fetch bandwidth contribute to the overall degradation. Most of these results would also apply to the other configurations of shared ICache, taking them even closer to zero degradation.

Section 4.3 evaluated a sharing model where the shared resource gets accessed evenly irrespective of the access needs of the individual cores. Instead, the control of a shared resource can be decided assertively based on the resource needs.

We explore *assertive ICache access* where, whenever there is an L1 miss, the other core can take control of the cache after miss detection. We assume that a miss can be detected and communicated to the other core in 3 cycles. Access would become shared again when the data returns. This does not incur any additional latency since the arrival cycle of the data is known well in advance of its return.

Figure 22 shows the results for *assertive ICache access*. We show results for eight threads, where contention is highest. We observe a 13.7% reduction in the degradation of integer workloads and an improvement of 22.5% for floating point workloads. Performance improvement is because of improved effective fetch bandwidth. These results are for eight threads, so there is no contribution from threads that are not sharing an ICache. A minor tweak to *assertive access* (for ICache as well as DCache and FPU) can ensure that the shared resource becomes a private resource when the other core of the conjoined pair is idle.

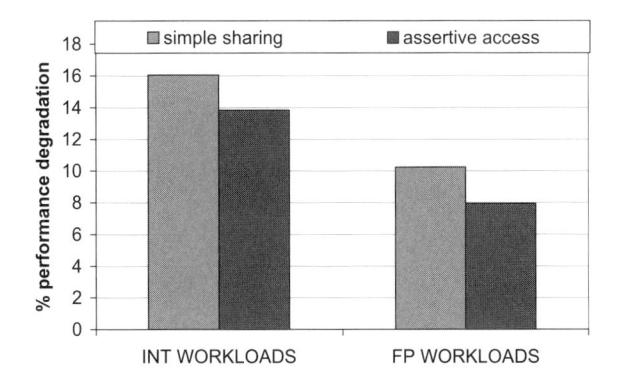

FIG. 22. ICache assertive access results when the original structure bandwidth is not doubled.

4.4.2 DCache Sharing

Performance loss due to DCache sharing is due to three factors—inter-thread conflict misses, reduced bandwidth, and potentially increased latency. We present two techniques for minimizing degradation due to DCache sharing.

Assertive access can also be used for the shared DCaches. Whenever there is an L1 miss on some data requested by a core, if the load is determined to be on the right path, the core relinquishes control over the shared DCache. There may be some delay between detection of an L1 miss and the determination that the load is on the right path. Once the core relinquishes control, the other core takes over full control and can then access the DCache whenever it wants. The timings are the same as with the ICache assertive access. This policy is still somewhat naive, assuming that the processor will stall for this load (recall, these are in-order cores) before another load is ready to issue—more sophisticated policies are possible.

Figure 23 shows the results. Assertive access leads to 29.6% reductions in the degradation for integer workloads and 23.7% improvements for floating point workloads. Improvements are due to improved data bandwidth.

The next technique that we study we call *static port assignment*. The DCache interface consists of two R/W ports. In the basic DCache sharing case, the DCache (and hence both the ports) can be accessed only every other cycle. Instead, one port can be statically assigned to each of the cores and that will make the DCache accessible every cycle.

Figure 23 shows the results comparing the baseline sharing policy against static port-to-core assignment. We observed a 33.6% reduction in degradation for integer workloads while the difference for FP workloads was only 3%. This outperforms the cycle-slicing mechanism for integer benchmarks, for the following reason: when

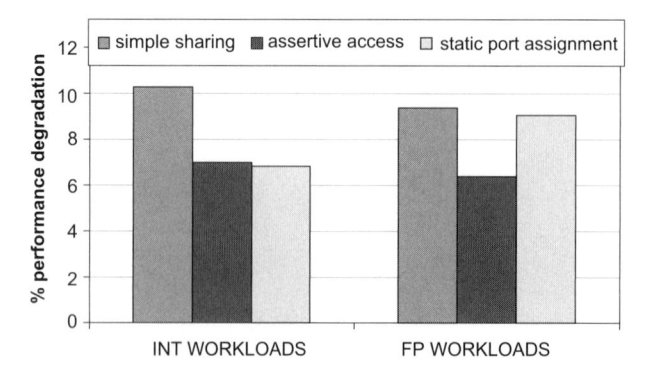

FIG. 23. Effect of assertive access and static assignment on the data cache.

load port utilization is not high, the likelihood (with port partitioning) of a port being available when a load becomes ready is high. However, with cycle-by-cycle slicing, the likelihood of a port being available that cycle is only 50%.

4.5 A Unified Conjoined-Core Architecture

We have studied various combinations of FPU, crossbar, ICache, and DCache sharing. We assumed a shared doubly-banked ICache with a fetch width of 16 bytes (similar to that used in Section 4.3.1), statically port-assigned shared DCache (similar to that used in Section 4.4.2), a fully-shared FPU and a shared crossbar input port for every conjoined-core pair. Access to the ICache, DCache, as well as the crossbar is assumed to incur a one cycle overhead, relative to the non-conjoined configuration. We assume that each shared structure can be *assertively accessed*. Assertive access for the statically port-assigned dual-ported DCache involves accessing the other port (the one not assigned to the core) assertively. Table IX shows the resulting area savings and performance for various sharing combinations. We map the applications to the cores such that "friendly" threads run on the conjoined cores where possible. All performance numbers are for the worst case when all cores are busy with threads.

The combination with all four types of sharing results in 38.1% core-area savings (excluding crossbar savings). In absolute terms, this is equivalent to the area occupied by 3.76 cores. If crossbar savings are included, then the total area saved is equivalent to *5.14 times the area of a core*. We observed an 11.9% degradation for integer workloads and 8.5% degradation for floating-point workloads. Note that the total performance degradation is significantly less than the sum of the individual degradation values that we observed for each kind of sharing. This is because a stall due to one bottleneck often either tolerates or obviates a stall due to some other bottleneck.

Thus, by applying conjoining to all of these structures, we can *more than double* the number of cores on the die, at a cost of less than 12% performance per core.

TABLE IX
RESULTS WITH MULTIPLE SHARINGS

Units shared	Perf. degradation		Core area savings (%)
	Int Aps (%)	FP Aps (%)	
Crossbar+FPU	0.97	1.2	23.1
Crossbar+FPU+ICache	4.7	3.9	33.0
Crossbar+FPU+DCache	6.1	6.8	45.2
ICache+DCache	11.4	7.6	32.0
Crossbar+FPU+ICache+DCache	11.9	8.5	55.1

5. Holistic Design of the Multi-Core Interconnect

This section examines the area, power, performance, and design issues for the on-chip interconnects on a chip multiprocessor, attempting to present a comprehensive view of a class of interconnect architectures. It shows that the design choices for the interconnect have significant effect on the rest of the chip, potentially consuming a significant fraction of the real estate and power budget. This research shows that designs that treat interconnect as an entity that can be independently architected and optimized ("multi-core oblivious") would not arrive at the best multi-core design. Several examples are presented showing the need for a holistic approach to design (e.g., careful co-design). For instance, increasing interconnect bandwidth requires area that then constrains the number of cores or cache sizes, and does not necessarily increase performance. Also, shared level-2 caches become significantly less attractive when the overhead of the resulting crossbar is accounted for. A hierarchical bus structure is examined which negates some of the performance costs of the assumed baseline architecture.

5.1 Interconnection Mechanisms

In this section, we detail three interconnection mechanisms that may serve particular roles in on-chip interconnect hierarchy—a shared bus fabric (SBF) that provides a shared connection to various modules that can source and sink coherence traffic, a point-to-point link (P2P link) that connects two SBFs in a system with multiple SBFs, and a crossbar interconnection system. In the subsequent sections, we will demonstrate the need for co-design using these mechanisms as our baseline.

Many different modules may be connected to these fabrics, which use them in different ways. But from the perspective of the core, an L2 miss goes out over the SBF to be serviced by higher levels of the memory hierarchy, another L2 on the same SBF, or possibly an L2 on another SBF connected to this one by a P2P link. If the core shares L2 cache with another core, there is a crossbar between the cores/L1 caches and the shared L2 banks. Our initial discussion of the SBF in this section assumes private L2 caches.

The results in this section are derived from a detailed model of a complex system, which are described in the next few sections. The casual reader may want to skim Sections 5.1 through 5.6 and get to the results in Section 5.7 more quickly.

5.2 Shared Bus Fabric

A Shared Bus Fabric is a high speed link needed to communicate data between processors, caches, IO, and memory within a CMP system in a coherent fashion. It is the on-chip equivalent of the system bus for snoop-based shared memory multiprocessors [37,94,68]. We model a MESI-like snoopy write-invalidate protocol with write-back L2s for this study [21,49]. Therefore, the SBF needs to support several coherence transactions (request, snoop, response, data transfer, invalidates, etc.) as well as arbitrate access to the corresponding buses. Due to large transfer distances on the chip and high wire delays, all buses must be pipelined, and therefore unidirectional. Thus, these buses appear in pairs; typically, a request traverses from the requester to the end of one bus, where it is queued up to be re-routed (possibly after some computation) across a broadcast bus that every node will eventually see, regardless of their position on the bus and distance from the origin. In the following discussion a bidirectional bus is really a combination of two unidirectional pipelined buses.

We are assuming, for this discussion, all cores have private L1 and L2 caches, and that the shared bus fabric connects the L2 caches (along with other units on the chip and off-chip links) to satisfy memory requests and maintain coherence. Below we describe a typical transaction on the fabric.

5.2.1 Typical Transaction on the SBF

A load that misses in the L2 cache will enter the shared bus fabric to be serviced. First, the requester (in this case, one of the cores) will signal the *central address arbiter* that it has a request. Upon being granted access, it sends the request over an *address bus* (AB in Fig. 24). Requests are taken off the end of the address bus and

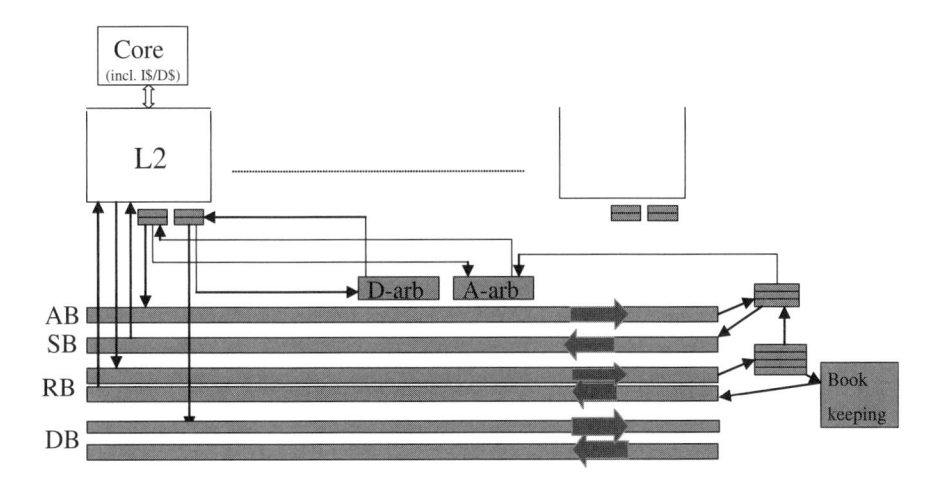

FIG. 24. The assumed shared bus fabric for our interconnection study.

placed in a snoop queue, awaiting access to the *snoop bus* (SB). Transactions placed on the snoop bus cause each snooping node to place a response on the *response bus* (RB). Logic and queues at the end of the response bus collect these responses and generate a broadcast message that goes back over the response bus identifying the action each involved party should take (e.g., source the data, change coherence state). Finally, the data is sent over a bidirectional *data bus* (DB) to the original requester. If there are multiple SBFs (e.g., connected by a P2P link), the address request will be broadcast to the other SBFs via that link, and a combined response from the remote SBF returned to the local one, to be merged with the local responses.

Note that the above transactions are quite standard for any shared memory multiprocessor implementing a snoopy write-invalidate coherence protocol [21].

5.2.2 Elements of the SBF

The composition of the SBF allows it to support all the coherence transactions mentioned above. We now discuss the primary buses, queues and logic that would typically be required for supporting these transactions. Figure 24 illustrates a typical SBF. Details of the modeled design are based heavily on the shared bus fabric in the Power5 multi-core architecture [50].

Each requester on the SBF interfaces with it via *request* and *data queues*. It takes at least one cycle to communicate information about the occupancy of the request queue to the requester. The request queue must then have at least two entries to maintain the throughput of one request every cycle. Similarly, all the units that can source data need to have data queues of at least two entries. Requesters connected to the SBF include cores, L2 and L3 caches, IO devices, memory controllers, and non-cacheable instruction units.

All requesters interface to the fabric through an arbiter for the address bus. The minimum latency through the arbiter depends on (1) the physical distance from the central arbiter to the most distant unit, and (2) the levels of arbitration. Caches are typically given higher priority than other units, so arbitration can take multiple levels based on priority. Distance is determined by the actual floorplan. Since the address bus is pipelined, the arbiter must account for the location of a requester on the bus in determining what cycle access is granted. Overhead of the arbiter includes control signals to/from the requesters, arbitration logic and some latches.

After receiving a grant from the central arbiter, the requester unit puts the address on the *address bus*. Each address request goes over the address bus and is then copied into multiple queues, corresponding to outgoing P2P links (discussed later) and to off-chip links. There is also a local *snoop queue* that queues up the requests and participates in the arbitration for the local *snoop bus*. Every queue in the fabric incurs at least one bus cycle of delay. The minimum size of each queue in the interconnect (there are typically queues associated with each bus) depends on the delay required for the arbiter to stall further address requests if the corresponding bus gets stalled. Thus it depends on the distance and communication protocol to the device or queue responsible for generating requests that are sinked in the queue, and the latency of requests already in transit on the bus. We therefore compute queue size based on floorplan and distance.

The snoop bus can be shared, for example by off-chip links and other SBFs, so it also must be accessed via an arbiter, with associated delay and area overhead. Since the snoop queue is at one end of the address bus, the snoop bus must run in the opposite direction of the address bus, as shown in Fig. 24. Each module connected to the snoop bus snoops the requests. Snooping involves comparing the request address with the address range allocated to that module (e.g., memory controllers) or checking the directory (tag array) for caches.

A response is generated after a predefined number of cycles by each snooper, and goes out over the *response bus*. The delay can be significant, because it can involve tag-array lookups by the caches, and we must account for possible conflicts with other accesses to the tag arrays. Logic at one end of the bidirectional response bus collects all responses and broadcasts a message to all nodes, directing their response to the access. This may involve sourcing the data, invalidating, changing coherence state, etc. Some responders can initiate a data transfer on a read request simultaneously with generating the snoop response, when the requested data is in appropriate coherence state. The responses are collected in queues. All units that can source data to the fabric need to be equipped with a data queue. A central arbiter interfacing with the data queues is needed to grant one of the sources access to the bus at a time.

Bidirectional data buses source data. They support two different data streams, one in either direction. Data bandwidth requirements are typically high.

It should be noted that designs are possible with fewer buses, and the various types of transactions multiplexed onto the same bus. However, that would require higher bandwidth (e.g., wider) buses to support the same level of traffic at the same performance, so the overheads are unlikely to change significantly. We assume for the purpose of this study that only the above queues, logic, and buses form a part of the SBF and contribute to the interconnection latency, power, and area overheads.

5.3 P2P Links

If there are multiple SBFs in the system, the connection between the SBFs is accomplished using P2P links. Multiple SBFs might be required to increase bandwidth, decrease signal latencies, or to ease floorplanning (all connections to a single SBF must be on a line). For example, if a processor has 16 cores as shown in Fig. 26, it becomes impossible to maintain die aspect ratio close to 1 unless there are two SBFs each supporting 8 cores.

Each P2P link should be capable of transferring all kinds of transactions (request/response/data) in both directions. Each P2P link is terminated with multiple queues at each end. There needs to be a queue and an arbiter for each kind of transaction described above.

5.4 Crossbar Interconnection System

The previous section assumed private L2 caches, with communication and coherence only occurring on L2 misses. However, if our architecture allows two or more cores to share L2 cache banks, a high bandwidth connection is required between the cores and the cache banks. This is typically accomplished by using a crossbar. It allows multiple core ports to launch operations to the L2 subsystem in the same cycle. Likewise, multiple L2 banks are able to return data or send invalidates to the various core ports in the same cycle.

The crossbar interconnection system consists of crossbar links and crossbar interface logic. A crossbar consists of address lines going from each core to all the banks (required for loads, stores, prefetches, TLB misses), data lines going from each core to the banks (required for writebacks) and data lines going from every bank to the cores (required for data reload as well as invalidate addresses). A typical implementation, shown in Fig. 25, consists of one address bus per core from which all the banks feed. Each bank has one outgoing data bus from which all the cores feed. Sim-

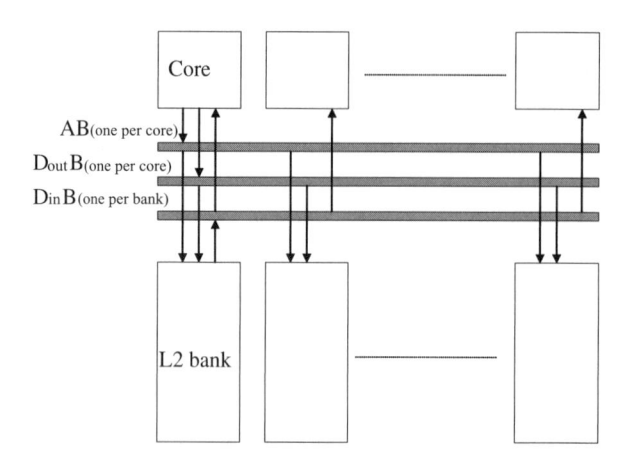

FIG. 25. A typical crossbar.

ilarly, corresponding to each write port of a core is an outgoing data bus that feeds all the banks.

Crossbar interface logic presents a simplified interface to the instruction fetch unit and the Load Store Unit in the cores. It typically consists of a load queue corresponding to each core sharing the L2. The load queue sends a request to the L2 bank appropriate to the request, where it is enqueued in a bank load queue (BLQ) (one per core for each bank to avoid conflict between cores accessing the same bank). The BLQs must arbitrate for the L2 tags and arrays, both among the BLQs, as well as with the snoop queue, the writeback queue, and the data reload queue—all of which may be trying to access the L2 at the same time. After L2 access (on a load request), the data goes through the reload queue, one per bank, and over the data bus back to the core. The above description of the crossbar interface logic is based on the crossbar implementation (also called core interface unit) in Power4 [49] and Power5 [50].

Note that even when the caches (or cache banks) are shared, an SBF is required to maintain coherence between various units in the CMP system.

5.5 Modeling Interconnect Area, Power, and Latency

Both wires and logic contribute to interconnect overhead. This section describes our methodology for computing various overheads for 65 nm technology. The scaling of these overheads with technology as well as other design parameters is discussed in more detail in [64].

5.5.1 Wiring Area Overhead

We address the area overheads of wires and logic separately.

The latency, area, and power overhead of a metal wire depends on the metal layer used for this wire. The technology that we consider facilitates 10 layers of metal, 4 layers in 1X plane and 2 layers in the higher planes (2X, 4X and 8X) [51]. The 1X metal layers are typically used for macro-level wiring [51]. Wiring tracks in higher layers of metal are very scarce and only used for time-critical signals running over a considerable distance (several millimeters of wire).

We evaluate crossbar implementations for 1X, 2X and 4X metal planes where both data and address lines use the same metal plane. For our SBF evaluations, the address bus, snoop bus, and control signals always use the 8X plane. Response buses preferably use the 8X plane, but can use the 4X plane. Data buses can be placed in the 4X plane (as they have more relaxed latency considerations). All buses for P2P links are routed in the 8X plane.

Methodological details for computing area overhead for a given number of wires can be found in [64]. Overheads depend on the pitch of wires, their metal layer,

TABLE X
DESIGN PARAMETERS FOR WIRES IN DIFFERENT METAL PLANES

Metal plane	Pitch (µm)	Signal wiring pitch (µm)	Repeater spacing (mm)	Repeater width (µm)	Latch spacing (mm)	Latch height (µm)	Channel leakage per repeater (µA)	Gate leakage per repeater (µA)
1X	0.2	0.5	0.4	0.4	1.5	120	10	2
2X	0.4	1.0	0.8	0.8	3.0	60	20	4
4X	0.8	2.0	1.6	1.6	5.0	30	40	8
8X	1.6	4.0	3.2	3.2	8.0	15	80	10

and the dimensions and spacing of the corresponding repeaters and latches. Table X shows the signal wiring pitch for wires in different metal planes for 65 nm. These pitch values are estimated by conforming to the considerations mentioned in [85]. The table also shows the minimum spacing for repeaters and latches as well as their heights for computing the corresponding area overheads. We model the height of the repeater macro to be 15 µm. The height of the latch macro given in the table includes the overhead of the local clock buffer and local clock wiring, but excludes the overhead of rebuffering the latch output which is counted separately. The values in Table X are for a bus frequency of 2.5 GHz and a bus voltage of 1.1 V.

5.5.2 Logic Area Overhead

Area overhead due to interconnection-related logic comes primarily from queues. Queues are assumed to be implemented using latches. We estimate the area of a 1-bit latch used for implementing the queues to be 115 µm^2 for 65 nm technology [93]. This size includes the local clock driver and the area overhead of local clock distribution. We also estimated that there is 30% overhead in area due to logic needed to maintain the queues (such as head and tail pointers, queue bypass, overflow signaling, request/grant logic, etc.) [28].

The interconnect architecture can typically be designed such that buses run over interconnection-related logic. The area taken up due to wiring is usually big enough that it (almost) subsumes the area taken up by the logic.

Because queues overwhelmingly dominate the logic area, we ignore the area (but not latency) of multiplexors and arbiters. It should be noted that the assumed overheads can be reduced by implementing queues using custom arrays instead of latches.

5.5.3 Power

Power overhead comes from wires, repeaters, and latches. For calculating dynamic dissipation in the wires, we optimistically estimate the capacitance per unit length of wire (for all planes) to be 0.2 pF/mm [48]. Repeater capacitance is assumed to be

30% of the wire capacitance [15]. The dynamic power per latch is estimated to be 0.05 mW per latch for 2.5 GHz at 65 nm [93]. This includes the power of the local clock buffer and the local clock distribution, but does not include rebuffering that typically follows latches.

Repeater leakage is computed using the parameters given in Table X. For latches, we estimate channel leakage to be 20 µA per bit in all planes (again not counting the repeaters following a latch). Gate leakage for a latch is estimated to be 2 µA per bit in all planes [15]. For computing dynamic and leakage power in the queues, we use the same assumptions as for the wiring latches.

More details of the power model and how it was derived can be found in [64].

5.5.4 Latency

The latency of a signal traveling through the interconnect is primarily due to wire latencies, wait time in the queues for access to a bus, arbitration latencies, and latching that is required between stages of interconnection logic. Latency of wires is determined by the spacing of latches. Spacing between latches for wires is given in Table X.

Arbitration can take place in multiple stages (where each stage involves arbitration among the same priority units) and latching needs to be done between every two stages. For 65 nm technology, we estimate that no more than four units can be arbitrated in a cycle. The latency of arbitration also comes from the travel of control between a central arbiter and the interfaces corresponding to request/data queues. Other than arbiters, every time a transaction has to be queued, there is at least a bus cycle of delay—additional delays depend on the utilization of the outbound bus.

5.6 Modeling the Cores

For this study, we consider a stripped version of out-of-order Power4-like cores [49]. We determine the area taken up by such a core at 65 nm to be 10 mm^2. The area and power determination methodology is similar to the one presented in [60]. The power taken up by the core is determined to be 10 W, including leakage.

For calculating on-chip memory sizes, we use the Power5 cache density, as measured from die photos [50], scaled to 65 nm. We determine it to be 1 bit per square micron, or 0.125 MB/mm^2. For the purpose of this study, we consider L2 caches as the only type of on-chip memory (besides the L1 caches associated with each core). We do not assume off-chip L3 cache, but in 65 nm systems, it is likely that L3 chips would be present as well (the number of L3 chips would be limited, however, due to the large number of pins that every L3 chip would require), but we account for that effect using somewhat optimistic estimates for effective bandwidth and memory latency. Off-chip bandwidth was modeled carefully based on pincount [15] and the number of memory channels (Rambus RDRAM interface was assumed).

We simulate a MESI-like [74,49] coherence protocol, and all transactions required by that protocol are faithfully modeled in our simulations. We also model weak consistency [33] for the multiprocessor, so there is no impact on CPI due to the latency of stores and writebacks.

Because our focus is on accurate modeling of the interconnect, including types of traffic not typically generated by processor simulations, we use a very different performance simulator than used in the previous sections, making heavy use of captured commercial traces. We use a combination of detailed functional simulation and queuing simulation tools [65]. The functional simulator is used for modeling the memory subsystem as well as the interconnection between modules. It takes instruction traces from a SMP system as input and generates coherence statistics for the modeled memory/interconnect sub-system. The queuing simulator takes as input the modeled subsystem, its latencies, coherence statistics, and the inherent CPI of the modeled core assuming perfect L2. It then generates the CPI of the entire system, accounting for real L2 miss rates and real interconnection latencies. Traffic due to syncs, speculation, and MPL (message passing library) effects is accounted for as well. The tools and our interconnection models have been validated against a real, implemented design.

The cache access times are calculated using assumptions similar to those made in CACTI [80]. Memory latency is set to 500 cycles. The average CPI of the modeled core over all the workloads that we use, assuming perfect L2, is measured to be 2.65. Core frequency is assumed to be 5 GHz for the 65 nm studies. Buses as well as the L2 are assumed to be clocked at half the CPU speed.

More details on the performance model can be found in [64].

5.6.1 Workload

All our performance evaluations have been done using commercial workloads, including TPC-C, TPC-W, TPC-H, Notesbench, and others further described in [65]. We use PowerPC instruction and data reference traces of the workloads running under AIX. The traces are taken in a non-intrusive manner by attaching a hardware monitor to a processor [34,65]. This enables the traces to be gathered while the system is fully loaded with the normal number of users, and captures the full effects of multitasking, data sharing, interrupts, etc. These traces even contain DMA instructions and non-cacheable accesses.

5.7 Shared Bus Fabric: Overheads and Design Issues

This section examines the various overheads of the shared bus fabric, and the implications this has for the entire multi-core architecture. We examine floorplans for several design points, and characterize the impact of the area, power, and latency overheads on the overall design and performance of the processor. This section demonstrates that the overheads of the SBF can be quite significant. It also illustrates the tension between the desire to have more cores, more cache, and more interconnect bandwidth, and how that plays out in total performance.

In this section, we assume private L2 caches and that all the L2s (along with NCUs, memory controllers, and IO Devices) are connected using a shared bus fabric. We consider architectures with 4, 8, and 16 cores. Total die area is assumed to be constant at 400 mm^2 due to yield considerations. Hence, the amount of L2 per core decreases with increasing number of cores. For 4, 8 and 16 cores, we evaluate multiple floorplans and choose those that maximized cache size per core while

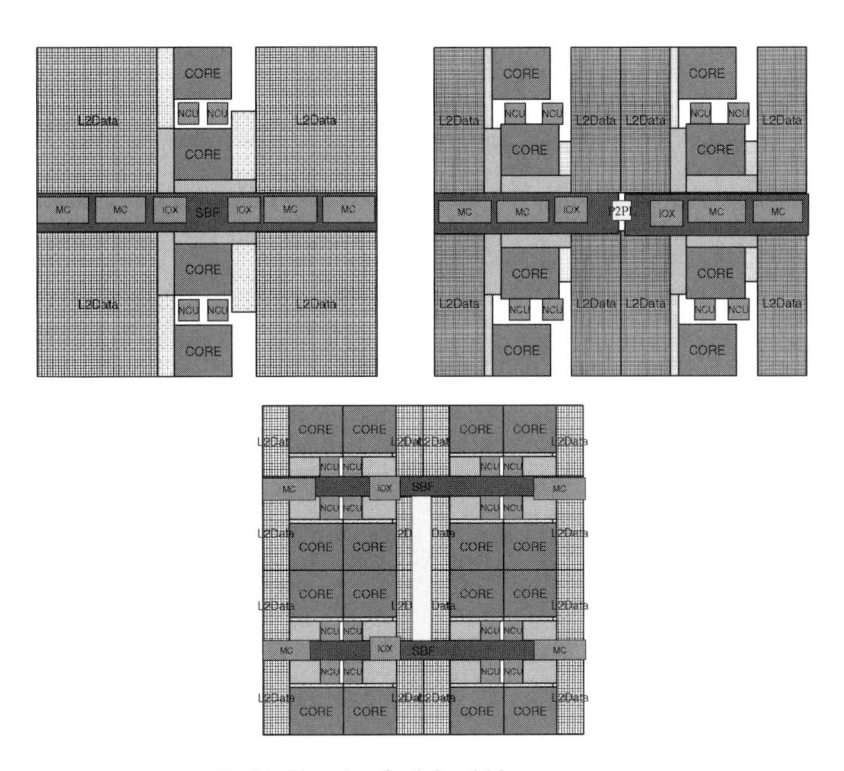

FIG. 26. Floorplans for 4, 8 and 16 core processors.

maintaining a die aspect ratio close to 1. In the default case, we consider the width of the address, snoop, response and data buses of the SBF to be 7, 12, 8, 38 (in each direction) bytes respectively—these widths are determined such that no more than 0.15 requests are queued up, on average, for the 8 core case. We also evaluate the effect of varying bandwidths. We can lay out 4 or 8 cores with a single SBF, but for 16 cores, we need two SBFs connected by a P2P link. In that case, we model two half-width SBFs and a 76 byte wide P2P link. Figure 26 shows the floorplans arrived at for the three cases. The amount of L2 cache per core is 8 MB, 3 MB and 0.5 MB for 4, 8 and 16 core processors, respectively. It must be mentioned that the 16-core configuration is somewhat unrealistic for this technology as it would result in inordinately high power consumption. However, we present the results here for completeness reasons.

Wires are slow and hence cannot be clocked at very high speeds without inserting an inordinately large number of latches. For our evaluations, the SBF buses are cycled at half the core frequency.

5.7.1 Area

The area consumed by the shared bus fabric comes from wiring and interconnection-related logic, as described in Section 5.5. Wiring overhead depends on the architected buses and the control wires that are required for flow control and arbitration. Control wires are needed for each multiplexor connected to the buses and signals to and from every arbiter. Flow control wires are needed from each queue to the units that control traffic to the queue.

Figure 27 shows the wiring area overhead for various processors. The graph shows the area overhead due to architected wires, control wires, and the total. We see that area overhead due to interconnections in a CMP environment can be significant. For the assumed die area of 400 mm^2, area overhead for the interconnect with 16 cores is 13%. Area overhead for 8 cores and 4 cores is 8.7% and 7.2% of the die area, respectively. Considering that each core is 10 mm^2, the area taken up by the SBF is sufficient to place 3–5 extra cores, or 4–6 MB of extra cache.

The graph also shows that area overhead increases quickly with the number of cores. This result assumes constant width architected buses, even when the number of cores is increased. If the effective bandwidth per core is kept constant, overhead would increase even faster.

The overhead due to control wires is high. Control takes up at least 37% of SBF area for 4 cores and at least 62.8% of the SBF area for 16 cores. This is because the number of control wires grows linearly with the number of connected units, in addition to the linear growth in the average length of the wires. Reducing SBF bandwidth does not reduce the control area overhead, thus it constrains how much area can be

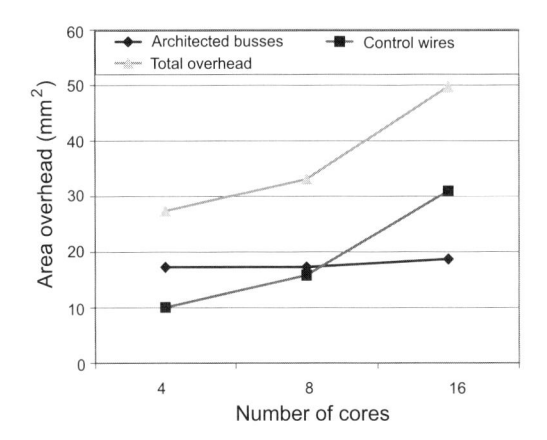

FIG. 27. Area overhead for shared bus fabric.

regained with narrower buses. Note that this argues against very lightweight (small, low performance) cores on this type of chip multiprocessor, because the lightweight core does not amortize the incremental cost to the interconnect of adding each core. Interconnect area due to logic is primarily due to the various queues, as described in Section 5.5. This overhead can often be comparable to that of SBF wires. Note, however, that the logic can typically be placed underneath the SBF wires. Thus, under these assumptions the SBF area is dominated by wires, but only by a small amount.

5.7.2 Power

The power dissipated by the interconnect is the sum of the power dissipated by wires and the logic. Figure 28 shows a breakdown of the total power dissipation by the interconnect.

The graph shows that total power due to the interconnect can be significant. The interconnect power overhead for the 16-core processor is more than the combined power of two cores. It is equal to the power dissipation of one full core even for the 8 core processor. Power increases superlinearly with the number of connected units. This is because of the (at least linear) increase in the number of control wires as well as the (at least linear) increase in the number of queuing latches. There is also a considerable increase in the bus traffic with the growing number of cores. Half the power due to wiring is leakage (mostly from repeaters).

Contrary to popular belief, interconnect power is not always dominated by the wires. The power due to logic can be, as in this case, more than the power due to wiring.

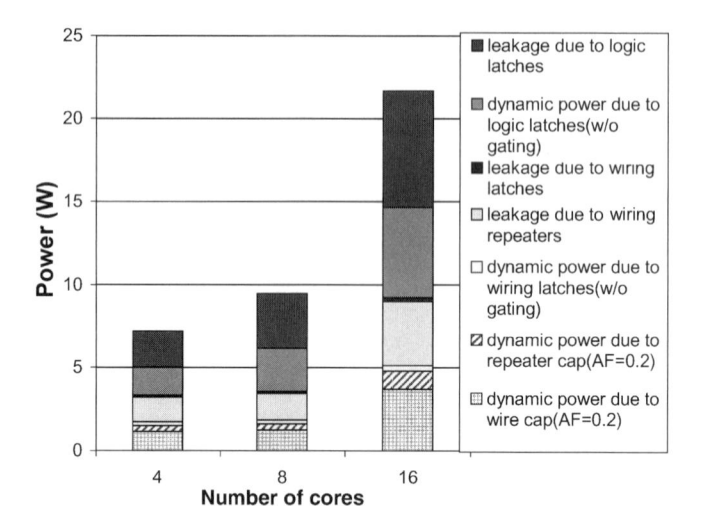

FIG. 28. Power overhead for shared bus fabric.

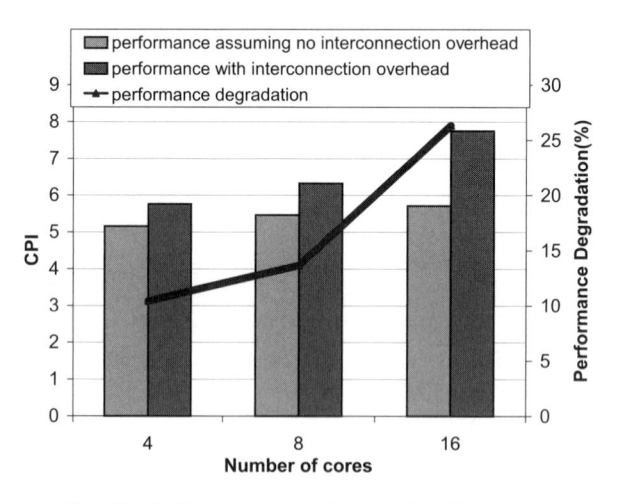

FIG. 29. Performance overhead due to shared bus fabric.

5.7.3 Performance

Figure 29 shows the per-core performance for 4, 8, and 16 core architectures both assuming no interconnection overhead (zero latency interconnection) and with interconnection overheads modeled carefully. Single-thread performance (even assuming

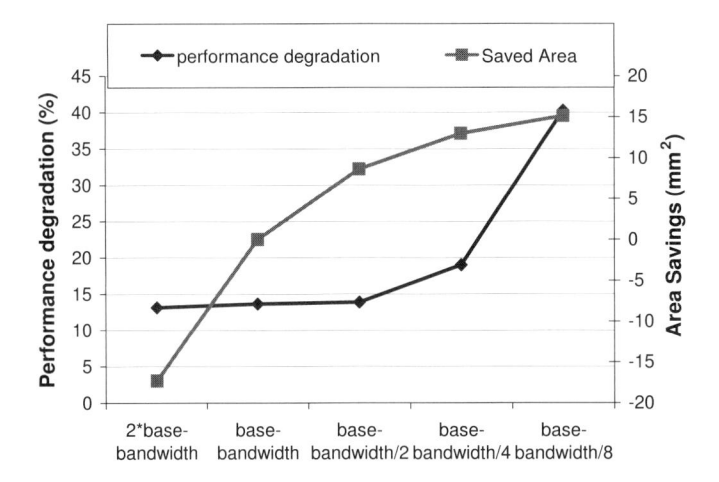

FIG. 30. Trading off interconnection bandwidth with area.

no interconnection overhead) goes down as the number of cores increases due to the reduced cache size per core. If interconnect overhead is considered, then the performance decreases much faster. In fact, performance overhead due to interconnection is more than 10% for 4 cores, more than 13% for 8 cores and more than 26% for 16 cores.

In results to this point, we keep bus bandwidth constant. In Fig. 30, we show the single-thread performance of a core in the 8 core processor case, when the width of the architected buses is varied by factors of two. The graph also shows the real estate saved compared to the baseline. In this figure, performance is relative to an ideal bus, so even the baseline sees a "degradation." We see that with wide buses, the area costs are significant, and the incremental performance is minimal—these interconnects are dominated by latency, not bandwidth. On the other hand, with narrow buses, the area saved by small changes in bandwidth is small, but the performance impact is significant.

There is definitely potential, then, to save real estate with a less aggressive interconnect. We could put that saved area to use. We ran simulations that assume that we put that area back into the caches. We find that over certain ranges, if the bandwidth is reduced by small factors, the performance degradation can be recovered using bigger caches. For example, decreasing the bandwidth by a factor of 2 decreases the performance by 0.57%. But it saves 8.64 mm^2. This can be used to increase the per-core cache size by 135 KB. When we ran simulations using new cache sizes, we observed a performance improvement of 0.675%. Thus, *we can decrease bus bandwidth and improve performance* (if only by small margins in this example), because the resulting bigger caches protect the interconnect from a commensurate increase in utilization. On the other hand, when bandwidth is decreased by a factor of 8, performance decreases by 31%, while the area it saves is 15.12 mm^2. The area savings is sufficient to increase per core cache size by only 240 KB. The increase in cache size was not sufficient to offset the performance loss in this case. Similarly, when doubling interconnect bandwidth over our baseline configuration, total performance decreased by 1.2% due to the reduced cache sizes.

This demonstrates the importance of co-designing the interconnect and memory hierarchy. It is neither true that the biggest caches nor the widest interconnect give the best performance; designing each of these subsystems independently is unlikely to result in the best design. This is another example of holistic design—not designing each component for maximum performance, but designing the components together for maximum whole processor performance.

5.8 Shared Caches and the Crossbar

The previous section presented evaluations with private L1 and L2 caches for each core, but many proposed chip multiprocessors have featured shared L2 caches, connected with crossbars. Shared caches allow the cache space to be partitioned dynamically rather than statically, typically improving overall hit rates. Also, shared data does not have to be duplicated. To fully understand the tradeoffs between private and shared L2 caches, however, we find that it is absolutely critical that we account for the impact of the interconnect.

5.8.1 Area and Power Overhead

The crossbar, shown in Fig. 25, connects cores (with L1 caches) to the shared L2 banks. The data buses are 32 bytes while the address bus is 5 bytes. Lower bandwidth solutions were found to adversely affect performance and render sharing highly unfruitful. In this section we focus on an 8-core processor with 8 cache banks, giving us the options of 2-way, 4-way, and full (8-way) sharing of cache banks. Crossbar wires can be implemented in the 1X, 2X or 4X plane. For a latency reduction of nearly a factor of two, the wire thickness doubles every time we go to a higher metal plane.

Figure 31 shows the area overhead for implementing different mechanisms of cache sharing. The area overhead is shown for two cases—one where the crossbar runs between cores and L2 and the other where the crossbar can be routed over L2 (there is one line for the latter case, because the overhead is independent of the degree of sharing above two). When the crossbar is placed between the L2 and the cores, interfacing is easy, but all wiring tracks result in area overhead. When the crossbar is

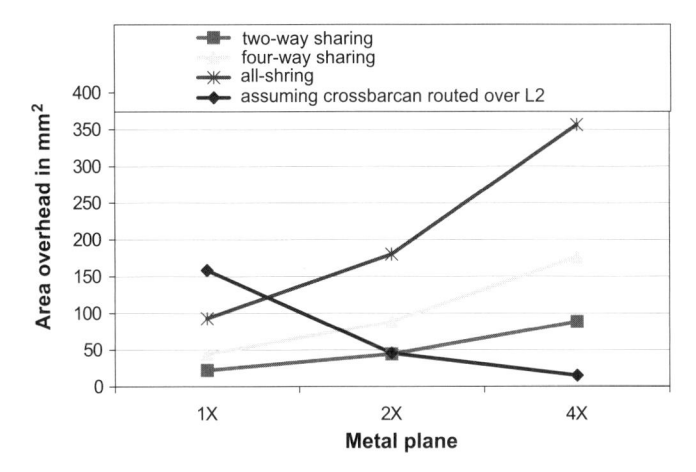

THE ARCHITECTURE OF EFFICIENT MULTI-CORE PROCESSORS 77

FIG. 31. Area overhead for cache sharing—results for crossbar routed over L2 assume uniform cache density.

routed over L2, area overhead is only due to reduced cache density to accommodate repeaters and latches. However, the implementation is relatively complex as vertical wires are needed to interface the core with the L2. We show the results assuming that the L2 density is kept uniform (i.e. even if repeaters/latches are dropped only over the top region of the cache, sub-arrays are displaced even in the other regions to maintain uniform density).

Cache sharing carries a heavy area overhead. If the total die area is around 400 mm^2, then the area overhead for an acceptable latency (2X) is 11.4% for 2-way sharing, 22.8% for four-way sharing and 46.8% for full sharing (nearly half the chip!). Overhead increases as we go to higher metal layers due to increasing signal pitch values. When we assume that the crossbar can be routed over L2, area overhead is still substantial; however, in that case it improves as we move up in metal layers. At low levels the number of repeater/latches, which must displace cache, is highest.

The point of sharing caches is that it gives the illusion of more cache space, as threads sharing a larger cache tend to share it more effectively than statically partitioning the space among the individual threads. However, in this case, the cores gain significant *real* cache space by foregoing sharing, raising doubts about whether sharing has any benefit.

The high area overhead again suggests that issues of interconnect/cache/core co-design must be considered. For crossbars sitting between cores and L2, just two-way sharing results in an area overhead equivalent to more than the area of two cores. Four-way sharing results in an area overhead of 4 cores. An 8-way sharing results in an area overhead of 9 cores. If the same area were devoted to caches, one could instead put 2.75, 5.5 and 11.6 MB of extra caches, respectively.

Similarly, the power overhead due to crossbars is very significant. The overhead can be more than the power taken up by three full cores for a completely shared cache and more than the power of one full core for 4-way sharing. Even for 2-way sharing, power overhead is more than half the power dissipation of a single core. Hence, even if power is the primary constraint, the benefits of the shared caches must be weighed against the possibility of more cores or significantly more cache.

5.8.2 Performance

Because of the high area overhead for cache sharing, the total amount of on-chip caches decreases with sharing. We performed our evaluations for the most tightly packed floorplans that we could find for 8-core processors with different levels of sharing. When the crossbar wires are assumed to be routed in the 2X plane between cores and L2, total cache size is 20, 14 and 4 MB respectively for 2-way, 4-way and full sharing. When the crossbar is assumed to be routed over L2 (and assuming uniform cache density), the total cache size was 22 MB for 4X and 18.2 MB for 2X. Figure 32 presents the results for a fixed die area and cache sizes varied accordingly (i.e. taking into account crossbar area overhead).

Figure 32, assumes a constant die area and considers interconnection area overhead. It shows that performance, even without considering the interconnection la-

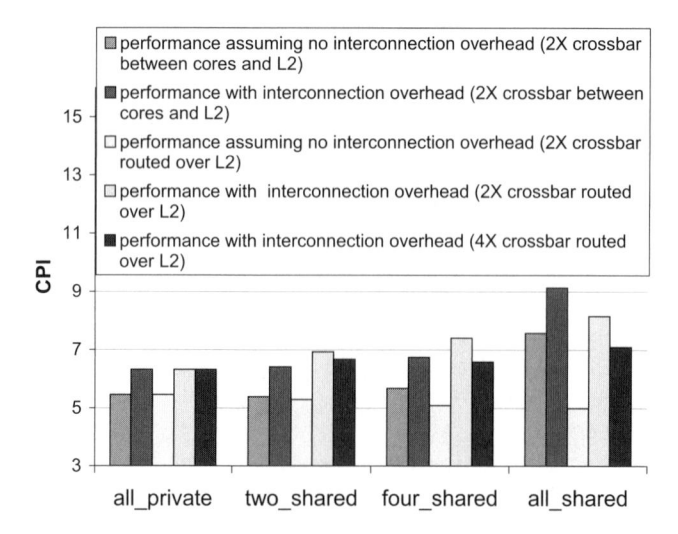

FIG. 32. Evaluating cache sharing for a fixed die area—area overhead taken into account.

tency overhead (and hence purely the effect of cache sharing), either does not improve or improves only by a slight margin. This is due to the reduced size of on-chip caches to accommodate the crossbar. If interconnect latencies are accounted for (higher sharing means longer crossbar latencies), sharing degrades performance even between two cores. Note that in this case, the conclusion reached ignoring interconnect area effects is opposite that reached when those effects are considered.

Note that performance loss due to increased L2 hit latency can possibly be mitigated by using L2 latency hiding techniques, like overlapping of L2 accesses or prefetching. However, our results definitely show that having shared caches becomes *significantly less desirable* than previously accepted if interconnection overheads are considered. We believe that the conclusion holds, in general, for uniform access time caches and calls for evaluation of caching strategies with careful consideration of interconnect overheads. Further analysis needs to be done for intelligent NUCA (non-uniform cache access) caches [53].

These results again demonstrate that a design that does not adequately account for interconnection costs in the architecture phase will not arrive at the best architecture.

5.9 An Example Holistic Approach to Interconnection

The intent of this section is to apply one lesson learned from the high volume of data gathered in this research. Our interconnect architectures to this point were highly driven by layout. The SBF spans the width of the chip, allowing us to connect as many units as possible in a straight line across the chip. However, the latency overheads of a long SBF encourage us to consider alternatives. This section describes a more hierarchical approach to interconnects, which can exploit shorter

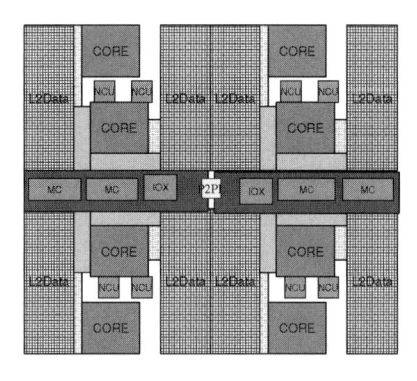

FIG. 33. Hierarchical approach (splitting SBFs).

buses with shorter latencies when traffic remains local. We will be considering the 8-core processor again.

The effectiveness of such an approach will depend on the probability that an L2 miss is serviced on a local cache (an L2 connected to the same SBF), rather than a cache on a remote SBF. We will refer to this probability as *"thread bias."* A workload with high thread bias means that we can identify and map "clusters" of threads that principally communicate with each other on the same SBF.

In this section, we split the single SBF that spanned the chip vertically into two SBFs, with a P2P link between them (see Fig. 33). Local accesses benefit from decreased distances. Remote accesses suffer because they travel the same distances as before, but see additional queuing and arbitration overheads between interconnects.

Figure 34 shows the performance of the split SBF for various thread bias levels. The SBF is split vertically into two, such that each SBF piece now supports 4 cores, 4 NCUs, 2 memory controllers and 1 IO Device. The X-axis shows the thread bias in terms of the fraction of misses satisfied by an L2 connected to the same SBF. A 25% thread bias means that one out of four L2 misses are satisfied by an L2 connected to the same SBF piece. These results are obtained through statistical simulation by synthetically skewing the traffic pattern.

The figure also shows the system performance for a single monolithic SBF (the one used in previous sections). As can be seen, if thread bias is more than 17%, the

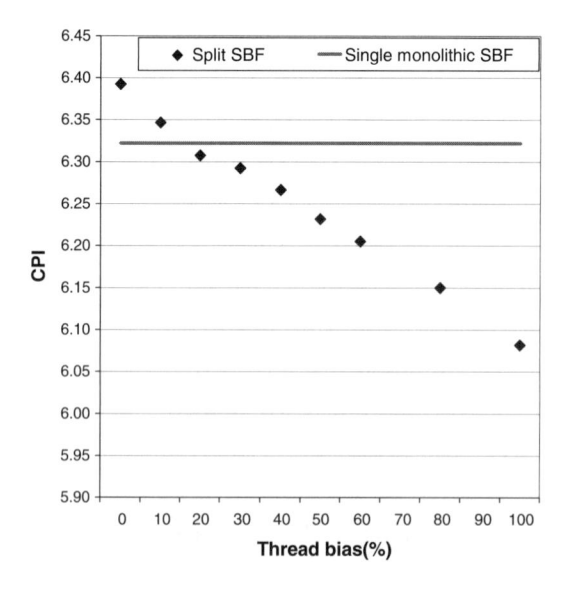

FIG. 34. Split vs Monolithic SBF.

performance of split SBF can overtake performance of a monolithic SBF. Note that 17% is lower than the statistical probability of locally satisfying an L2 miss assuming uniform distribution (3/7). Hence, the split SBF, in this case, is always a good idea.

6. Summary and Conclusions

The decreasing marginal utility of transistors and the increasing complexity of processor design has led to the advent of chip multiprocessors. Like other technology shifts, the move to multi-core architectures creates an opportunity to change the way we do, or even think about, architecture. This chapter makes the case that significant gains in performance and efficiency can be achieved by designing the processor holistically—designing the components of the processor (cores, caches, interconnect) to be part of an effective system, rather than designing each component to be effective on its own. A multi-core oblivious design methodology is one where the processor subsystems are designed and optimized without any cognizance of the overall chip multiprocessing systems that they would become parts of. This results in processors that are inefficient in terms of area and power. The paper shows that this inefficiency is due to the inability of such processors to react to workload diversity, processor overprovisioning, and the high cost of connecting cores and caches. This paper recommends a holistic approach to multi-core design where the processor subsystems are designed from the ground up to be parts of a chip multiprocessing system.

Specifically, we discuss *single-ISA heterogeneous multi-core architectures* for adapting to workload diversity. These architectures consists of multiple types of processing cores on the same die. These cores can all execute the same ISA, but represent different points in the power-performance continuum. Applications are mapped to cores in a way that the resource demands of an application match the resources provided by the corresponding core. This results in increased computational efficiency and hence higher throughput for a given area and/or power budget. A throughput improvement of up to 63% and energy savings of up to four-fold are shown. We also look at the design of cores for a heterogeneous CMP. We show that the best way to design a heterogeneous CMP is not to find individual cores that are well suited for the entire universe of applications, but rather to tune the cores to different classes of applications. We find that customizing cores to subsets of the workload results in processors that have greater performance and power benefits than heterogeneous designs with ordered set of cores. An example such design outperformed the best homogeneous CMP design by 15.4% and the best fully-customized monotonic design by 7.5%.

We also present *conjoined-core chip multiprocessing architectures*. These architectures consist of multiple cores on the die where the adjacent cores share the large, overprovisioned structures. Sharing results in reduced area requirement for such processors at a minimal cost in performance. Reduced area, in turn, results in higher yield, lower leakage, and potentially higher overall throughput per unit area. Up to 56% area savings with 10-12% loss in performance are shown.

This chapter also presents conventional interconnection mechanisms for multi-core architectures and demonstrates that the interconnection overheads are significant enough that they affect the number, size, and design of cores and caches. It shows the need to co-design the cores, caches, and the interconnects, and also presents an example holistic approach to interconnection design for multi-core architectures.

ACKNOWLEDGEMENTS

Most of the research described in this chapter was carried out by the authors with the help of various collaborators, without whom this work would not have been possible. Those collaborators include Norm Jouppi, Partha Ranganathan, Keith Farkas, and Victor Zyuban. Research described in this chapter was funded in part by HP Labs, IBM, Intel, NSF, and the Semiconductor Research Corporation.

REFERENCES

[1] http://www.amd.com/us-en/Processors/ProductInformation/0,,30_118_13909,00.html.
[2] http://www.amd.com/us-en/processors/productinformation/0,,30_118_8825,00.html.
[3] http://www.amd.com/us-en/Processors/ProductInformation/0,,30_118_9485_9484,00.html.
[4] http://www.arm.com/products/cpus/arm11mpcoremultiprocessor.html.
[5] http://www.broadcom.com/products/enterprise-small-office/communications-processors.
[6] http://www.cavium.com/octeon_mips64.html.
[7] http://www.geek.com/procspec/hp/pa8800.htm.
[8] http://www.intel.com/pressroom/kits/quickreffam.htm.
[9] http://www.intel.com/products/processor/coreduo/.
[10] http://www.intel.com/products/processor/pentium_d/index.htm.
[11] http://www.intel.com/products/processor/pentiumxe/index.htm.
[12] http://www.intel.com/products/processor/xeon/index.htm.
[13] http://www.razamicroelectronics.com/products/xlr.htm.
[14] http://www.xbox.com/en-us/hardware/xbox360/powerplay.htm.
[15] International Technology Roadmap for Semiconductors 2003, http://public.itrs.net.

[16] Digital Equipment Corporation, *Alpha 21064 and Alpha 21064A: Hardware Reference Manual*, Digital Equipment Corporation, 1992.
[17] Digital Equipment Corporation, *Alpha 21164 Microprocessor: Hardware Reference Manual*, Digital Equipment Corporation, 1998.
[18] Compaq Corporation, *Alpha 21264/EV6 Microprocessor: Hardware Reference Manual*, Compaq Corporation, 1998.
[19] Intel Corporation, *Measuring Processor Performance with SPEC2000—A White Paper*, Intel Corporation, 2002.
[20] Annavaram M., Grochowski E., Shen J., "Mitigating Amdahl's Law through EPI throttling", in: *International Symposium on Computer Architecture*, 2005.
[21] Archibald J., Baer J.-L., "Cache coherence protocols: Evaluation using a multiprocessor simulation model", *ACM Trans. Comput. Syst.* **4** (4) (1986) 273–298.
[22] Balakrishnan S., Rajwar R., Upton M., Lai K., "The impact of performance asymmetry in emerging multicore architectures", in: *International Symposium on Computer Architecture*, June 2005.
[23] Barroso L., Gharachorloo K., McNamara R., Nowatzyk A., Qadeer S., Sano B., Smith S., Stets R., Verghese B., "Piranha: A scalable architecture based on single-chip multiprocessing", in: *The 27th Annual International Symposium on Computer Architecture*, June 2000.
[24] Bowhill W., "A 300-MHz 64-b quad-issue CMOS microprocessor", in: *ISSCC Digest of Technical Papers*, February 1995.
[25] Brooks D., Tiwari V., Martonosi M., "Wattch: A framework for architectural-level power analysis and optimizations", in: *International Symposium on Computer Architecture*, June 2000.
[26] Burns J., Gaudiot J.-L., "Area and system clock effects on SMT/CMP processors", in: *The 2001 International Conference on Parallel Architectures and Compilation Techniques*, IEEE Computer Society, 2001, p. 211.
[27] Burns J., Gaudiot J.-L., "SMT layout overhead and scalability", *IEEE Transactions on Parallel and Distributed Systems* **13** (2) (February 2002).
[28] Clabes J., Friedrich J., Sweet M., DiLullo J., Chu S., Plass D., Dawson J., Muench P., Powell L., Floyd M., Sinharoy B., Lee M., Goulet M., Wagoner J., Schwartz N., Runyon S., Gorman G., Restle P., Kalla R., McGill J., Dodson S., "Design and implementation of the Power5 microprocessor", in: *International Solid-State Circuits Conference*, 2004.
[29] Collins J., Tullsen D., "Clustered multithreaded architectures—pursuing both IPC and cycle time" in: *International Parallel and Distributed Processing Symposium*, April 2004.
[30] Diefendorff K., "Compaq chooses SMT for Alpha", *Microprocessor Report* **13** (16) (December 1999).
[31] Dobberpuhl D.W., "A 200-MHz 64-b dual-issue CMOS microprocessor", *IEEE Journal of Solid-State Circuits* **27** (11) (November 1992).
[32] Dolbeau R., Seznec A., "CASH: Revisiting hardware sharing in single-chip parallel processor", IRISA Report 1491, November 2002.
[33] Dubois M., Scheurich C., Briggs F., "Synchronization, coherence, and event ordering in multiprocessors", *IEEE Computer* **21** (2) (1988).

[34] Eickemeyer R.J., Johnson R.E., Kunkel S.R., Squillante M.S., Liu S., "Evaluation of multithreaded uniprocessors for commercial application environments", in: *International Symposium on Computer Architecture*, 1996.
[35] Emer J., "EV8: The post-ultimate Alpha", in: *Conference on Parallel Architectures and Computing Technologies*, September 2001.
[36] Espasa R., Ardanaz F., Emer J., Felix S., Gago J., Gramunt R., Hernandez I., Juan T., Lowney G., Mattina M., Seznec A., "Tarantula: A vector extension to the alpha architecture", in: *International Symposium on Computer Architecture*, May 2002.
[37] Frank S.J., "Tightly coupled multiprocessor systems speed memory access times", in: *Electron*, January 1984.
[38] Ghiasi S., Grunwald D., "Aide de camp: Asymmetric dual core design for power and energy reduction", University of Colorado Technical Report CU-CS-964-03, 2003.
[39] Ghiasi S., Keller T., Rawson F., "Scheduling for heterogeneous processors in server systems", in: *Computing Frontiers*, 2005.
[40] Gieseke B., "A 600-MHz superscalar RISC microprocessor with out-of-order execution", in: *ISSCC Digest of Technical Papers*, February 1997.
[41] Grochowski E., Ronen R., Shen J., Wang H., "Best of both latency and throughput", in: *IEEE International Conference on Computer Design*, 2004.
[42] Gupta S., Keckler S., Burger D., "Technology independent area and delay estimates for microprocessor building blocks", University of Texas at Austin Technical Report TR-00-05, 1998.
[43] Hammond L., Nayfeh B.A., Olukotun K., "A single-chip multiprocessor", *IEEE Computer* **30** (9) (1997).
[44] Hammond L., Willey M., Olukotun K., "Data speculation support for a chip multiprocessor", in: *The Eighth International Conference on Architectural Support for Programming Languages and Operating Systems*, October 1998.
[45] Hennessy J., Patterson D., *Computer Architecture: A Quantitative Approach*, Morgan Kaufmann Publishers, Inc., 2002.
[46] Hennessy J.L., Jouppi N.P., "Computer technology and architecture: An evolving interaction", *Computer* **24** (9) (September 1991) 18–29.
[47] Horowitz M., Alon E., Patil D., Naffziger S., Kumar R., Bernstein K., "Scaling, power, and the future of CMOS", in: *IEEE International Electron Devices Meeting*, December 2005.
[48] Horowitz M., Ho R., Mai K., "The future of wires", Invited Workshop Paper for SRC Conference, May 1999.
[49] IBM, Power4, http://www.research.ibm.com/power4.
[50] IBM, "Power5: Presentation at microprocessor forum", 2003.
[51] Kaanta C., Cote W., Cronin J., Holland K., Lee P., Wright T., "Submicron wiring technology with tungsten and planarization", in: *Fifth VLSI Multilevel Interconnection Conference*, 1988.
[52] Kahle J.A., Day M.N., Hofstee H.P., Johns C.R., Maeurer T.R., Shippy D., "Introduction to the Cell multiprocessor", *IBM Journal of Research and Development* (September 2005).

[53] Kim C., Burger D., Keckler S., "An adaptive, non-uniform cache structure for wire-delay dominated on-chip caches", in: *International Conference on Architectural Support for Programming Languages and Operating Systems*, 2002.
[54] Klauser A., "Trends in high-performance microprocessor design", in: *Telematik-2001*, 2001.
[55] Kongetira P., Aingaran K., Olukotun K., "Niagara: A 32-way multithreaded Sparc processor" in: *IEEE MICRO Magazine*, March 2005.
[56] Kotla R., Devgan A., Ghiasi S., Keller T., Rawson F., "Characterizing the impact of different memory-intensity levels", in: *IEEE 7th Annual Workshop on Workload Characterization*, 2004.
[57] Kowaleski J., "Implementation of an Alpha microprocessor in SOI", in: *ISSCC Digest of Technical Papers*, February 2003.
[58] Krishnan V., Torrellas J., "A clustered approach to multithreaded processors", in: *International Parallel Processing Symposium*, March 1998, pp. 627–634.
[59] Kumar A., "The HP PA-8000 RISC CPU", in: *Hot Chips VIII*, August 1996.
[60] Kumar R., Farkas K.I., Jouppi N.P., Ranganathan P., Tullsen D.M., "Single-ISA heterogeneous multi-core architectures: The potential for processor power reduction", in: *International Symposium on Microarchitecture*, December 2003.
[61] Kumar R., Jouppi N.P., Tullsen D.M., "Conjoined-core chip multiprocessing", in: *International Symposium on Microarchitecture*, December 2004.
[62] Kumar R., Tullsen D.M., Jouppi N.P., "Core architecture optimization for heterogeneous chip multiprocessors", in: *15th International Symposium on Parallel Architectures and Compilation Techniques*, September 2006.
[63] Kumar R., Tullsen D.M., Ranganathan P., Jouppi N.P., Farkas K.I., "Single-ISA heterogeneous multi-core architectures for multithreaded workload performance", in: *International Symposium on Computer Architecture*, June 2004.
[64] Kumar R., Zyuban V., Tullsen D.M., "Interconnections in multi-core architectures: Understanding mechanisms, overheads and scaling", in: *Proceedings of International Symposium on Computer Architecture*, June 2005.
[65] Kunkel S., Eickemeyer R., Lipasti M., Mullins T., Krafka B., Rosenberg H., VanderWiel S., Vitale P., Whitley L., "A performance methodology for commercial servers", *IBM Journal of R&D* (November 2000).
[66] Laudon J., "Performance/watt the new server focus", in: *The First Workshop on Design, Architecture, and Simulation of Chip-Multiprocessors*, November 2005.
[67] Li J., Martinez J., "Power-performance implications of thread-level parallelism in chip multiprocessors", in: *Proceedings of International Symposium on Performance Analysis of Systems and Software*, 2005.
[68] Lovett T., Thakkar S., "The symmetry multiprocessor system", in: *International Conference on Parallel Processing*, August 1988.
[69] Merritt R., "Designers cut fresh paths to parallelism", in: *EE Times*, October 1999.
[70] Moore G., "Cramming more components onto integrated circuits", *Electronics* **38** (8) (1965).
[71] Morad T., Weiser U., Kolodny A., "ACCMP—asymmetric cluster chip-multiprocessing", CCIT Technical Report 488, 2004.

[72] Morad T.Y., Weiser U.C., Kolodny A., Valero M., Ayguade E., "Performance, power efficiency and scalability of asymmetric cluster chip multiprocessors", *Computer Architecture Letters* **4** (2005).
[73] Mulder J.M., Quach N.T., Flynn M.J., "An area model for on-chip memories and its applications", *IEEE Journal of Solid State Circuits* **26** (2) (February 1991).
[74] Papamarcos M., Patel J., "A low overhead coherence solution for multiprocessors with private cache memories", in: *International Symposium on Computer Architecture*, 1988.
[75] Rabaey J.M., "The quest for ultra-low energy computation opportunities for architectures exploiting low-current devices", April 2000.
[76] Sankaralingam K., Nagarajan R., Liu H., Kim C., Huh J., Burger D., Keckler S.W., Moore C.R., "Exploiting ILP, TLP, and DLP with the polymorphous TRIPS architecture", in: *International Symposium on Computer Architecture*, June 2003.
[77] Sherwood T., Perelman E., Hamerly G., Calder B., "Automatically characterizing large scale program behavior", in: *Tenth International Conference on Architectural Support for Programming Languages and Operating Systems*, October 2002.
[78] Sherwood T., Perelman E., Hamerly G., Sair S., Calder B., "Discovering and exploiting program phases", in: *IEEE Micro: Micro's Top Picks from Computer Architecture Conferences*, December 2003.
[79] Sherwood T., Perelman E., Hamerly G., Calder B., "Automatically characterizing large-scale program behavior", in: *International Conference on Architectural Support for Programming Languages and Operating Systems*, October 2002.
[80] Shivakumar P., Jouppi N., "CACTI 3.0: An integrated cache timing, power and area model", Technical Report 2001/2, Compaq Computer Corporation, August 2001.
[81] Snavely A., Tullsen D., "Symbiotic jobscheduling for a simultaneous multithreading architecture", in: *Eighth International Conference on Architectural Support for Programming Languages and Operating Systems*, November 2000.
[82] SPEC, Spec cpu2000 documentation, http://www.spec.org/osg/cpu2000/docs/.
[83] Sun, UltrasparcIV, http://siliconvalley.internet.com/news/print.php/3090801.
[84] Swanson S., Michelson K., Schwerin A., Oskin M., "Wavescalar", in: *The 36th Annual IEEE/ACM International Symposium on Microarchitecture, Washington, DC, USA*, IEEE Computer Society, 2003, p. 291.
[85] Theis T.N., "The future of interconnection technology", *IBM Journal of R&D* (May 2000).
[86] Tremblay M., "Majc-5200: A VLIW convergent mpsoc", in: *Microprocessor Forum*, October 1999.
[87] Tullsen D., "Simulation and modeling of a simultaneous multithreading processor", in: *22nd Annual Computer Measurement Group Conference*, December 1996.
[88] Tullsen D., Brown J., "Handling long-latency loads in a simultaneous multithreading processor", in: *34th International Symposium on Microarchitecture*, December 2001.
[89] Tullsen D., Eggers S., Emer J., Levy H., Lo J., Stamm R., "Exploiting choice: Instruction fetch and issue on an implementable simultaneous multithreading processor", in: *23rd Annual International Symposium on Computer Architecture*, May 1996.
[90] Tullsen D., Eggers S., Levy H., "Simultaneous multithreading: Maximizing on-chip parallelism", in: *22nd Annual International Symposium on Computer Architecture*, June 1995.

[91] Waingold E., Taylor M., Srikrishna D., Sarkar V., Lee W., Lee V., Kim J., Frank M., Finch P., Barua R., Babb J., Amarasinghe S., Agarwal A., "Baring it all to software: Raw machines", *Computer* **30** (9) (1997) 86–93.

[92] Wall D., "Limits of instruction-level parallelism", in: *International Symposium on Architectural Support for Programming Languages and Operating Systems*, April 1991, pp. 176–188.

[93] Warnock J., Keaty J., Petrovick J., Clabes J., Kircher C., Krauter B., Restle P., Zoric B., Anderson C., "The circuit and physical design of the Power4 microprocessor", *IBM Journal of R&D* (January 2002).

[94] Wilson A., "Hierarchical cache/bus architecture for shared memory multiprocessors", in: *International Symposium on Computer Architecture*, June 1987.

[95] Zyuban V., "Unified architecture level energy-efficiency metric", in: *2002 Great Lakes Symposium on VLSI*, April 2002.

Designing Computational Clusters for Performance and Power

KIRK W. CAMERON, RONG GE, AND XIZHOU FENG

Computer Science
212 Knowledge Works II Building
Corporate Research Center
Virginia Tech
Blackburg, VA 24061
USA
cameron@vt.edu
ge@cs.vt.edu
fengx@cs.vt.edu

Abstract

Power consumption in computational clusters has reached critical levels. High-end cluster performance improves exponentially while the power consumed and heat dissipated increase operational costs and failure rates. Yet, the demand for more powerful machines continues to grow. In this chapter, we motivate the need to reconsider the traditional performance-at-any-cost cluster design approach. We propose designs where power *and* performance are considered critical constraints. We describe power-aware and low power techniques to reduce the power profiles of parallel applications and mitigate the impact on performance.

1. Introduction	90
1.1. Cluster Design PARADIGM Shift	91
2. Background	91
2.1. Computational Clusters	92
2.2. Performance	92
2.3. Power	93
2.4. Power-Aware Computing	93
2.5. Energy	94
2.6. Power-Performance Tradeoffs	95
3. Single Processor System Profiling	96
3.1. Simulator-Based Power Estimation	96
3.2. Direct Measurements	97

3.3. Event-Based Estimation 98
3.4. Power Reduction and Energy Conservation 98
4. Computational Cluster Power Profiling 98
    4.1. A Cluster-Wide Power Measurement System 99
    4.2. Cluster Power Profiles 103
5. Low Power Computational Clusters 112
    5.1. Argus: Low Power Cluster Computer 112
6. Power-Aware Computational Clusters 125
    6.1. Using DVS in High-Performance Clusters 126
    6.2. Distributed DVS Scheduling Strategies 128
    6.3. Experimental Framework 131
    6.4. Analyzing an Energy-Conscious Cluster Design 137
    6.5. Lessons from Power-Aware Cluster Design 148
7. Conclusions 149
    References 149

1. Introduction

High-end computing systems are a crucial source for scientific discovery and technological revolution. The unmatched level of computational capability provided by high-end computers enables scientists to solve challenging problems that are insolvable by traditional means and to make breakthroughs in a wide spectrum of fields such as nanoscience, fusion, climate modeling and astrophysics [40,63].

The designed peak performance for high-end computing systems has increased rapidly in the last two decades. For example, the peak performance of the No. 1 supercomputer in 1993 was below 100 Gflops. This value increased 2800 times within 13 years to 280 TFlops in 2006 [65].

Two facts primarily contribute to the increase in peak performance of high-end computers. The first is increasing microprocessor speed. The operating frequency of a microprocessor almost doubled every 2 years in the 1990s [10]. The second is the increasing size of high-end computers. The No. 1 supercomputer in the 1990s consists of about 1000 processors; today's No. 1 supercomputer, BlueGene/L, is about 130 times larger, consisting of 131,072 processors [1].

There is an increasing gap between achieved "sustained" performance and the designed peak performance. Empirical data indicates that the sustained performance achieved by average scientific applications is about 10–15% of the peak performance. Gordon Bell prize winning applications [2,59,61] sustain 35 to 65% of peak performance. Such performance requires the efforts of a team of experts working collaboratively for years. LINPACK [25], arguably the most scalable and optimized benchmark code suite, averages about 67% of the designed peak performance on TOP500 machines in the past decade [24].

The power consumption of high-end computers is enormous and increases exponentially. Most high-end systems use tens of thousands of cutting edge components in clusters of SMPs,[1] and the power dissipation of these components increases by 2.7 times every two years [10]. Earth Simulator requires 12 megawatts of power. Future petaflop systems may require 100 megawatts of power [4], nearly the output of a small power plant (300 megawatts). High power consumption causes intolerable operating cost and failure rates. For example, a petaflop system will cost $10,000 per hour at $100 per megawatt excluding the additional cost of dedicated cooling. Considering commodity components fail at an annual rate of 2–3% [41], this system with 12,000 nodes will sustain hardware failure once every twenty-four hours. The mean time between failures (MTBF) [67] is 6.5 hours for LANL ASCI Q, and 5.0 hours for LLNL ASCI white [23].

1.1 Cluster Design PARADIGM Shift

The traditional performance-at-any-cost cluster design approach produces systems that make inefficient use of power and energy. Power reduction usually results in performance degradation, which is undesirable for high-end computing. The challenge is to reduce power consumption without sacrificing cluster performance. Two categories of approaches are used to reduce power for embedded and mobile systems: low power and power-aware. The low power approach uses low power components to reduce power consumption with or without a performance constraint, and the power-aware approach uses power-aware components to maximize performance subject to a power budget. We describe the effects of both of these approaches on computational cluster performance in this chapter.

2. Background

In this section, we provide a brief review of some terms and metrics used in evaluating the effects of power and performance in computational clusters.

---

[1] SMP stands for Symmetric Multi-Processing, a computer architecture that provides fast performance by making multiple CPUs available to complete individual processes simultaneously. SMP uses a single operating system and shares common memory and disk input/output resources.

2.1 Computational Clusters

In this chapter, we use the term *computational cluster* to refer to any collection of machines (often SMPs) designed to support parallel scientific applications. Such clusters differ from commercial server farms that primarily support embarrassingly parallel client-server applications. Server farms include clusters such as those used by Google to process web queries. Each of these queries is independent of any other allowing power-aware process scheduling to leverage this independence. The workload on these machines often varies with time, e.g. demand is highest during late afternoon and lowest in early morning hours.

Computational clusters are designed to accelerate simulation of natural phenomena such as weather modeling or the spread of infectious diseases. These applications are not typically embarrassingly parallel, that is there are often dependences among the processing tasks required by the parallel application. These dependencies imply power reduction techniques for server farms that exploit process independence may not be suitable for computational clusters. Computational cluster workloads are batch scheduled for full utilization 24 hours a day, 7 days per week.

2.2 Performance

An ultimate measure of system performance is the execution time T or delay D for one or a set of representative applications [62]. The execution time for an application is determined by the CPU speed, memory hierarchy and application execution pattern.

The sequential execution time $T(1)$ for a program on a single processor consists of two parts: the time that the processor is busy executing instructions T_{comp}, and the time that the process waits for data from the local memory system $T_{memoryaccess}$ [21], i.e.,

$$T(1) = T_{comp}(1) + T_{memoryaccess}(1). \tag{1}$$

Memory access is expensive: the latency for a single memory access is almost the same as the time for the CPU to execute one hundred instructions. The term $T_{memoryaccess}$ can consume up to 50% of execution time for an application whose data accesses reside in cache 99% of the time.

The parallel execution time on n processors $T(n)$ includes three other components as parallel overhead: the synchronization time due to load imbalance and serialization $T_{sync}(n)$; the communication time $T_{comm}(n)$ that the processor is stalled for data to be communicated from or to remote processing node; and the time that the processor is busy executing extra work $T_{extrawork}(n)$ due to decomposition and task assignment. The parallel execution time can be written as

$$T(n) = T_{comp}(n) + T_{memoryaccess}(n) + T_{sync}(n) + T_{comm}(n) \\ + T_{extrawork}(n). \quad (2)$$

Parallel overhead $T_{sync}(n)$, $T_{comm}(n)$ and $T_{extrawork}(n)$ are quite expensive. For example, the communication time for a single piece of data can be as large as the computation time for thousands of instructions. Moreover, parallel overhead tends to increase with the number of processing nodes.

The ratio of sequential execution time to parallel execution time on n processors is the parallel speedup, i.e.,

$$speedup(n) = \frac{T(1)}{T(n)}. \quad (3)$$

Ideally, the speedup on n processors is equal to n for a fixed-size problem, or the speedup grows linearly with the number of processors. However, the achieved speedup for real applications is typically sub-linear due to parallel overhead.

2.3 Power

The power consumption of CMOS logic circuits [58] such as processor and cache logic is approximated by

$$P = ACV^2 f + P_{short} + P_{leak}. \quad (4)$$

The power consumption of CMOS logic consists of three components: dynamic power $P_d = ACV^2 f$ which is caused by signal line switching; short circuit power P_{short} which is caused by through-type current within the cell; and leak power P_{leak} which is caused by leakage current. Here f is the operating frequency, A is the activity of the gates in the system, C is the total capacitance seen by the gate outputs, and V is the supply voltage. Of these three components, dynamic power dominates and accounts for 70% or more, P_{short} accounts for 10–30%, and P_{leak} accounts for about 1% [51]. Therefore, CMOS circuit power consumption is approximately proportional to the operating frequency and the square of supply voltage when ignoring the effects of short circuit power and leak power.

2.4 Power-Aware Computing

Power-aware computing describes the use of *power-aware* components to save energy. Power-aware components come with a set of power-performance modes. A high performance mode consumes more power than a low performance mode but provides better performance. By scheduling the power-aware components among different power-performance modes according to the processing needs, a power-aware system can reduce the power consumption while delivering the performance required by an application.

Power aware components, including processor, memory, disk, and network controller were first available to battery-powered mobile and embedded systems. Similar technologies have recently emerged in high end server products.

In this chapter, we focus on power-aware computing using power-aware processors. Several approaches are available for CPU power control. A DVFS (dynamic voltage frequency scaling) capable processor is equipped with several performance modes, or operating points. Each operating point is specified by a frequency and core voltage pair. An operating point with higher frequency provides higher peak performance but consumes more power. Many current server processors support DVFS. For example, Intel Xeon implements SpeedStep, and AMD Opteron supports PowerNow. SpeedStep and PowerNow are trademarked by Intel and AMD respectively; this marketing language labels a specific DVFS implementation.

For DVFS capable processors, scaling down voltage reduces power quadratically. However, scaling down the supply voltage often decreases the operating frequency and causes performance degradation. The maximum operating frequency of the CPU is roughly linear to its core voltage V, as described by the following equation [58]:

$$f_{max} \propto (V - V_{threshold})^2 / V. \tag{5}$$

Since operating frequency f is usually correlated to execution time of an application, reducing operating frequency will increase the computation time linearly when the CPU is busy.

However, the effective sustained performance for most applications is not simply determined by the CPU speed (i.e. operating frequency). Both application execution patterns and system hardware characteristics affect performance. For some codes, the effective performance may be insensitive to CPU speed. Therefore, scaling down the supply voltage and the operating frequency could reduce power consumption significantly without incurring noticeable additional execution time. Hence, the opportunity for power aware computing lies in appropriate DVFS scheduling which switches CPU speed to match the application performance characteristics.

2.5 Energy

While power (P) describes consumption at a discrete point in time, energy (E) specifies the number of joules used for time interval (t_1, t_2) as a product of the average power and the delay ($D = t_2 - t_1$):

$$E = \int_{t_1}^{t_2} P\, dt = P_{avg} \times (t_2 - t_1) = P_{avg} \times D. \tag{6}$$

Equation (6) specifies the relation between power, delay and energy. To save energy, we need to reduce the delay, the average power, or both. Performance improvements such as code transformation, memory remapping and communication optimization may decrease the delay. Clever system scheduling among various power-performance modes may effectively reduce average power without affecting delay.

In the context of parallel processing, by increasing the number of processors, we can speedup the application but also increase the total power consumption. Depending on the parallel scalability of the application, the energy consumed by an application may be constant, grow slowly or grow very quickly with the number of processors.

In power aware cluster computing, both the number of processors and the CPU speed configuration of each processor affect the power-performance efficiency of the application.

2.6 Power-Performance Tradeoffs

As discussed earlier, power and performance often conflict with one another. Some relation between power and performance is needed to define optimal in this context. To this end, some product forms of delay D (i.e. execution time T) and power P are used to quantify power-performance efficiency. Smaller products represent better efficiency. Commonly used metrics include PDP (the $P \times D$ product, i.e. energy E), PD2P (the $P \times D^2$ product), and PD3P (the $P \times D^3$ product), respectively. These metrics can also be represented in the forms of energy and delay products such as EDP and ED2P.

These metrics put different emphasis on power and performance, and are appropriate for evaluating power-performance efficiency for different systems. PDP or energy is appropriate for low power portable systems where battery life is the major concern. PD2P [19] metrics emphasize both performance and power; this metric is appropriate for systems which need to save energy with some allowable performance loss. PD3P [12] emphasizes performance; this metric is appropriate for high-end systems where performance is the major concern but energy conservation is desirable.

3. Single Processor System Profiling

Three primary approaches: simulators, direct measurements and performance counter based models, are used to profile power of systems and components.

3.1 Simulator-Based Power Estimation

In this discussion, we focus on architecture level simulators and categorize them across system components, i.e. microprocessor and memory, disk and network. These power simulators are largely built upon or used in conjunction with performance simulators that provide resource usage counts, and estimate energy consumption using resource power models.

Microprocessor power simulators. Wattch [11] is a microprocessor power simulator interfaced with a performance simulator, SimpleScalar [13]. Wattch models power consumption using an analytical formula $P_d = CV_{dd}^2 af$ for CMOS chips, where C is the load capacitance, V_{dd} is the supply voltage, f is the clock frequency, and a is the activity factor between 0 and 1. Parameters V_{dd}, f and a are identified using empirical data. The load capacitance C is estimated using the circuit and the transistor sizes in four categories: array structure (i.e. caches and register files), CAM structures (e.g. TLBs), complex logic blocks, and clocking. When the application is simulated on SimpleScalar, the cycle-accurate hardware access counts are used as input to the power models to estimate energy consumption.

SimplePower [68] is another microprocessor power simulator built upon SimpleScalar. It estimates both microprocessor and memory power consumption. Unlike Wattch which estimates circuit and transistor capacitance using their sizes, SimplePower uses a capacitance lookup table indexed by input vector transition. SimplePower differs with Wattch in two ways. First, it integrates rather than interfaces with SimpleScalar. Second, it uses the capacitance lookup table rather than empirical estimation of capacitance. The capacitance lookup table could lead to more accurate power simulation. However, this accuracy comes at the expense of flexibility as any change in circuit and transistor would require changes in the capacitance lookup table.

TEM^{2}P^{2}EST [22] and the Cai–Lim model [14] are similar. They both build upon the SimpleScalar toolset. These two approaches add complexity in power models and functional unit classification, and differ from Wattch. First these two models use an empirical mode and an analytical mode. Second, they model both dynamic and leakage power. Third, they include a temperature model using power dissipation.

Network power simulators. Orion [69] is an interconnection network power simulator at the architectural-level based on the performance simulator LSE [66]. It models power analytically for CMOS chips using architectural-level parameters, thus reducing simulation time compared to circuit-level simulators while providing reasonable accuracy.

System power simulators. Softwatt [39] is a complete system power simulator that models the microprocessor, memory systems and disk based on SimOS [60]. Softwatt calculates the power values for microprocessor and memory systems using analytical power models and the simulation data from the log-files. The disk energy consumption is measured during simulation based on assumptions that full power is consumed if any of the ports of a unit is accessed, otherwise no power is consumed.

Powerscope [34] is a tool for profiling the energy usage of mobile applications. Powerscope consists of three components: the system monitor samples system activity by periodically recording the program counter (PC) and process identifier (PID) of the currently executing process; the energy monitor collects and stores current samples; and the energy analyzer maps the energy to specific processes and procedures.

3.2 Direct Measurements

There are two basic approaches to measure processor power directly. The first approach [7,50] inserts a precision resistor into the power supply line using a multimeter to measure its voltage drop. The power dissipation by the processor is the product of power supply voltage and current flow, which is equal to the voltage drop over the resistor divided by its resistance. The second approach [48,64] uses an ammeter to measure the current flow of the power supply line directly. This approach is less intrusive as it does not need to cut wires in the circuits.

Tiwari et al. [64] used ammeters to measure current drawn by a processor while running programs on an embedded system and developed a power model to estimate power cost. Isci et al. [48] used ammeters to measure the power for P4 processors to derive their event-count based power model. Bellosa et al. [7] derived CPU power by measuring current on a precision resistor inserted between the power line and supply for a Pentium II CPU; they used this power to validate their event-count based power model and save energy. Joseph et al. [50] used a precision resistor to measure power for a Pentium Pro processor.

These approaches can be extended to measure single processor system power. Flinn et al. [34] used a multimeter to sample the current being drawn by a laptop from its external power source.

3.3 Event-Based Estimation

Most high-end CPUs have a set of hardware counters to count performance events such as cache hit/miss, memory load, etc. If power is mainly dissipated by these performance events, power can be estimated based on performance counters. Isci et al. [48] developed a runtime power monitoring model which correlates performance event counts with CPU subunit power dissipation on real machines. CASTLE [50] did similar work on performance simulators (SimpleScalar) instead of real machines. Joule Watcher [7] also correlates power with performance events, the difference is that it measures the energy consumption for a single event such as a floating point operation, L2 cache access, and uses this energy consumption for energy-aware scheduling.

3.4 Power Reduction and Energy Conservation

Power reduction and energy conservation has been studied for decades, mostly in the area of energy-constrained, low power, real time and mobile systems [38,54,55, 71]. Generally, this work exploits the multiple performance/power modes available on components such as processor [38,54,71], memory [27,28], disk [17], and network card [18]. When any component is not fully utilized, it can be set to a lower power mode or turned off to save energy. The challenge is to sustain application performance and meet a task deadline in spite of mode switching overhead.

4. Computational Cluster Power Profiling

Previous studies of power consumption on high performance clusters focus on building-wide power usage [53]. Such studies do not separate measurements by individual clusters, nodes or components. Other attempts to estimate power consumption for systems such as ASC Terascale facilities use rule-of-thumb estimates (e.g. 20% peak power) [4]. Based on past experience, this approach could be completely inaccurate for future systems as power usage increases exponentially for some components.

There are two compelling reasons for in-depth study of the power usage of cluster applications. First, there is need for a scientific approach to quantify the energy cost of typical high-performance systems. Such cost estimates could be used to accurately estimate future machine operation costs for common application types. Second, a component-level study may reveal opportunities for power and energy savings. For example, component-level profiles could suggest schedules for powering down equipment not being used over time.

Profiling power directly in a distributed system at various granularities is challenging. First, we must determine a methodology for separating component power after conversion from AC to DC current in the power supply for a typical server. Next, we must address the physical limitations of measuring the large number of nodes found in typical clusters. Third, we must consider storing and filtering the enormous data sets that result from polling. Fourth, we must synchronize the polling data for parallel programs to analyze parallel power profiles.

Our measurement system addresses these challenges and provides the capability to automatically measure power consumption at component level synchronized with application phases for power-performance analysis of clusters and applications. Though we do make some simplifying assumptions in our implementation (e.g. the type of multimeter), our tools are built to be portable and require only a small amount of retooling for portability.

4.1 A Cluster-Wide Power Measurement System

Figure 1 shows the prototype system we created for power-performance profiling. We measure the power consumption of the major computing resources (i.e.

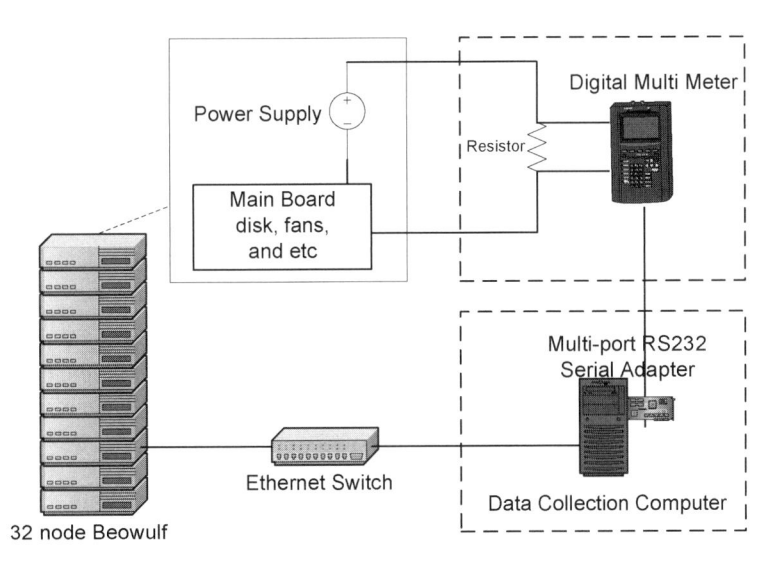

FIG. 1. Our system prototype enables measurement of cluster power at component granularity. For scalability, we assume the nodes are homogeneous. Thus, one node is profiled and software is used to remap applications when workloads are non-uniform. A separate PC collects data directly from the multimeters and uses time stamps to synchronize measured data to an application.

CPU, memory, disk, and NIC) on the slave nodes in a 32-node Beowulf. Each slave node has one 933 MHz Intel Pentium III processor, 4 256M SDRAM modules, one 15.3 GB IBM DTLA-307015 DeskStar hard drive, and one Intel 82559 Ethernet Pro 100 onboard Ethernet controller.

ATX extension cables connect the tested node to a group of 0.1 ohm sensor resistors on a circuit board. The voltage on each resistor is measured with one RadioShack 46-range digital multi meter 22-812 that has been attached to a multi port RS232 serial adapter plugged into a data collection computer running Linux. We measure 10 power points using 10 independent multi meters between the power supply and components simultaneously.

The meters broadcast live measurements to the data collection computer for data logging and processing through their RS232 connections. Each meter sends 4 samples per second to the data collection computer.

Currently, this system measures one slave node at a time. The power consumed by a parallel application requires summation of the power consumption on all nodes used by the application. Therefore, we first measure a second node to confirm that power measurements are nearly identical across like systems, and then use node remapping to study the effective power properties of different nodes in the cluster without requiring additional equipment. To ensure confidence in our results, we complete each experiment at least 5 times based on our observations of variability.

Node remapping works as follows. Suppose we are running a parallel workload on M nodes, we fix the measurement equipment to one physical node (e.g. node #1) and repeatedly run the same workload M times. Each time we map the tested physical node to a different virtual node. Since all slave nodes are identical (as they should be and we experimentally confirmed), we use the M independent measurements on one node to emulate one measurement on M nodes.

4.1.1 Isolating Power by Component

For parallel applications, a cluster can be abstracted as a group of identical nodes consisting of CPU, memory, disk, and network interface. The power consumed by a parallel application is computed by equations presented in Section 2 with direct or derived power measurement for each component.

In our prototype system, the mother board and disk on each slave node are connected to a 250 W ATX power supply through one ATX main power connector and one ATX peripheral power connector respectively. We experimentally deduce the correspondence between ATX power connectors and node components.

Since disk is connected to a peripheral power connection independently, its power consumption can be directly measured through +12VDC and +5VDC pins on the peripheral power connect. To map the component on the motherboard with the pins on the main power connector, we observe the current changes on all non-COM pins by adding/removing components and running different micro benchmarks which access isolated components over time. Finally, we are able to conclude that the CPU is powered through four +5VDC pins; memory, NIC and others are supplied through +3.3VDC pins; the +12VDC feeds the CPU fan; and other pins are constant and small (or zero) current. The CPU power consumption is obtained by measuring all +5VDC pins directly.

The idle part of memory system power consumption is measured by extrapolation. Each slave node in the prototype has four 256 MB memory modules. We measure the power consumptions of the slave node configured with 1, 2, 3, and 4 memory modules separately, then estimate the idle power consumed by the whole memory system.

The slave nodes in the prototype are configured with onboard NIC. It is hard to separate its power consumption from other components directly. After, observing that the total system power consumption changes slightly when we disable the NIC or pull out the network cable and consulting the documentation of the NIC (Intel 82559 Ethernet Pro 100), we approximate it with constant value of 0.41 W.

For further verification, we compared our measured power consumption for CPU and disk with the specifications provided by Intel and IBM separately and they matched well. Also by running memory access micro benchmarks, we observed that if accessed data size is located within L1/L2 cache, the memory power consumption does not change; while once main memory is accessed, the memory power consumption we measured increases correspondingly.

4.1.2 Automating Cluster Power Profiling and Analysis

To automate the entire profiling process we require enough multimeters to measure directly, in real-time, a single node—10 in our system. Under this constraint, we fully automate data profiling, measurement and analysis by creating a tool suite named *PowerPack*. PowerPack consists of utilities, benchmarks and libraries for controlling, recording and processing power measurements in clusters. PowerPack's profiling software structure is shown in Fig. 2.

In PowerPack, the PowerMeter control thread reads data samples coming from a group of meter readers which are controlled by globally shared variables. The control thread modifies the shared variables according to messages received from applications running on the cluster. Applications trigger message operations through a set of application level library calls that synchronize the live profiling process with the application source code. These application level library calls can be inserted into the source code of the profiled applications. The commonly used subset of the power profile library API includes:

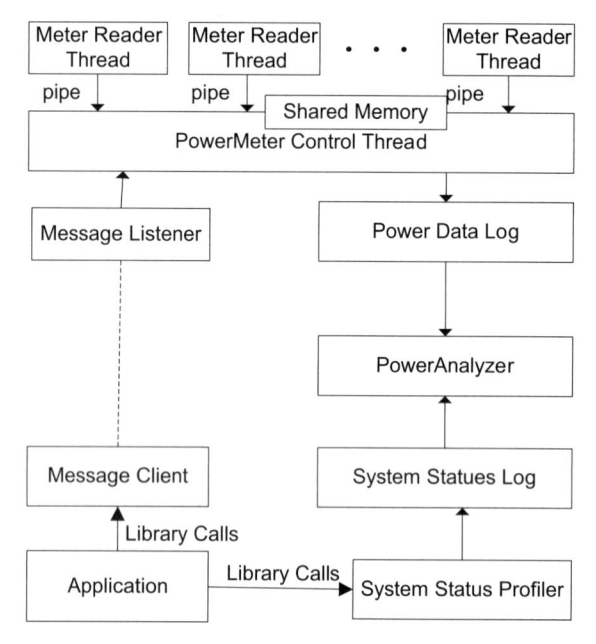

FIG. 2. Automation with software. We created scalable, multi-threaded software to collect and analyze power meter data in real time. An application programmer interface was created to control (i.e. start/stop/init) multimeters and to enable synchronization with parallel codes.

```
pmeter_init (char *ip_address, int *port);
pmeter_log  (char *log_file, int *option);
pmeter_start_session  ( char * lable);
pmeter_finalize  ( );
psyslog_start_session (char *label, int
*interval);
psyslog_pausr ( );
```

The power profile log and the system status log are processed with the PowerAnalyzer, a software module that implements functions such as converting DC current to power, interpolating between sampling points, decomposing pins power to component power, computing power and energy consumed by applications and system, and performing related statistical calculations.

4.2 Cluster Power Profiles

4.2.1 Single Node Measurements

To better understand the power consumption of distributed applications and systems, we first profile the power consumption of a single slave node. Figure 3 provides power profiles for system idle (Fig. 3(a)) and system under load (Fig. 3(b)) for the 171.swim benchmark included in SPEC CPU2000 [44].

From this figure, we make the following observations.

Whether system is idle or busy, the power supply and cooling fans always consume ~20 W of power; about 1/2 system power when idle and 1/3 system power when busy. This means optimal design for power supply and cooling fans could lead to considerable power savings. This is interesting but beyond the scope of this work, so in our graphs we typically ignore this power.

During idle time, CPU, memory, disk and other chipset components consume about 17 W of power in total. When system is under load, CPU power dominates (e.g. for 171.swim, it is 35% of system power; for 164.gzip, it is 48%).

Additionally, the power consumed by each component varies under different workloads. Figure 4 illustrates the power consumption of four representative workloads. Each workload is bounded by the performance of a single component. For our prototype, the CPU power consumption ranges from 6 to 28 W; the memory system power consumption ranges from 3.6 to 9.4 W; the disk power consumption ranges from 4.2 to 10.8 W. Figure 4 indicates component use affects total power consump-

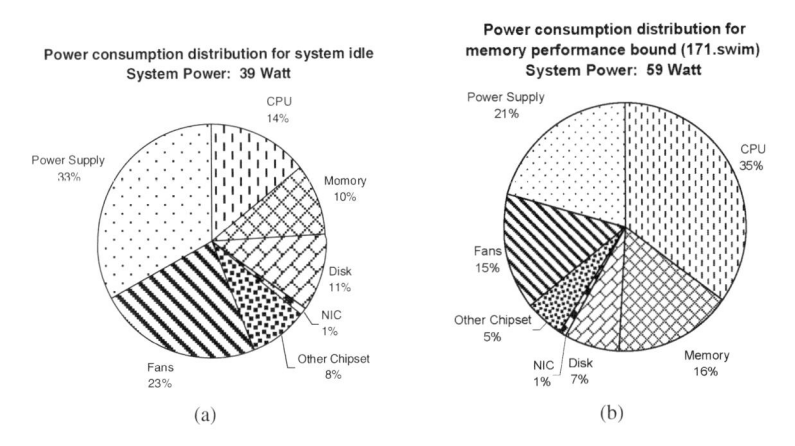

FIG. 3. Power profiles for a single node (a) during idle operation, and (b) under load. As the load increases, CPU and memory power dominate total system power.

Power Consumption Distribution for Different Workloads

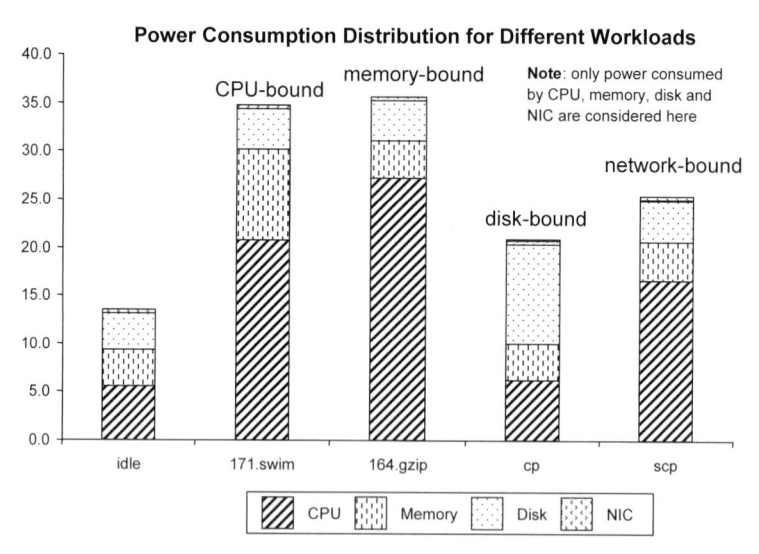

FIG. 4. Different applications stress different components in a system. Component usage is reflected in power profiles. When the system is not idle, it is unlikely that the CPU is 100% utilized. During such periods, reducing power can impact total power consumption significantly. Power-aware techniques (e.g. DVS) must be studied in clusters to determine if power savings techniques impact performance significantly.

tion yet it may be possible to conserve power in non-idle cases when the CPU or memory is not fully utilized.

4.2.2 Cluster-Wide Measurements

We continue illustrating the use of our prototype system by profiling the power-energy consumption of the NAS parallel benchmarks (Version 2.4.1) on the 32-node Beowulf cluster. The NAS parallel benchmarks [5] consist of 5 kernels and 3 pseudo-applications that mimic the computation and data movement characteristics of large scale CFD applications. We measured CPU, memory, NIC and disk power consumption over time for different applications in the benchmarks at different operating points. We ignore power consumed by the power supply and the cooling system because they are constant and machine dependent as mentioned.

4.2.2.1 Nodal Power Profiles Over Time.
Figure 5(a) shows the power profile of NPB FT benchmark (class B) during the first 200 seconds of a run on 4 nodes. The profile starts with a warm up phase and an initialization phase followed by N iterations (for class A, $N = 6$; for class B, $N = 20$). The power profiles

Power Profile of FT Benchmark (class B, NP=4)

(a)

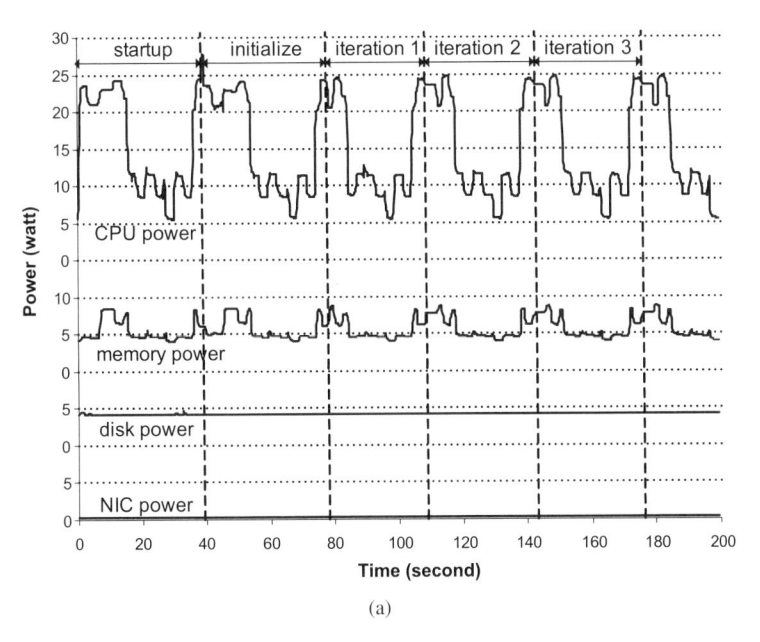

FIG. 5. FT power profiles. (a) The first 200 seconds of power use on one node of four for the FT benchmark, class B workload. Note component results are overlaid along the y-axis for ease of presentation. Power use for CPU and memory dominate and closely reflect system performance. (b) An expanded view of the power profile of FT during a single iteration of computation followed by communication.

are identical for all iterations in which spikes and valleys occur with regular patterns coinciding with the characteristics of different computation stages. The CPU power consumption varies from 25 W in the computation stage to 6 W in the communication stage. The memory power consumption varies from 9 W in the computation stage to 4 W in the communication stage. Power trends in the memory during computation are often the inverse of CPU power. Additionally, the disk uses near constant power since FT rarely accesses the file system. NIC power probably varies with communication, but as discussed, we emulate it as a constant since the maximum usage is quite low (0.4 W) compared to all other components. For simplification, we ignore the disk and NIC power consumption in succeeding discussions and figures where they do not change, focusing on CPU and memory behavior. An in-depth view of the power profile during one (computation + communication) iteration is presented in Fig. 5(b).

Expanded View of Power Profile of FT(class B, NP=4)

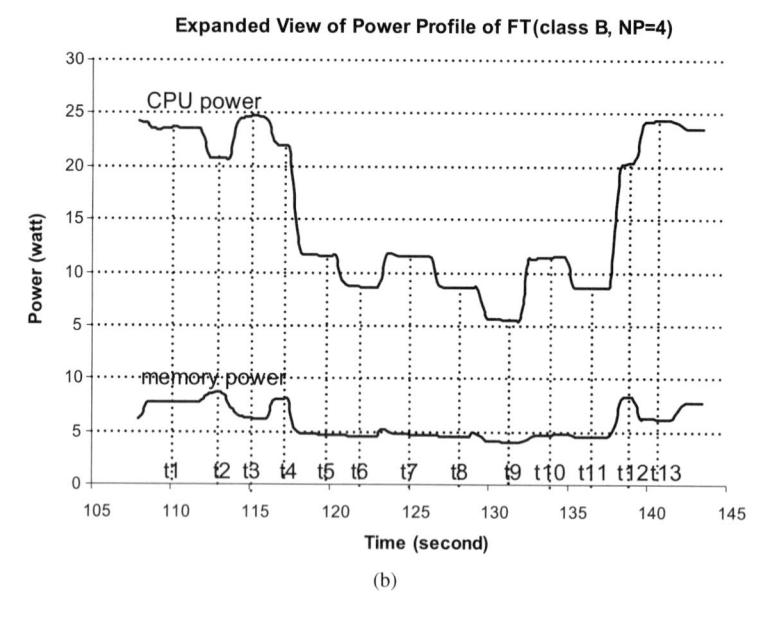

FIG. 5. *(continued)*

4.2.2.2 *Power Profiles for Varying Problem Sizes.*

Figure 6(a) shows the power profile of the FT benchmark (using the smaller class A workload) during the first 50 seconds of a run on 4 nodes. FT has similar patterns for different problem sizes (see Fig. 5(a)). However, iterations are shorter in duration for the smaller (class A) problem set making peaks and values more pronounced; this is effectively a reduction in the communication to computation ratio when the number of nodes is fixed.

4.2.2.3 *Power Profiles for Heterogeneous Workloads.*

For the FT benchmark, workload is distributed evenly across all working nodes. We use our node remapping technique to provide power profiles for all nodes in the cluster (in this case just 4 nodes). For FT, there are no significant differences. However, Fig. 6(b) shows a counter example snapshot for a 10 second interval of SP synchronized across nodes. For the SP benchmark, class A problem sizes running on 4 nodes result in varied power profiles for each node.

4.2.2.4 *Power Profiles for Varying Node Counts.*

The power profile of parallel applications also varies with the number of nodes used in the execution if we fix problem size (i.e. strong scaling). We have profiled the power consump-

Power profile of FT benchmark (class A, NP=4)

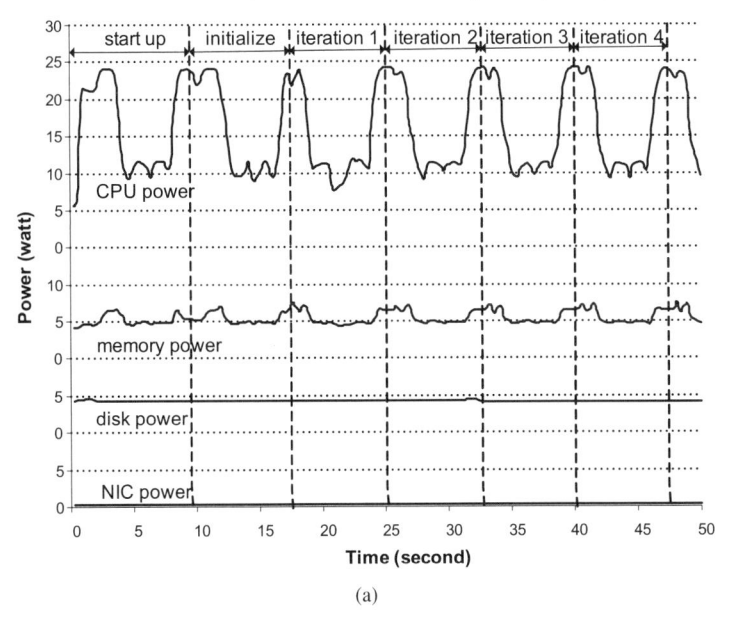

(a)

Power Profile of SP on Different Node (class A, NP=4)

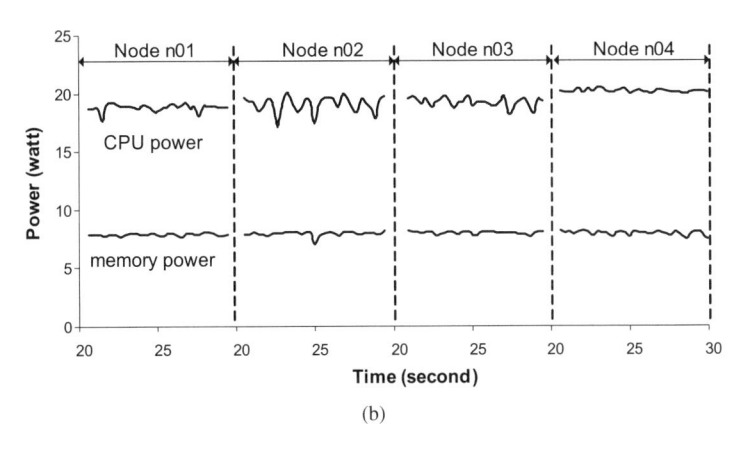

(b)

FIG. 6. (a) The first 50 seconds of power use on one node of four for the FT benchmark, class A workload. For smaller workloads running this application, trends are the same while data points are slightly more pronounced since communication to computation ratios have changed significantly with the change in workload. (b) Power use for code SP that exhibits heterogeneous performance and power behavior across nodes. Note: x-axis is overlaid for ease of presentation—repeats 20–30 second time interval for each node.

tion for all the NPB benchmarks on all execution nodes with different numbers of processors (up to 32) and several classes of problem sizes. Figure 7(a–c) provides an overview of the profile variations on different system scales for benchmarks FT, EP, and MG. These figures show segments of synchronized power profiles for different number of nodes; all the power profiles correspond to the same computing phase in the application on the same node.

These snapshots illustrate profile results for distributed benchmarks using various numbers of nodes under class A workload. Due to space limitations in a single graph, here we focus on power amplitude only, so each time interval is simply a fixed length snapshot (though the x-axis does not appear to scale). For FT and MG, the profiles are similar for different system scale except the average power decreases with the number of execution nodes; for EP, the power profile is identical for all execution nodes.

4.2.3 Cluster Energy-Performance Efficiency

For parallel systems and applications, we would like to use E (see Eq. (6)) to reflect energy efficiency, and use D to reflect performance efficiency. To compare the energy-performance behavior of different parallel applications such as NPB benchmarks, we use two metrics: (1) *normalized delay* or the speedup (from Eq. (3)) defined as $D_{\#\text{ of node}=1}/D_{\#\text{ of node}=n}$; and (2) *normalized system energy*, or the ratio of

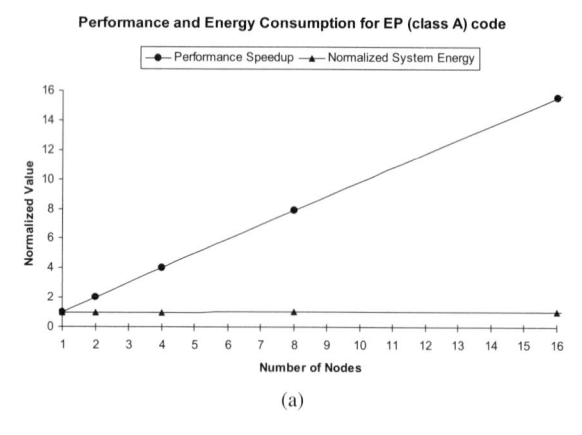

FIG. 7. Energy performance efficiency. These graphs use normalized values for performance (i.e. speedup) and energy. Energy reflects total system energy. (a) EP shows linear performance improvement with no change in total energy consumption. (b) MG is capable of some speedup with the number of nodes with a corresponding increase in the amount of total system energy necessary. (c) FT shows only minor improvements in performance for significant increases in total system energy.

Performance and Energy Consumption for MG (class A) code

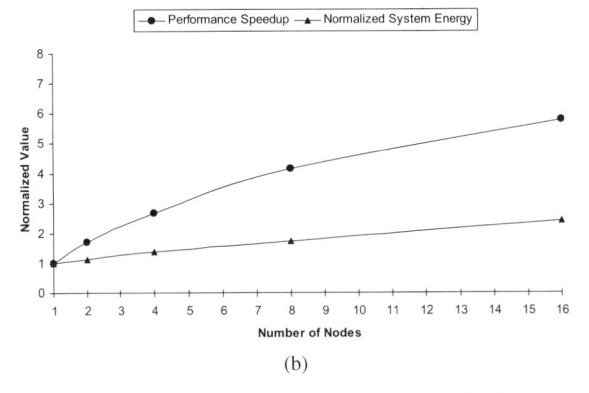

(b)

Performance and Energy Consumption for FT (class A) code

(c)

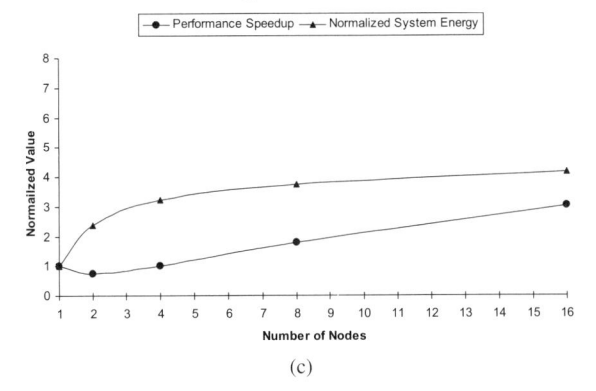

FIG. 7. (*continued*)

multi-node to single-node energy consumption, defined as $E_{\text{\# of node}=n}/E_{\text{\# of node}=1}$. Plotting these two metrics on the same graph with x-axis as the number of nodes, we identify 3 energy-performance categories for the codes measured.

Type I: energy remains constant or approximately constant while performance increases linearly. EP, SP, LU and BT belong to this type (see Fig. 7(a)).

Type II: both energy and performance increase linearly but performance increases faster. MG and CG belong to this type (see Fig. 7(b)).

Type III: both energy and performance increase linearly but energy consumption increases faster. FT and IS belong to this type. For small problem sizes, the IS benchmark gains little in performance speedup using more nodes but consumes much more energy (see Fig. 7(c)).

Since average total system power increases linearly (or approximately linearly) with the number of nodes, we can express energy efficiency as a function of the number of nodes and the performance efficiency:

$$\frac{E_n}{E_1} = \frac{\overline{P}_n \cdot D_n}{\overline{P}_1 \cdot D_1} = \frac{\overline{P}_n}{\overline{P}_1} \cdot \frac{D_n}{D_1} \approx \frac{n \cdot D_n}{D_1}. \qquad (7)$$

In this equation, the subscript refers to the number of nodes used by the application. Equation (7) shows that energy efficiency of parallel applications on clusters is strongly tied to parallel speedup (D_1/D_n). In other words, as parallel programs increase in efficiency with the number of nodes (i.e. improved speedup) they make more efficient use of the additional energy.

4.2.4 Application Characteristics

The power profiles observed are regular and coincide with the computation and communication characteristics of the codes measured. Patterns may vary by node, application, component and workload, but the interaction or interdependency among CPU, memory, disk and NIC have definite patterns. This is particularly obvious in the FT code illustrated in the t1 through t13 labels in Fig. 5(b). FT phases include computation (t1), reduce communication (t2), computation (t3:t4) and all-to-all communication (t5:t11). More generally, we also observe the following for all codes:

1. CPU power consumption decreases when memory power increases. This reflects the classic memory wall problem where access to memory is slow, inevitably causing stalls (low power operations) on the CPU.
2. Both CPU power and memory power decrease with message communication. This is analogous to the memory wall problem where the CPU stalls while waiting on communication. This can be alleviated by non-blocking messages, but this was not observed in the Ethernet-based system under study.
3. For all the codes studied (except EP), the normalized energy consumption increases as the number of nodes increases. In other words, while performance is gained from application speedup, there is a considerable price paid in increased total system energy.
4. Communication distance and message size affect the power profile patterns. For example, LU has short and shallow power profiles while FT phases are significantly longer. This highlights possible opportunities for power and energy savings (discussed next).

4.2.5 Resource Scheduling

We mentioned an application's energy efficiency is dependent on its speedup or parallel efficiency. For certain applications such as FT and MG, we can achieve speedup by running on more processors while increasing total energy consumption. The subjective question remains as to whether the performance gain was *worth* the additional resources. Our measurements indicate there are tradeoffs between power, energy, and performance that should be considered to determine the best resource "operating points" or the best configuration in number of nodes (NP) based on the user's needs.

For performance-constrained systems, the best operating points will be those that minimize delay (D). For power-constrained systems, the best operating points will be those that minimize power (P) or energy (E). For systems where power-performance must be balanced, the choice of appropriate metric is subjective. The energy-delay product (see Section 2.6) is commonly used as a single metric to weigh the effects of power and performance.

Figure 8 presents the relationships between four metrics (normalized D and E, EDP, and ED2P) and the number of nodes for the MG benchmark (class A). To minimize energy (E), the system should schedule only one node to run the application which corresponds in this case to the worst performance. To minimize delay (D), the system should schedule 32 nodes to run the application or 6 times speedup for more than 4 times the energy. For power-performance efficiency, a scheduler using the EDP metric would recommend 8 nodes for a speedup of 2.7 and an energy cost

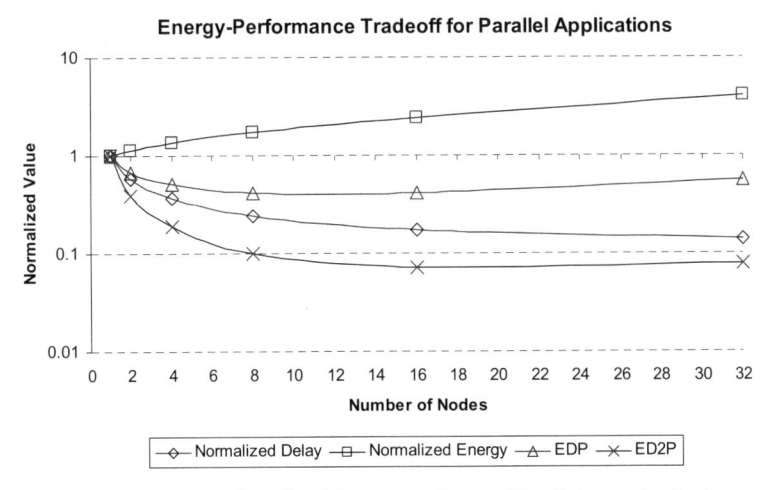

FIG. 8. To determine the number of nodal resources that provides the best rate of return on energy usage is a subjective process. Different metrics recommend different configurations. For the MG code shown here, minimizing delay means using 32 processors; minimizing energy means using 1 processor; the EDP metric recommends 8 processors while the ED2P metric recommends 16 processors. Note: y-axis in log.

of 1.7 times the energy of 1 node. Using the ED2P metric a smart scheduler would recommend 16 nodes for a speedup of 4.1 and an energy cost of 2.4 times the energy of 1 node. For fairness, the average delay and energy consumption obtained from multiple runs are used in Fig. 8.

For existing cluster systems, power-conscious resource allocation can lead to significant energy savings with controllable impact on performance. Of course, there are more details to consider including how to provide the scheduler with application-specific information. This is the subject of ongoing research in power-aware cluster computing.

5. Low Power Computational Clusters

To address operating cost and reliability concerns, large-scale systems are being developed with low power components. This strategy, used in construction of Green Destiny [70] and IBM BlueGene/L [8], requires changes in architectural design to improve performance. For example, Green Destiny relies on driving the Transmeta Crusoe processor [52] development while BlueGene/L uses a version of the embedded PowerPC chip modified with additional floating point support. In essence, the resulting high-end machines are no longer strictly composed of commodity parts—making this approach very expensive to sustain.

To illustrate the pros and cons of a low power computational cluster, we developed the Argus prototype, a high density, low power supercomputer built from an IXIA network analyzer chassis and load modules. The prototype is configured as a diskless cluster scalable to 128 processors in a single 9U chassis. The entire system has a footprint of $1/4$ m^2 (2.5 ft^2), a volume of 0.09 m^3 (3.3 ft^3) and maximum power consumption of less than 2200 W. In this section, we compare and contrast the characteristics of Argus against various machines including our 32-node Beowulf and Green Destiny.

5.1 Argus: Low Power Cluster Computer

Computing resources may be special purpose (e.g. Earth Simulator) or general purpose (e.g. network of workstations). While these high-end systems often provide unmatched computing power, they are extremely expensive, requiring special cooling systems, enormous amounts of power and dedicated building space to ensure reliability. It is common for a supercomputing resource to encompass an entire building and consume tens of megawatts of power.

In contrast low-power, high-throughput, high-density systems are typically designed for a single task (e.g. image processing). These machines offer exceptional speed (and often guaranteed performance) for certain applications. However, design constraints including performance, power, and space make them expensive to develop and difficult to migrate to future generation systems.

We propose an alternative approach augmenting a specialized system (i.e. an Ixia network analyzer) that is designed for a commodity marketplace under performance, power, and space constraints. Though the original Ixia machine is designed for a single task, we have created a configuration that provides general-purpose high-end parallel processing in a Linux environment. Our system provides computational power surpassing Green Destiny [30,70] (another low-power supercomputer) while decreasing volume by a factor of 3.

5.1.1 System Design

Figure 9 is a detailed diagram of the prototype architecture we call Argus. This architecture consists of four sets of separate components: the IXIA chassis, the IXIA Load Modules, the multi port fast Ethernet switch and an NFS server.

The chassis contains a power supply and distribution unit, cooling system, and runs windows and proprietary software (IX server and IX router). Multiple (up to 16) Load Modules plug into the chassis and communicate with the chassis and each other via an IX Bus (mainly used for system management, much too slow for message transfer). Each Load Module provides up to 8 RISC processors (called port proces-

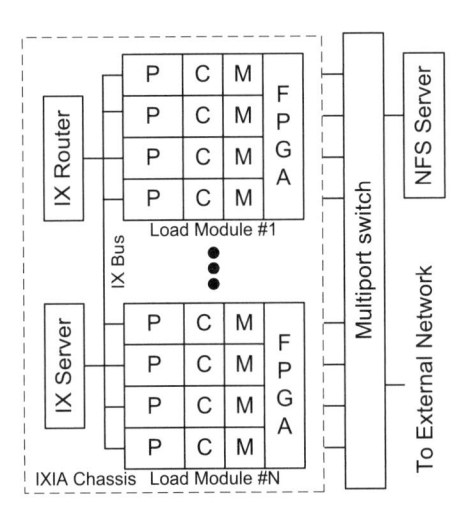

FIG. 9. The hardware architecture Argus. Up to 16 Load Modules are supported in a single IXIA 1600T chassis. A single bus interconnects modules and the chassis PC while external disks and the cluster front-end are connected via an Ethernet switch. P = processor, C = cache, M = memory.

sors) in a dense form factor and each processor has its own operating system, cache (L1 and L2), main memory and network interface. Additional FPGA elements on each Load Module aid real-time analysis of network traffic. Though the performance abilities of these FPGAs have merit, we omit them from consideration for two reasons: (1) reprogramming is difficult and time consuming, and (2) it is likely FPGA elements will not appear in succeeding generation Load Modules to reduce unit cost.

There is no disk on each Load Module. We allocate a small portion of memory at each port to store an embedded version of the Linux OS kernel and application downloaded from the IX Server. An external Linux machine running NFS file server is used to provide external storage for each node. A possible improvement is to use networked memory as secondary storage but we did not attempt this in the initial prototype. Due to cost considerations, although the Load Modules support 1000 Mbps Ethernet on copper, we used a readily available switch operating at 100 Mbps.

The first version of the Argus prototype is implemented with one IXIA 1600T chassis and 4 LM1000TXS4 Load Modules (see http://www.ixiacom.com/library/ catalog/ for specification) [20] configured as a 16-node distributed memory system, i.e., each port processor is considered as an individual node.

Another option is to configure each Load Module as an SMP node. This option requires use of the IxBus between Load Modules. The IxBus bus (and the PowerPC 750CXe processor) does not maintain cache coherence and has limited bandwidth. Thus, early on we eliminated this option from consideration since software-driven cache coherence will limit performance drastically. We opted to communicate data between all processors through the Ethernet connection. Hence one recommendation for future implementations is to significantly increase the performance and capabilities of the IX Bus. This could result in a cluster of SMPs architecture allowing hybrid communications for improved performance.

Each LM1000TXS4 Load Module provides four 1392 MIPS PowerPC 750CXe RISC processors [45] and each processor has one 128 MB memory module and one network port with auto-negotiating 10/100/1000 Mbps Copper Ethernet interface. The 1392 MIPS PowerPC 750CXe CPU employs 0.18 micrometer CMOS copper technology, running at 600 MHz with 6.0 W typical power dissipation. This CPU has independent on-chip 32 K bytes, eight-way set associative, physically addressed caches for instructions and data. The 256 KB L2 cache is implemented with on-chip, two-way set associative memories and synchronous SRAM for data storage. The external SRAM are accessed through a dedicated L2 cache port. The PowerPC 750CXe processor can complete two instructions per CPU cycle. It incorporates 6 execution units including one floating-point unit, one branch processing unit, one system register unit, one load/store unit and two integer units. Therefore, the theoretical peak performance of the PowerPC 750CXe is 1200 MIPS for integer operations and 600 MFLOPS for floating-point operations.

In Argus, message passing (i.e. MPI) is chosen as the model of parallel computation. We ported gcc3.2.2 and glib for PowerPC 750 CXe to provide a useful development environment. MPICH 1.2.5 (the MPI implementation from Argonne National Lab and Michigan State University) and a series of benchmarks have been built and installed on Argus. Following our augmentation, Argus resembles a standard Linux-based cluster running existing software packages and compiling new applications.

5.1.2 Low Power Cluster Metrics

According to design priorities, general-purpose clusters can be classified into four categories:

Performance: These are traditional high-performance systems (e.g. Earth Simulator) where performance (GFLOPS) is the absolute priority.

Cost: These are systems built to maximize the performance/cost ratio (GFLOPS/$) using commercial-off-the-shelf components (e.g. Beowulf).

Power: These systems are designed for reduced power (GFLOPS/W) to improve reliability (e.g. Green Destiny) using low-power components.

Density: These systems have specific space constraints requiring integration of components in a dense form factor with specially designed size and shape (e.g. Green Destiny) for a high performance/volume ratio (GFLOPS/ft^3).

Though high performance systems are still a majority in the HPC community; low cost, low power, low profile and high density systems are emerging. Blue Gene/L (IBM) [1] and Green Destiny (LANL) are two examples designed under cost, power and space constraints.

Argus is most comparable to Green Destiny. Green Destiny prioritizes reliability (i.e. power consumption) though this results in a relatively small form factor. In contrast, the Argus design prioritizes space providing general-purpose functionality not typical in space-constrained systems. Both Green Destiny and Argus rely on system components targeted at commodity markets.

Green Destiny uses the Transmeta Crusoe TM5600 CPU for low power and high density. Each blade of Green Destiny combines server hardware, such as CPU, memory, and the network controller into a single expansion card. Argus uses the PowerPC 750CXe embedded microprocessor which consumes less power but matches the sustained performance of the Transmeta Crusoe TM5600. Argus' density comes at the expense of mechanical parts (namely local disk) and multiple processors on each load module (or blade). For perspective, 240 nodes in Green Destiny fill a single rack (about 25 ft^3); Argus can fit 128 nodes in 3.3 ft^3. This diskless design makes Argus more dense and mobile yet less suitable for applications requiring significant storage.

5.1.2.1 TCO Metrics.
As Argus and Green Destiny are similar, we use the total cost of ownership (TCO) metrics proposed by Feng et al. [30] as the basis of evaluation. For completeness, we also evaluate our system using traditional performance benchmarks.

$$\text{TCO} = \text{AC} + \text{OC}, \tag{8}$$

$$\text{AC} = \text{HWC} + \text{SWC}, \tag{9}$$

$$\text{OC} = \text{SAC} + \text{PCC} + \text{SCC} + \text{DTC}. \tag{10}$$

TCO refers to all expenses related to acquisition, maintaining and operating the computing system within an organization. Equations (8)–(10) provide TCO components including acquisition cost (AC), operations cost (OC), hardware cost (HWC), software cost (SWC), system-administration cost (SAC), power-consumption cost (PCC), space-consumption cost (SCC) and downtime cost (DTC). The ratio of total cost of ownership (TCO) and the performance (GFLOPS) is designed to quantify the effective cost of a distributed system.

According to a formula derived from Arrhenius' Law,[2] component life expectancy decreases 50% for every 10 °C (18 °F) temperature increase. Since system operating temperature is roughly proportional to its power consumption, lower power consumption implies longer component life expectancy and lower system failure rate. Since both Argus and Green Destiny use low power processors, the performance to power ratio (GFLOPS/W) can be used to quantify power efficiency. A high GFLOPS/W implies lower power consumption for the same number of computations, and hence lower system working temperature and higher system reliability (i.e. lower component failure rate).

Since both Argus and Green Destiny provide small form factors relative to traditional high-end systems, the performance to space ratio (GFLOPS/ft^2 for footprint and GFLOPS/ft^3 for volume) can be used to quantify computing density. Feng et al. propose the footprint as the metric of computing density [30]. While Argus performs well in this regard for a very large system, we argue it is more precise to compare volume. We provide both measurements in our results.

5.1.2.2 Benchmarks.
We use an iterative benchmarking process to determine the system performance characteristics of the Argus prototype for general comparison to a performance/cost design (i.e. Beowulf) and to target future design improvements. Benchmarking is performed at two levels:

---

[2] Reaction rate equation of Swedish physical chemist and Nobel Laureate Svante Arrhenius (1859–1927) is used to derive time to failure as a function of $e^{-E_a/KT}$, where E_a is activation energy (eV), K is Boltzmann's constant, and T is absolute temperature in Kelvin.

Micro-benchmarks: Using several micro benchmarks such as LMBENCH [57], MPPTEST [37], NSIEVE [49] and Livermore LOOPS [56], we provide detailed performance measurements of the core components of the prototype CPU, memory subsystem and communication subsystem.

Kernel application benchmarks: We use LINPACK [25] and the NAS Parallel Benchmarks [5] to quantify performance of key application kernels in high performance scientific computing. Performance bottlenecks in these applications may be explained by measurements at the micro-benchmark level.

For direct performance comparisons, we use an on-site 32-node Beowulf cluster called DANIEL. Each node on DANIEL is a 933 MHz Pentium III processor with 1 Gigabyte memory running Red Hat Linux 8.0. The head node and all slave nodes are connected with two 100M Ethernet switches. We expect DANIEL to out-perform Argus generally, though our results normalized for clock rate (i.e. using machine clock cycles instead of seconds) show performance is comparable given DANIEL is designed for performance/cost and Argus for performance/space.

For direct measurements, we use standard UNIX system calls and timers when applicable as well as hardware counters if available. Whenever possible, we use existing, widely-used tools (e.g. LMBENCH) to obtain measurements. All measurements are the average or minimum results over multiple runs at various times of day to avoid outliers due to local and machine-wide perturbations.

5.1.3 Analyzing a Low Power Cluster Design

5.1.3.1 Measured Cost, Power and Space Metrics.
We make direct comparisons between Argus, Green Destiny and DANIEL based on the aforementioned metrics. The results are given in Table I. Two Argus systems are considered: Argus64 and Argus128. Argus64 is the 64-node update of our current prototype with the same Load Module. Argus128 is the 128-node update with the more advanced IXIA Application Load Module (ALM1000T8) currently available [20]. Each ALM1000T8 load module has eight 1856 MIPS PowerPC processors with Gigabit Ethernet interface and 1 GB memory per processor. Space efficiency is calculated by mounting 4 chassis' in a single 36U rack (excluding I/O node and Ethernet switches to be comparable to Green Destiny). The LINPACK performance of Argus64 is extrapolated from direct measurements on 16-nodes and the performance of Argus128 is predicted using techniques similar to Feng et al. as 2×1.3 times the performance of Argus64.

All data on the 32-node Beowulf, DANIEL is obtained from direct measurements. There is no direct measurement of LINPACK performance for Green Destiny in the literature. We use both the Tree Code performance as reported [70] and the estimated

TABLE I
COST, POWER, AND SPACE EFFICIENCY

Machine	DANIEL	Green Destiny	ARGUS64	ARGUS128
CPUs	32	240	64	128
Performance (GFLOPS)	17	39 (101)	13	34
Area (foot^2)	12	6	2.5	2.5
TCO ($K)	100	350	100~150	100~200
Volume (foot^3)	50	30	3.3	3.3
Power (kW)	2	5.2	1	2
GFLOPS/proc	0.53	0.16 (0.42)	0.20	0.27
GFLOPS per chassis	0.53	3.9	13	34
TCO efficiency (GFLOPS/K$)	0.17	0.11 (0.29)	0.08~0.13	0.17~0.34
Computing density (GFLOPS/foot^3)	0.34	1.3 (3.3)	3.9	10.3
Space efficiency (GFLOPS/foot^2)	1.4	6.5 (16.8)	20.8	54.4
Power efficiency (GFLOPS/foot^3)	8.5	7.5 (19.4)	13	17

Notes. For Green Destiny, the first value corresponds to its Tree Code performance; the second value in parenthesis is its estimated LINPACK performance. All other systems use LINPACK performance. The downtime cost of DANIEL is not included when computing its TCO since it is a research system and often purposely rebooted before and after experiments. The TCO of the 240-node Green Destiny is estimated based on the data of its 24-node system.

LINPACK performance by Feng [29] for comparison denoted with parenthesis in Table I.

We estimated the acquisition cost of Argus using prices published by IBM in June 2003 and industry practice. Each PowerPC 750Cxe costs less than $50. Considering memory and other components, each ALM Load Module will cost less than $1000. Including software and system design cost, each Load Module could sell for $5000–$10,000. Assuming the chassis costs another $10,000, the 128-node Argus may cost $90K–170K in acquisition cost (AC). Following the same method proposed by Feng et al., the operating cost (OC) of Argus is less than $10K. Therefore, we estimate the TCO of Argus128 is below $200K. The downtime cost of DANIEL is not included when computing its TCO since it is a research system and often purposely rebooted before and after experiments. The TCO of the 240-node Green Destiny is estimated based on the data of its 24-node system.

Though TCO is suggested as a better metric than acquisition cost, the estimation of downtime cost (DTC) is subjective while the acquisition cost is the largest component of TCO. Though, these three systems have similar TCO performance, Green

Destiny and Argus have larger acquisition cost than DANIEL due to their initial system design cost. System design cost is high in both cases since the design cost has not been amortized over the market size—which would effectively occur as production matures.

The Argus128 is built with a single IXIA 1600T chassis with 16 blades where each blade contains 8 CPUs. The chassis occupies $44.5 \times 39.9 \times 52$ cm^3 (about 0.09 m^3 or 3.3 ft^3). Green Destiny consists of 10 chassis; each chassis contains 10 blades; and each blade has only one CPU. DANIEL includes 32 rack-dense server nodes and each node has one CPU.

Due to the large difference in system footprints (50 ft^3 for DANIEL, 30 ft^3 for Green Destiny and 3.3 ft^3 for Argus) and relatively small difference in single processor performance (711 MFLOPS for DANIEL, 600 MFLOPS for Green Destiny and 300 MFLOPS for Argus), Argus has the highest computing density, 30 times higher than DANIEL, and 3 times higher than Green Destiny.

Table I shows Argus128 is twice as efficient as DANIEL and about the same as Green Destiny. This observation contradicts our expectations that Argus should fair better against Green Destiny in power efficiency. However upon further investigation we suspect either (1) the Argus cooling system is less efficient (or works harder given the processor density), (2) our use of peak power consumption on Argus compared to average consumption on Green Destiny is unfair, (3) the Green Destiny LINPACK projections (not measured directly) provided in the literature are overly optimistic, or (4) some combination thereof. In any case, our results indicate power efficiency should be revisited in succeeding designs though the results are respectable particularly given the processor density.

5.1.3.2 Performance Results.
A single RLX System 324 chassis with 24 blades from Green Destiny delivers 3.6 GFLOPS computing capability for the Tree Code benchmark. A single IXIA 1600T with 16 Load Modules achieves 34 GFLOPS for the LINPACK benchmark. Due to the varying number of processors in each system, the performance per chassis and performance per processor are used in our performance comparisons. Table I shows DANIEL achieves the best performance per processor and Argus achieves the worst. Argus has poor performance on double MUL operation (discussed in the next section) which dominates operations in LINPACK. Argus performs better for integer and single precision float operations. Green Destiny outperforms Argus on multiply operations since designers were able to work with Transmeta engineers to optimize the floating point translation of the Transmeta processor.

Memory hierarchy performance (latency and bandwidth) is measured using the lat_mem_rd and bw_mem_xx tools in the LMBENCH suite. The results are summarized in Table II. DANIEL, using its high-power, high-profile off-the-shelf Intel

TABLE II
MEMORY SYSTEM PERFORMANCE

Parameters	ARGUS	DANIEL
CPU Clock Rate	600 MHz	922 MHz
Clock Cycle Time	1.667 ns	1.085 ns
L1 Data Cache Size	32 KB	16 KB
L1 Data Cache Latency	3.37 ns ≈ 2 cycles	3.26 ns ≈ 3 cycles
L2 Data Cache Size	256 KB	256 KB
L2 Data Cache Latency	19.3 ns ≈ 12 cycles	7.6 ns ≈ 7 cycles
Memory Size	128 MB	1 GB
Memory Latency	220 ns ≈ 132 cycles	153 ns ≈ 141 cycles
Memory Read Bandwidth	146 ~ 2340 MB/s	514 ~ 3580 MB/s
Memory Write Bandwidth	98 ~ 2375 MB/s	162 ~ 3366 MB/s

technology, outperforms Argus at each level in the memory hierarchy in raw performance (time). Normalizing with respect to cycles however, shows how clock rate partially explains the disparity. The resulting "relative performance" between DANIEL and Argus is more promising. Argus performs 50% better than Daniel at the L1 level, 6% better at the main memory level, but much worse at the L2 level. Increasing the clock rate of the PowerPC processor and the L2 implementation in Argus would improve raw performance considerably.

IPC is the number of *instructions* executed per *clock cycle*. Throughput is the number of instructions executed per second (or *IPC* × *clock_cycle*). Peak throughput is the maximum throughput possible on a given architecture. Peak throughput is only attained when ideal IPC (optimal instruction-level parallelism) is sustained on the processor. Memory accesses, data dependencies, branching, and other code characteristics limit the achieved throughput on the processor. Using microbenchmarks, we measured the peak throughput for various instruction types on the machines under study.

Table III shows the results of our throughput experiments. Integer performance typically outperforms floating point performance on Argus. For DANIEL (the Intel architecture) floating point (F-xxx in Table III) and double (D-xxx in Table III) performances are comparable for ADD, MUL, and DIV, respectively. This is not true for Argus where F-MUL and D-MUL are significantly different as observed in our LINPACK measurements. We expect the modified version of the PowerPC architecture (with an additional floating point unit) present in IBM BlueGene/L will equalize the performance difference with the Intel architecture in future systems. CPU throughput measurements normalized for clock rates (MIPS) show Argus performs better than DANIEL for integer ADD/DIV/MOD, float ADD/MUL and double ADD instructions, but worse for integer MUL and double DIV instructions.

TABLE III
INSTRUCTION PERFORMANCE

Instruction	ARGUS			DANIEL		
	Cycles	IPC	MIPS	Cycles	IPC	MIPS
I-BIT	1	1.5	900	1	1.93	1771
I-ADD	1	2.0	1200	1	1.56	1393
I-MUL	2	1.0	300	4	3.81	880
I-DIV	20	1.0	30	39	1.08	36
I-MOD	24	1.0	25	42	1.08	24
F-ADD	3	3.0	600	3	2.50	764
F-MUL	3	3.0	600	5	2.50	460
F-DIV	18	1.0	33	23.6	1.08	42
D-ADD	3	3.0	600	3	2.50	764
D-MUL	4	2.0	300	5	2.50	460
D-DIV	32	1.0	19	23.6	1.08	42

Notes. IPC: instructions per clock, MIPS: millions of instructions per second, I: integer, F: single precision floating point; D: double precision floating point.

The performance of message communication is critical to overall parallel system performance. We measured memory communication latency and bandwidth with the MPPTEST tool available in the MPICH distribution. Results show that Argus performance is slightly worse yet comparable to DANIEL. MPI point-to-point latency is 104 ns (about 62 CPU cycles) on Argus and 87 ns (about 80 CPU cycles) on DANIEL. Both systems use 10/100 Mbps Ethernet so this is somewhat expected. However, we observed larger message injection overhead on Argus as message size approaches typical packet size. This is most likely due to the memory hierarchy disparity already described.

For further comparison, we measured the performance of two additional sequential benchmarks: NSIEVE and Livermore Loops. NSIEVE is a sieve of Eratosthenes program that varies array sizes to quantify the performance of integer operations. Livermore loops is a set of 24 DO-loops extracted from operational code used at Lawrence Livermore National Laboratory.

The NSIEVE benchmark results show that for small array size, Argus has a higher MIPS rating (980) than DANIEL (945). However, as array sizes increase, the relative performance of Argus decreases versus DANIEL. This reflects the differences in L2 cache performance between Argus and DANIEL.

The performance results from Livermore loops are summarized in Table IV. We observe DANIEL achieves 1.5–2 times higher MFLOPS rating than Argus for most statistical values, Argus achieves the best, worst-case execution time for this benchmark. For instance, in real time codes where worst-case performance must be as-

TABLE IV
LIVERMORE LOOPS

	ARGUS		DANIEL	
	MFLOPS	NORM.	MFLOPS	NORM.
Maximum Rate	731.5	1.22	1281.9	1.37
Quartile Q3	225.0	0.38	377.6	0.40
Average Rate	174.5	0.29	278.9	0.30
Geometric Mean	135.5	0.23	207.2	0.22
Median Q2	141.6	0.24	222.2	0.24
Harmonic Mean	106.6	0.18	133.6	0.14
Quartile Q1	66.4	0.11	132.6	0.14
Minimum Rate	46.2	0.08	20.0	0.02
Standard Dev	133.8	0.22	208.5	0.22
Average Efficiency	18.52%		16.16%	
Mean Precision (digits)	6.24		6.35	

Notes. NORM: normalized performance, obtained by dividing MFLOPS by CPU clock rate

sumed, Argus may be a better choice. However, examining performance normalized for clock rates (NORM) on this benchmark, the two systems perform similarly.

The Argus prototype architecture can execute both commercial and scientific applications. In this paper, we focus on scientific applications and provide results for two benchmark suites: LINPACK [25] and the NAS parallel benchmarks [5]. Since we already established the performance difference between Argus and DANIEL for single node (see previous section), here we only discuss the parallel performance of Argus.

LINPACK is arguably the most widely used benchmark for scientific applications and its measurements form the basis for the Top500 list [63] of fastest supercomputers in the world. Our measurements use HPL, a parallel version of the linear algebra subroutines in LINPACK that solve a (random) dense linear system in double precision (64-bit) arithmetic on distributed-memory computers. HPL provides the ability to scale workloads for better performance by adjusting array sizes. To ensure good performance, we compiled and installed the BLAS libraries with the aid of ATLAS (Automatically Tuned Linear Algebra Software). Table V shows the LINPACK benchmark results on the 16-node Argus prototype. The prototype achieves 3.4 GFLOPS, about 210 MFLOPS each node or 70% peak throughput of "double MUL" operations.

The NAS Parallel Benchmark (NPB) is a set of 8 programs designed to help evaluate the performance of parallel supercomputers. This benchmark suite consists of five application kernels and three pseudo-applications derived from computational fluid dynamics applications. These benchmarks are characterized with different compu-

TABLE V
LINPACK RESULTS ON ARGUS

NP	Problem size	GFLOPS	GFLOPS/proc	Speedup
1	3000	0.297	0.297	1.00
2	3000	0.496	0.248	1.67
4	5000	0.876	0.219	2.95
8	8000	1.757	0.221	5.91
16	12,000	3.393	0.212	11.42

TABLE VI
PARALLEL BENCHMARK RESULTS ON ARGUS

CODE	Performance (MOP/s)		
	$NP = 1$	$NP = 4$	$NP = 16$
CG	19.61	46.04	88.12
EP	1.69	6.75	24.08
IS	4.06	3.62	18.02
LU	48.66	188.24	674.62
MG	45.50	84.51	233.36
BT	40.04	131.76	436.29
SP	28.72	90.99	299.71

tation/communication ratios described in [5]. The raw performance of NPB 2.4.1 with a problem size of W on Argus is shown in Table VI. To better identify the performance trends, Fig. 10 provides the scalability of Argus under strong scaling (i.e. fixed problem size and increasing number of processors).

For 16 nodes, EP and LU show the best scalability. The embarrassingly-parallel code (EP) should achieve linear speedup since little communication is present. LU achieves super-linear speedup that appears to be levelling off. As working set size remains fixed with increases in the number of processors, communication is minimal (i.e. strong or fixed-size scaling). Super-linear performance is achieved as the working set gets smaller and smaller on a per node basis.

The curve of IS initially drops but then grows with the number of nodes. These codes stress communication performance. The levelling off of performance indicates the communication costs are not saturating the Ethernet interconnect up to 16 nodes.

The other four curves (SP, BT, CG and MG) have similar trends but different slopes. The performance of these codes reflects the communication to computation ratio. EP and LU are dominated by computation. IS and FT are dominated by communication. These codes sit somewhere in between. Trends here are similar (though less pronounced) than the communication-bound codes. These codes (SP, BT, CG,

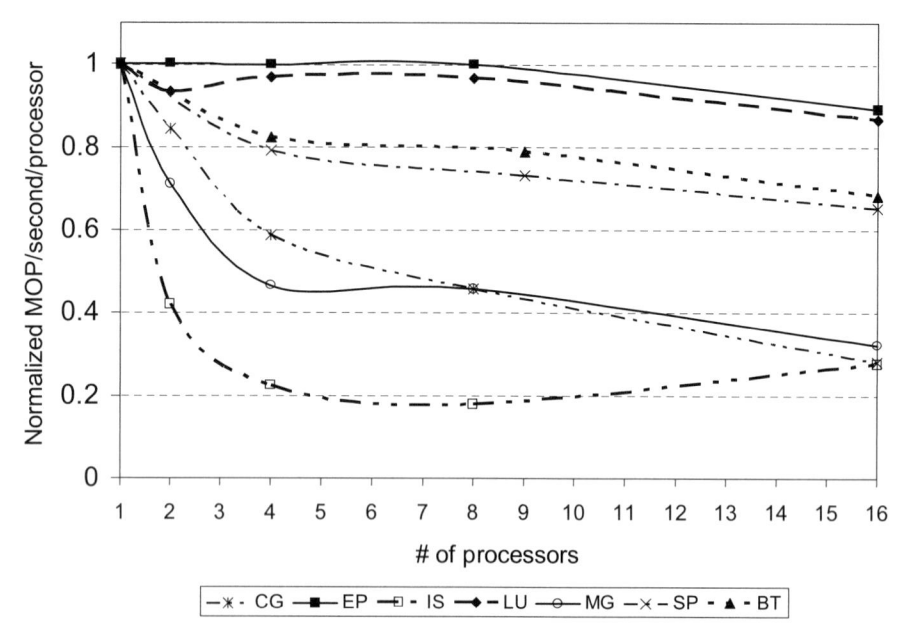

FIG. 10. Argus scalability. These curves show the varying scalability of parallel benchmarks from the NPB 2.4.1 release (class W). Main limitations on performance in the prototype include memory bandwidth and FP operation throughput. However, the result is a low power cluster capable of computing scientific applications.

and MG) are more sensitive to the number of nodes as it affects the number of communications. Performance then is likely to move downward with the number of nodes until a plateau is reached prior to network saturation (i.e. similar to the plateau in IS and FT performance). At some later point all of these codes will reach the limits of either the input data set size (Amdahl's Law) or the interconnect technology (saturation) where performance will drop drastically again. Our system is too small to observe these types of problems, so this is the subject of future work.

5.1.4 Lessons From a Low Power Cluster Design

Argus exemplifies an architectural design with trade-offs between performance, cost, space and power. The Argus prototype is a new approach to cluster computing that uses the aggregate processing elements on network analysis Load Modules for parallel computing. The analysis shows this architecture has advantages such as high scalability, small volumetric footprint, reduced power, high availability, and ultra-high processor density.

Argus achieves higher computing efficiency than Green Destiny, a comparable system with similar power efficiency. Argus is capable of packing more processors per blade than Green Destiny at present though future versions of both machines will undoubtedly address this.

The benchmarking measurements and comparisons with DANIEL indicate that the current Argus prototype has two major performance limitations due to the architectural characteristics of the embedded PowerPC processor: L2 cache latency and hardware support for double precision. Also the communication overhead on the processing node should and could be improved through system-specific hardware and software tuning of MPI. Furthermore, results from a larger prototype with a faster interconnect would allow more comprehensive scalability analyses.

There are two concerns with the low power cluster design approach highlighted by our experiments with Argus. First, performance is not considered a critical design constraint. In all low power approaches including Argus, Green Destiny and IBM BlueGene/L, performance is sacrificed to reduce the power consumption of the machine. BlueGene/L has been the most successful at overcoming this constraint by (1) redesign of the PowerPC embedded architecture to support double precision floating point operations, and (2) custom design of a 130,000+ processor system.

Second, the low power approach is limited since it assumes all applications suffer from poor power efficiency. This contradicts our earlier findings that the power needs of applications vary significantly over time. Together, these observations motivate the need for power-conscious approaches in high-performance that adapt to the performance phases of applications.

6. Power-Aware Computational Clusters

Recently, power has become a critical issue for large data centers. Studies of power consumption and energy conservation on commercial servers have emerged. Bohrer et al. [9] found dynamic voltage and frequency scaling (DVFS) for processors is effective for saving energy in web servers. Carrera et al. [17] found multi-speed disks can save energy up to 23% for network servers. Zhu et al. [72,73] combines several techniques including multi-speed disks and data migration to reduce energy consumption while meeting performance goals.

Power reduction becomes critical in high-performance computing to ensure reliability and limit operating cost. Two kinds of systems are now built to accommodate these goals: systems with low power components (discussed in the previous section [8,31,32]) and systems with power-aware components [43]. Energy reduction using power-aware technologies had not been exploited in high-performance computing until our research was launched in 2004 [15].

In contrast to Argus, Green Destiny and BlueGene/L, our power-aware approach for HPC uses off-the-shelf DVS technologies[3] in server-class systems to exploit scientific workload characteristics. Power-aware clusters include components that have multiple power/performance modes (e.g. CPU's with DVS). The time spent within and transitioning to/from these power/performance modes determines the delay (cost in performance) and energy (cost in power, heat, etc.).

There are two ways to schedule DVS transitions. *External* distributed DVS scheduling techniques are autonomous and control DVS power/performance modes in a cluster as processes separate from the application under study. External schedulers may be system-driven (e.g. the cpuspeed daemon) or user-driven (e.g. setting DVS from the command line). *Internal* distributed DVS scheduling techniques use source-level performance profiling to direct DVS power/performance modes with source-code instrumentation.

6.1 Using DVS in High-Performance Clusters

Dynamic Voltage Scaling (DVS) is a technology now common in high-performance microprocessors [3,46]. DVS works on a very simple principal: decreasing the supply voltage to the CPU consumes less power.

As shown in Eq. (4), the dynamic power consumption (P) of a CMOS-based microprocessor is proportional to the product of total capacitance load (C), frequency (f), the square of the supply voltage (V^2), and percentage of active gates (A) or $P \approx ACV^2 f$. As shown in Eq. (6), energy consumption (measured in joules) is reduced by lowering the average power (P_{avg}) consumed for some duration or delay (D) or $E = P_{avg} \times D$.

There are two reasons for using DVS to conserve energy in HPC server clusters. The first reason is to exploit the dominance of CPU power consumption on the system node (and thus cluster) power consumption. Figure 3(b) shows the breakdown of system node power obtained using direct measurement [33]. This figure shows percentage of total system power for a Pentium III CPU (35%) under load. This percentage is lower (15%) but still significant when the CPU is idle. While the Pentium III can consume nearly 45 W, recent processors such as Itanium 2 consume over 100 W and a growing percentage of total system power. Reducing the average power consumed by the CPU can result in significant server energy savings magnified in cluster systems.

The second reason for using DVS to conserve energy in HPC server clusters is to save energy without increasing execution time. DVS provides the ability to adaptively reduce power consumption. By reducing CPU power when peak speed is not

---

[3] DVS capabilities are now available in server-class architectures including Intel Xeon (SE7520 chipset) and AMD Opteron (Tyan s2882 board) dual processor nodes.

Normalized Delay and Energy for Swim

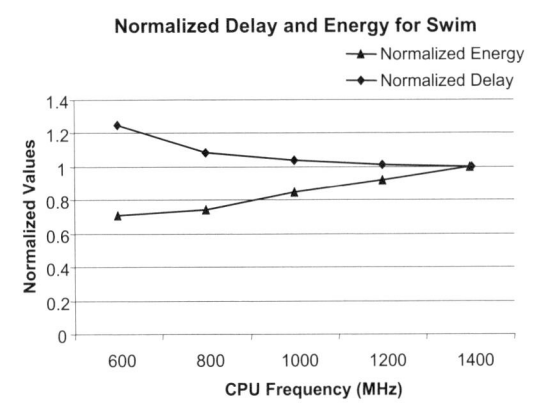

FIG. 11. The energy-delay crescendo for swim shows the effect of application-dependent CPU slackness on (node) energy and performance measured at a single NEMO node. For swim, energy conservation can be achieved with (at times) reasonable performance loss.

needed (e.g. idle or slack periods) a DVS scheduling algorithm can reduce energy consumption. To minimize the impact on execution time we must ensure (1) supply voltage is reduced during CPU times when peak speed is not necessary, and (2) period duration outweighs voltage state transition costs.[4]

Figure 11 shows the use of DVS on a single node to exploit CPU slack due to memory stalls. In this example we run swim from the SPEC 2000 benchmark suite on a DVS-enabled node at various fixed voltages shown as the resulting frequency[5] on the x-axis in increasing order. Lower frequency (i.e. lower voltage) means lower CPU performance. The values plotted on the y-axis are normalized to the highest (i.e. fastest) frequency respectively for energy and delay (execution time). This energy-delay "crescendo" for swim shows when reducing CPU frequency (from right to left) the delay (execution time) increase varies from almost no increase at 1200 MHz to about 25% increase at 600 MHz. The corresponding total system energy decreases steadily with lower frequencies. Simply put, the memory stalls in swim produce enough slack periods for DVS to save energy (e.g. 8% at 1200 MHz) with almost no impact on execution time (<1%).

In the rest of this section, we will analyze the tradeoffs of various DVS scheduling techniques designed to exploit CPU slack time in computational clusters. For parallel

---

[4] In our AMD Opteron-based systems transition costs vary from 20–30 μs. Manufacturers set lower bounds (∼10 μs) to achieve system stability following mode transitions.

[5] To be precise, DVS affects both voltage and frequency. Voltage and frequency are not independent as shown in Table I. However for ease of discussion, we describe measurements in terms of the resulting frequency.

codes, idle CPU periods will be workload dependent and result from both memory and remote communication stalls.

6.2 Distributed DVS Scheduling Strategies

Now that we have established DVS as a viable approach to conserving energy while maintaining performance for HPC applications, we turn our attention to describing several approaches to schedule DVS transitions over the duration of a parallel code. Our goal in this section is not to explore every possible alternative in distributed DVS scheduling, but to provide detail on three techniques that differ in complexity and efficiency.

The scheduling techniques studied can be characterized as follows: (1) user- or system-driven, (2) internal or external to the application. The techniques can be evaluated by directly measuring the amount of total system energy consumed and the amount of execution time required to solution. Figure 12 provides an overview of the three scheduling methods studied.

6.2.1 CPUSPEED Daemon

6.2.1.1 Strategy #1: Using the CPUSPEED Daemon.
System-driven, external control of distributed DVS scheduling can be implemented as a system process or Daemon. The daemon we study is the CPUSPEED program included in the latest Fedora Core releases.[6] CPUSPEED schedules the DVS modes of a single node according to the CPU utilization information recorded by the system in the /proc directory of Linux. Other operating systems (e.g. Windows

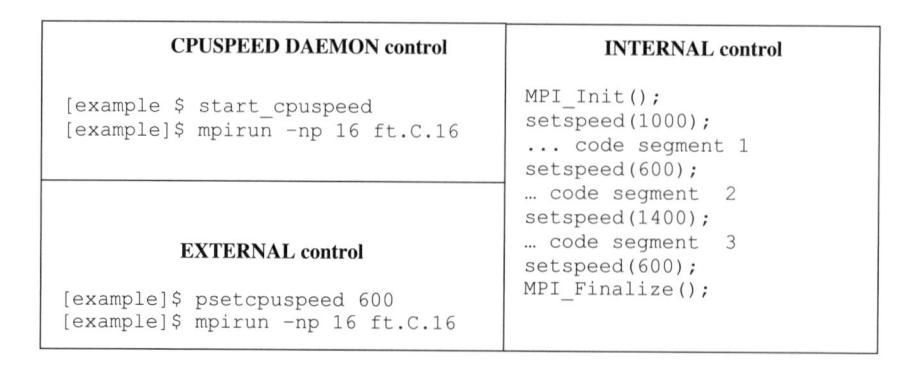

CPUSPEED DAEMON control	INTERNAL control

[example $ start_cpuspeed
[example]$ mpirun -np 16 ft.C.16
	MPI_Init();
	setspeed(1000);
	... code segment 1
	setspeed(600);

EXTERNAL control
 	
[example]$ psetcpuspeed 600
[example]$ mpirun -np 16 ft.C.16
	... code segment 2
	setspeed(1400);
	... code segment 3
	setspeed(600);
	MPI_Finalize();

FIG. 12. Illustrations of the usage of three distributed DVS control strategies.

[6] See http://carlthompson.net/Software/CPUSpeed.

running on a laptop) provide comparable daemons for scheduling CPU, disk, and monitor power modes when the system is underutilized. These processes run autonomously and typically use saturation-based counters (or thresholds) and simple history-based information (e.g. CPU utilization) to migrate components between power modes.

Assuming a power aware node supports **m** operating points (voltage steps or frequencies) and the current operating point is **S**, the basic algorithm for CPUSPEED is as follows:

```
while( true )
{
  poll %CPU-usage from ''/proc/stat''
  if %CPU-usage < minimum-threshold
      S = 0
  else if %CPU-usage > maximum-threshold
      S = m
  else if %CPU-usage < CPU-usage-threshold
      S = max( S-1, 0)
  else
      S = min( S+1, m)
  set-cpu-speed ( speed[S] )
  sleep (interval)
}
```

6.2.2 EXTERNAL

6.2.2.1 Strategy #2: Scheduling from the Command-Line.
User-driven, external control can be implemented as a program invocation from the command line or as a system-call from a process external to the application. This is the approach used to save energy on a single node for the swim code shown in Fig. 11. In the distributed version of this approach, the user synchronizes and sets the frequency for each node statically[7] prior to executing the application. For distributed applications that are memory/communication bound or imbalanced applications with large amounts of CPU slack or idle time, this approach is simple and effective. Performance profiling can be used to determine the amount of time the application spends stalled on the CPU for a given node configuration. Individual nodes can then be set to different DVS speeds appropriate to their share of the workload.

---

[7] Dynamic settings are more appropriate for internal control from within the application (discussed next).

The process of DVS scheduling using external control is as follows: First we run a series of microbenchmarks to determine the effect of DVS on common types of code blocks including computationally intensive, memory intensive and communication intensive sequences. Next, we profile the performance of the application under study as a black box. We then determine which power mode settings are appropriate for the entire application running on a single node. Prior to execution, we set the individual nodes accordingly.

6.2.3 INTERNAL

6.2.3.1 Strategy #3: Scheduling within Application.
User-driven, internal control can be implemented using an API designed to interface with the power-aware component in the node. By controlling DVS from within an application, we can control the granularity of DVS mode transitions. The appropriate level of granularity depends on the amount of slack time and the overhead for mode transitions. For some codes with intensive loop-based computation, transitions between power modes around basic blocks may be appropriate. For other codes, function-level granularity may be more useful. At the extreme end, we can resort to external scheduling at node granularity.

Application-level control requires an API. We created such an API as part of our PowerPack framework discussed earlier. The insertion of API DVS control commands can be implemented by a compiler, middleware, or manually.

The process of DVS scheduling using internal API control is as follows: First we run a series of microbenchmarks to determine the effect of DVS on common types of code blocks including computationally intensive, memory intensive and communication intensive sequences. Next, we profile the performance of the application under study at a fine granularity identifying code sequences that share characteristics with our microbenchmarks. We then determine which power mode settings are appropriate for a given code sequence and insert the appropriate API calls around the code blocks. For now we do this manually. As part of future work we plan to integrate this into a compiler or run-time tool.

Figure 12 provides an example using each of the three strategies described. In the rest of this chapter, we use CPUSPEED DAEMON to refer to strategy #1, EXTERNAL to refer to strategy #2, and INTERNAL to refer to strategy #3. Using CPUSPEED DAEMON, users execute their application after the daemon is running. Using EXTERNAL, users determine a suitable operating frequency and set all the nodes to this operating point[8] (such as 600 MHz in the example in Fig. 12) before

---

[8] For now we focus on a homogeneous implementation of the EXTERNAL scheduling algorithm. Heterogeneous (different nodes at different speeds) is straightforward but requires further profiling which is actually accomplished by the INTERNAL approach.

executing the application. Using INTERNAL, users insert DVS function calls into the source code, and execute the re-compiled application. When either external or internal scheduling is used, CPUSPEED must be turned off.

6.3 Experimental Framework

Our experimental framework is composed of five components: experimental platform, performance and energy profiling tools, data collection and analysis software, microbenchmarks, and metrics for analyzing system power-performance.

6.3.1 NEMO: Power-Aware Cluster

To better understand the impact of DVS technologies on future high performance computing platforms with DVS, we built a scalable cluster of high-performance, general purpose CPU's with DVS capabilities. Our experimental framework is shown in Fig. 13. It interacts with NEMO, a 16-node DVS-enabled cluster,[9] Baytech power management modules and a data workstation.

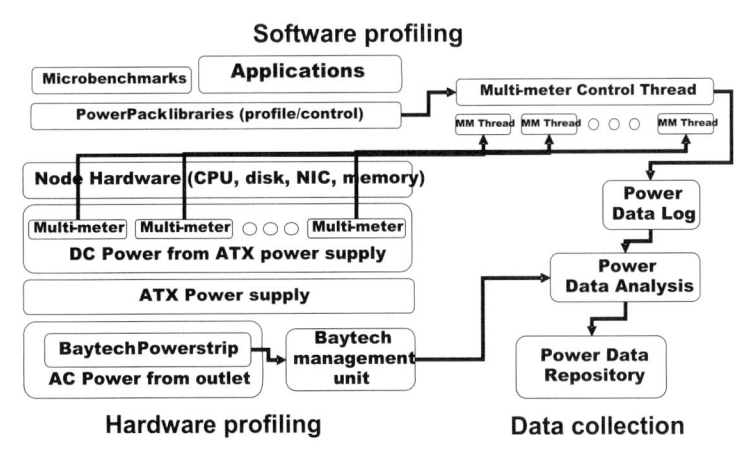

FIG. 13. The PowerPack framework. This software framework is used to measure, profile, and control several power-aware clusters including the cluster under study. Measurements are primarily obtained from the ACPI interface to the batteries of each node in the cluster and the Baytech Powerstrips for redundancy. The PowerPack libraries provide an API to control the power modes of each CPU from within the applications. Data is collected and analyzed from the Baytech equipment and the ACPI interface.

---

[9] We use this system prototype to compare and contrast the scheduling policies discussed. Our techniques are general and equally applicable to emergent server-based clusters with DVS enabled dual AMD Opterons and Intel Xeon processors. This cluster was constructed prior to the availability of such nodes to the general public.

TABLE VII
PENTIUM M 1.4 GHz OPERATING POINTS

Frequency	Supply voltage
1.4 GHz	1.484 V
1.2 GHz	1.436 V
1.0 GHz	1.308 V
800 MHz	1.180 V
600 MHz	0.956 V

NEMO is constructed with 16 Dell Inspiron 8600 laptops connected by 100M Cisco System Catalyst 2950 switch. Each node is equipped with a 1.4 GHz Intel Pentium M processor using Centrino mobile technology to provide high-performance with reduced power consumption. The processor includes on-die 32 KB L1 data cache, on-die 1 MB L2 cache, and each node has 1 GB DDR SDRAM. Enhanced Intel SpeedStep technology allows the system to dynamically adjust the processor among five supply voltage and clock frequency settings given by Table VII. The lower bound on SpeedStep transition latency is approximately 10 microseconds according to the manufacturer [47].

We installed open-source Linux Fedora Core 2 (version 2.6) and MPICH (version 1.2.5) on each node and use MPI (message passing interface) for communication. We use CPUFreq as the interface for application-level control of the operating frequency and supply voltage of each node.

6.3.2 Power, Energy and Performance Profiling on NEMO

For redundancy and to ensure correctness, we use two independent techniques to directly measure energy consumption.

The first direct power measurement technique is to poll the battery attached to the laptop for power consumption information using ACPI. An ACPI smart battery records battery states to report remaining capacity in mWh (1 mWh = 3.6 J). This technique provides polling data updated every 15–20 seconds. The energy consumed by an application is the difference of remaining capacity between execution beginning and ending when the system is running on DC battery power. To ensure reproducibility in our experiments, we do the following operations prior to all power measurements:

1. Fully charge all batteries in the cluster.
2. Disconnect (automatically) all laptops from wall outlet power remotely.
3. Discharge batteries for approximately 5 min to ensure accurate measurements.
4. Run parallel applications and record polling data.

The second direct power measurement technique uses specialized remote management hardware available from Bay Technical (Baytech) Associates in Bay St. Louis, MS. With Baytech proprietary hardware and software (GPML50), power related polling data is updated each minute for all outlets. Data is reported to a management unit using the SNMP protocol. We additionally use this equipment to connect and disconnect building power from the machines as described in the first technique.

To correlate the energy and performance profile, we also generate profiles of tested applications automatically by using an instrumented version of MPICH. We do application performance and energy profiling separately due to the overhead incurred by event tracing and recording.

6.3.3 PowerPack Software Enhancements

While direct measurement techniques are collectively quite useful, it was necessary to overcome two inherent problems to use them effectively. First, these tools may produce large amounts of data for typical scientific application runs. This data must be collected efficiently and analyzed automatically (if possible). Second, we must coordinate power profiling across nodes and account for hardware polling rates within a single application. As in the original PowerPack suite of applications, this includes correlating energy data to source code.

To overcome these difficulties, we enhanced our PowerPack tool suite. As we discussed earlier in this chapter, PowerPack automates power measurement data collection and analysis in clusters. We added portable libraries for (low-overhead) timestamp-driven coordination of energy measurement data and DVS control at the application-level using system calls. ACPI data is also obtained and collated using this new library (libbattery.a). Lastly, we created software to filter and align data sets from individual nodes for use in energy and performance analysis and optimization. The data shown herein is primarily obtained using our ACPI-related libraries; however all data is verified using the Baytech hardware.

6.3.4 Energy-Performance Microbenchmarks

We measure and analyze results for a series of microbenchmark codes (part of our PowerPack tool suite) to profile the memory, CPU, and network interface energy behavior at various static DVS operating points. These microbenchmarks are grouped into three categories:

I. Memory-Bound Microbenchmark
II. CPU-bound microbenchmark
III. Communication-bound microbenchmark

Normalized energy and delay of memory access

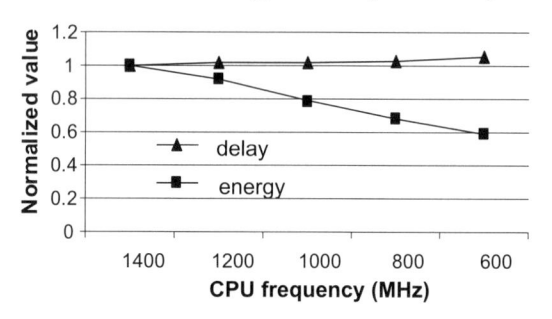

FIG. 14. Normalized energy and delay for a memory bound microbenchmark. Memory bound code phases provide opportunities for energy savings without impacting performance.

We leave disk-bounded microbenchmarks for future study, though disk-bound applications will provide more opportunities for energy saving.

I. Memory-Bound Microbenchmark. Figure 14 presents the energy consumption and delay of memory accesses under different CPU frequencies. The measured code reads and writes elements from a 32 MB buffer with stride of 128 Bytes, which assures each data reference is fetched from main memory. At 1.4 GHz, the energy consumption is at its maximum, while execution time is minimal. The energy consumption decreases with operating frequency, and it drops to 59.3% at the lowest operating point 600 MHz. However, execution time is only minimally affected by the decreases in CPU frequency; the worst performance at 600 MHz shows a decrease of only 5.4% in performance. The conclusion is memory-bound applications offer good opportunity for energy savings since memory stalls reduce CPU efficiency.

Using our weighted power-performance efficiency metrics (EDP), we can further explain this phenomenon. The best energy operating point is 600 MHz which is 40.7% more efficient than the fastest operating point (1.4 GHz). More pointedly, in our context this memory behavior explains the single node behavior of codes such as the swim benchmark (see Fig. 11).

II. CPU-Bound Microbenchmark. Figure 15 shows energy consumption and delay using DVS for a CPU-intensive micro benchmark. This benchmark reads and writes elements in a buffer of size 256 KB with stride of 128 B, where each calculation is has an L2 cache access. Since L2 cache is on-die, we can consider this code CPU-intensive. The energy consumption for a CPU-intensive phase is dramatically different from a memory bound code phase in that the CPU is always busy and involved in computation.

Normalized energy and delay for L 2 cache access

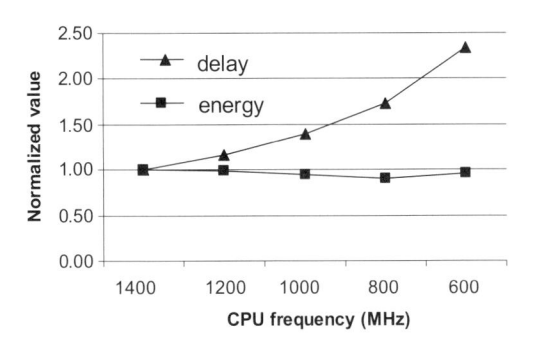

FIG. 15. Normalized energy and delay for a CPU bound microbenchmark. CPU bound code phases DO NOT provide opportunities for energy savings. To maximize performance, maximum CPU speed is needed.

As we expect, the results in Fig. 15 do not favor energy conservation. Delay increases with CPU frequency near-linearly. At the lowest operating point, the performance loss can be 134%. On the other hand, energy consumption decreases initially, and then goes up. Minimum energy consumption occurs at 800 MHz (10% decrease). Energy consumption then actually increases at 600 MHz. The dramatic decrease in performance by the slow down to 600 MHz cancels out the reduced power consumption. That is, while average power may decrease, the increased execution time causes total energy to increase. If we limit memory accesses to registers thereby eliminating the latency associated with L2 hits the results are even more striking. The lowest operating point consumes the most energy and takes 245% longer to complete. The computationally bound SPEC code mgrid exhibits behavior that reflects this data. For the parallel benchmarks we studied we rarely observe this exact behavior. Computational phases for parallel codes are normally bound to some degree by memory accesses.

III. Communication-Bound Microbenchmark. Figure 16 shows the normalized energy and execution time for MPI primitives. Figure 16(a) is the round trip time for sending and receiving 256 KB of data. Figure 16(b) is the round trip time for a 4 KB message with stride of 64 B. Compared to memory load latency of 110 ns, simple communication primitives MPI_Send and MPI_Recv take dozens of microseconds, and collective communication takes several hundreds of microseconds for two nodes, both present more CPU slack time than memory access.

Normalized energy and delay for 256K round trip

(a)

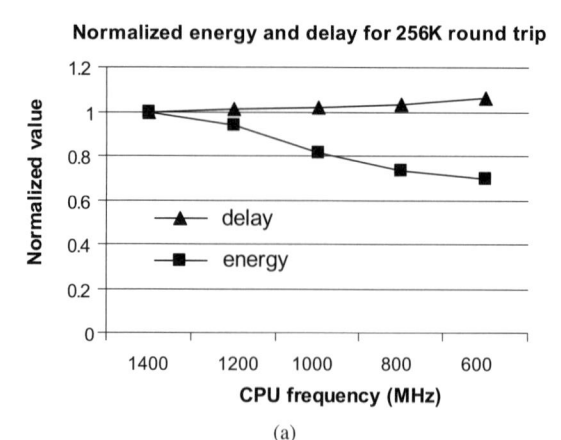

Normalized energy and delay for strided message

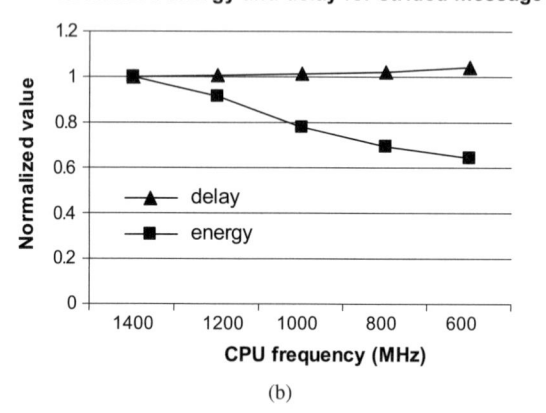

(b)

FIG. 16. Normalized energy and delay for a communication bound microbenchmark. Round trip delay is measured for (a) 256 KB non-strided message, and (b) 4 KB message with 64 B stride. Communication bound code phases provide opportunities for energy savings.

As we expect, the crescendos in Fig. 16(a) and (b) are favorable to energy conservation for both types of communication. For the 256K round trip, energy consumption at 600 MHz decreases 30.1% and execution time increases 6%. For 4 KB message with stride of 64 B, at 600 MHz the energy consumption decreases 36% and execution time increases 4%.

The energy gains apparent during communications are related to the communication to computation ratio of the application. As this ratio decreases, so should the impact of communication on the effectiveness of DVS strategies.

6.3.5 Energy-Performance Efficiency Metrics

When different operating points (i.e. frequency) are used, both energy and delay vary even for the same benchmark. A fused metric is required to quantify the energy-performance efficiency. In this section, we use ED2P ($E \times D^2$) and ED3P ($E \times D^3$) to choose "optimal" operating points (i.e., the CPU frequency that has the minimum ED2P or ED3P value for given benchmarks) in DVS scheduling for power-aware clusters. ED2P is proportional to J/MIPS^2, and ED3P is proportional to J/MIPS^3. Since the ED3P metric emphasizes performance, smaller performance loss is expected for scheduling with ED3P in contrast to scheduling with ED2P. As before, both energy and delay are normalized with the values at the highest frequencies.

6.4 Analyzing an Energy-Conscious Cluster Design

This section presents our experimental results for the NAS parallel benchmarks (NPB) [6] using three DVS scheduling strategies. The benchmarks, which are derived from computational fluid applications, consist of five parallel kernels (EP, MG, CG, FT and IS) and three pseudo-applications (LU, SP and BT). These eight benchmarks feature different communication patterns and communication to computation ratios. We note experiments as XX.S.# where XX refers to the code name, S refers to the problem size, and # refers to the number of nodes. For example, LU.C.8 is the LU code run using the C sized workload on 8 nodes. In all our figures, energy and delay values are normalized to the highest CPU speed (i.e. 1400 MHz). This corresponds to energy and delay values without any DVS activity.

To ensure accuracy in our energy measurements using ACPI, we collected data for program durations measured in minutes. In some cases we used larger problem sizes to ensure application run length was long enough to obtain accurate measurements. In other cases we iterate application execution. This ensures the relatively slow ACPI refresh rates (e.g. 15–20 s) accurately record the energy consumption of the battery for each node. Additionally, we repeated each experiment at least 3 times or more to identify outliers.

6.4.1 CPUSPEED Daemon Scheduling

Figure 17 shows NAS PB results using CPUSPEED daemon to control DVS scheduling on our distributed power-aware cluster. We evaluate the effect of two versions of CPUSPEED: one is version 1.1 included in Fedora 2 and the other is version 1.2.1 included in Fedora 3. In version 1.1, the default minimum CPU speed transition interval value is 0.1 second; in version 1.2.1, the default interval value has been changed to 2 seconds. Since we have observed that CPUSPEED version 1.1 al-

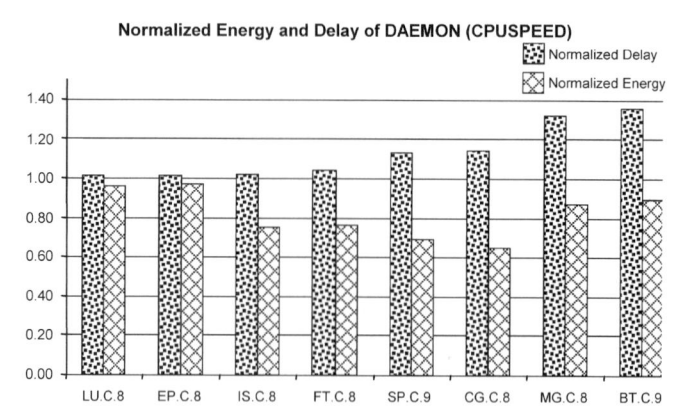

Normalized Energy and Delay of DAEMON (CPUSPEED)

FIG. 17. Energy-performance efficiency of NPB codes using CPUSPEED version 1.2.1. The results are sorted by normalized delay. Normalized delay is total application execution time with DVS divided by total application execution time without DVS. Values < 1 indicate performance loss. Normalized energy is total system energy with DVS divided by total system energy without DVS. Values < 1 indicate energy savings.

ways chooses the highest CPU speed for most NPB codes without significant energy savings [36], only results of the improved CPUSPEED 1.2.1 are shown in Fig. 17.

The effects of CPUSPEED vary with different codes. For LU and EP, it saves 3~4% energy with 1~2% delay increase in execution time. For IS and FT, it reduces 25% energy with 1~4% delay. For SP and CG, it reduces 31~35% energy with 13~14% delay increase. However, for MG and BT, it reduces 21% and 23% energy at the cost of 32% and 36% delay increase.

The original version of CPUSPEED 1.1 was equivalent to no DVS (our 1400 MHz base data point) since threshold values were never achieved. CPUSPEED version 1.2.1 improves energy-performance efficiency for scientific codes significantly by adjusting the thresholds. We intend to study the affects of varying thresholds for applications that perform poorly even under the improved version in future work.

Overall, CPUSPEED 1.2.1 does a reasonable job of conserving energy. However, for energy conservation of significance (>25%) 10% or larger increases in execution time are necessary, which is not acceptable to the HPC community. The history-based prediction of CPUSPEED is the main weakness of the CPUSPEED DAEMON scheduling approach. This motivates a study of scheduling techniques that incorporate application performance profiling in the prediction of slack states.

6.4.2 External Scheduling

We now examine coarse-grain, user-driven external control which assumes users know the overall energy-performance behavior of an application but treat the internals of the application as a black box.

We previously described the steps necessary to create a database of microbenchmark information for use in identifying DVS settings appropriate to an application. Applications with communication/computation or memory/computation ratios that match micro-benchmark characteristics allow a priori selection of DVS settings. Here, our goal is to analyze this DVS scheduling approach for the power mode settings in our system.

Table VIII gives raw figures for energy and delay for all the frequency operating points available on our system over all the codes in the NAS PB suite. As is evident, such numbers are a bit overwhelming to the reader. Furthermore, selecting a "good" frequency operating point is a subjective endeavor. For instance, BT at 1200 MHz has 2% additional execution time (delay) with 7% energy savings. Is this "better" or

TABLE VIII
ENERGY-PERFORMANCE PROFILES OF NPB BENCHMARKS

Code	CPU speed					
	Auto	600 MHz	800 MHz	1000 MHz	1200 MHz	1400 MHz
BT.C.9	1.36	1.52	1.27	1.14	1.05	1.00
	0.89	0.79	0.82	0.87	0.96	1.00
CG.C.8	1.14	1.14	1.08	1.04	1.02	1.00
	0.65	0.65	0.72	0.80	0.93	1.00
EP.C.8	1.01	2.35	1.75	1.40	1.17	1.00
	0.97	1.15	1.03	1.02	1.03	1.00
FT.C.8	1.04	1.13	1.07	1.04	1.02	1.00
	0.76	0.62	0.70	0.80	0.93	1.00
IS.C.8	1.02	1.04	1.01	0.91	1.03	1.00
	0.75	0.68	0.73	0.75	0.94	1.00
LU.C.8	1.01	1.58	1.32	1.18	1.07	1.00
	0.96	0.79	0.82	0.88	0.95	1.00
MG.C.8	1.32	1.39	1.21	1.10	1.04	1.00
	0.87	0.76	0.79	0.85	0.97	1.00
SP.C.9	1.13	1.18	1.08	1.03	0.99	1.00
	0.69	0.67	0.74	0.81	0.91	1.00

Notes. Only partial results are shown here. In each cell, the number on the top is the normalized delay and the number at the bottom is the normalized energy. The column "auto" means scheduling using CPUSPEED. The columns "XXX MHz" refer to the static external setting of processor frequency.

Normalized Delay and Energy Using EXTERNAL control (ED3P)

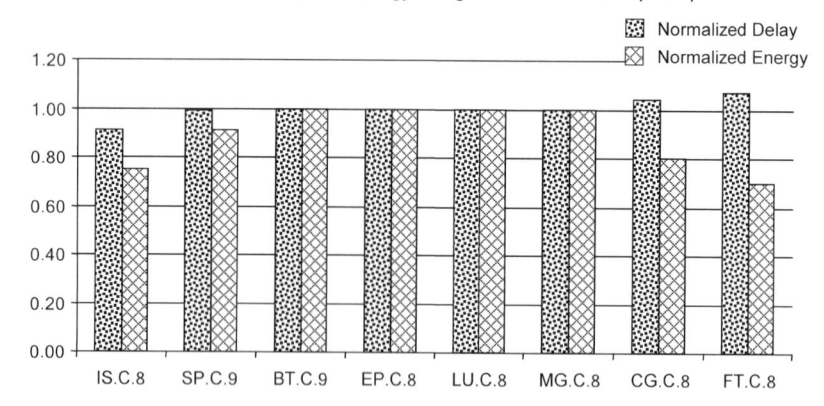

FIG. 18. Energy-performance efficiency of NPB codes using EXTERNAL DVS control. ED3P is chosen as the energy-performance metric in this figure. The results are sorted by normalized delay.

"worse" than BT at 1000 MHz with 4% additional execution time and 20% energy savings? Such comparisons require a metric to evaluate.

Figure 18 shows the energy-performance efficiency of NPB benchmarks using external control with the ED3P (ED^3) metric to weight performance significantly more than energy savings. This figure is obtained as follows: For each benchmark, compute the ED^3 value at each operating point using corresponding normalized delay and normalized energy, and use the operating point which has the smallest ED^3 value as the scheduling point thereafter. If two points have the same ED^3 value, choose the point with best performance. External DVS scheduling shown reduces energy with minimum execution time increase and selects an operating frequency that is application dependent—thus overcoming the weakness of CPUSPEED.

The effects of external DVS scheduling can be classified in three categories:

- Energy reduction with minimal performance impact. For FT, EXTERNAL saves 30% energy with 7% delay increase in execution time. For CG, EXTERNAL saves 20% energy with 4% delay increase in execution time.
- Energy reduction and performance improvement.[10] For SP, EXTERNAL saves 9% energy and also improves execution time by 1%. For IS, EXTERNAL saves 25% energy with 9% performance improvement.

    No energy savings and no performance loss. BT, EP, LU, MG fall into this category.

[10] These results are repeatable. Similar phenomena have been observed by other researchers. Our explanation is message communication is not sensitive to frequency above a certain threshold. Higher communication frequency (common to IS and SP) increases the probability of traffic collisions and longer waiting times for retransmission.

Normalized Delay and Energy Using EXTERNAL Control (ED2P)

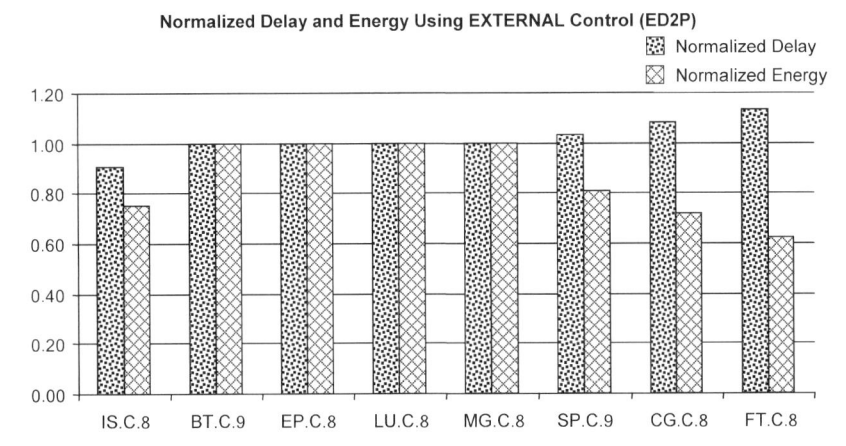

FIG. 19. Energy-performance efficiency of NPB codes using EXTERNAL control. ED2P is chosen as the energy-performance metric in this figure. The results are sorted by normalized delay.

If users allow slightly larger performance impact for more energy saving, ED2P (ED^2) or EDP (ED) can be used as the energy-performance metric. Figure 19 shows the effects of ED2P metrics on external DVS scheduling. The trend is the same as Fig. 18, but the metric may recommend frequency operating points where energy savings have slightly more weight than execution time delays. For example, ED2P would recommend different operating points for FT corresponding to energy savings of 38% with 13% delay increase; for CG, it selects 28% energy with 8% delay increase. For SP, it selects 19% energy with 3% delay increase.

We can use energy-delay crescendos to observe the effects on delay and energy visually for comparison to our microbenchmark results (Fig. 20). These figures indicate we can classify the NPB benchmarks as follows:

Type I (EP): near zero energy benefit, linear performance decrease when scaling down CPU speed. This is similar to the observed effects of CPU bound codes. The EP code performs very little communication and is basically computationally bound to the performance of any given node. Thus, reducing the CPU speed hurts performance and energy conservation for HPC is unlikely.

Type II (BT, MG and LU): near linear energy reduction and near linear delay increase, the rate of delay increase and energy reduction is about same. The results for these codes fall between CPU bound and memory or communication bound. The effects overall can lead to some energy savings, but EXTERNAL control means phases cannot adapt to changes in communication to computation ratio. In this case the overall effect is performance loss for energy savings, not acceptable in HPC.

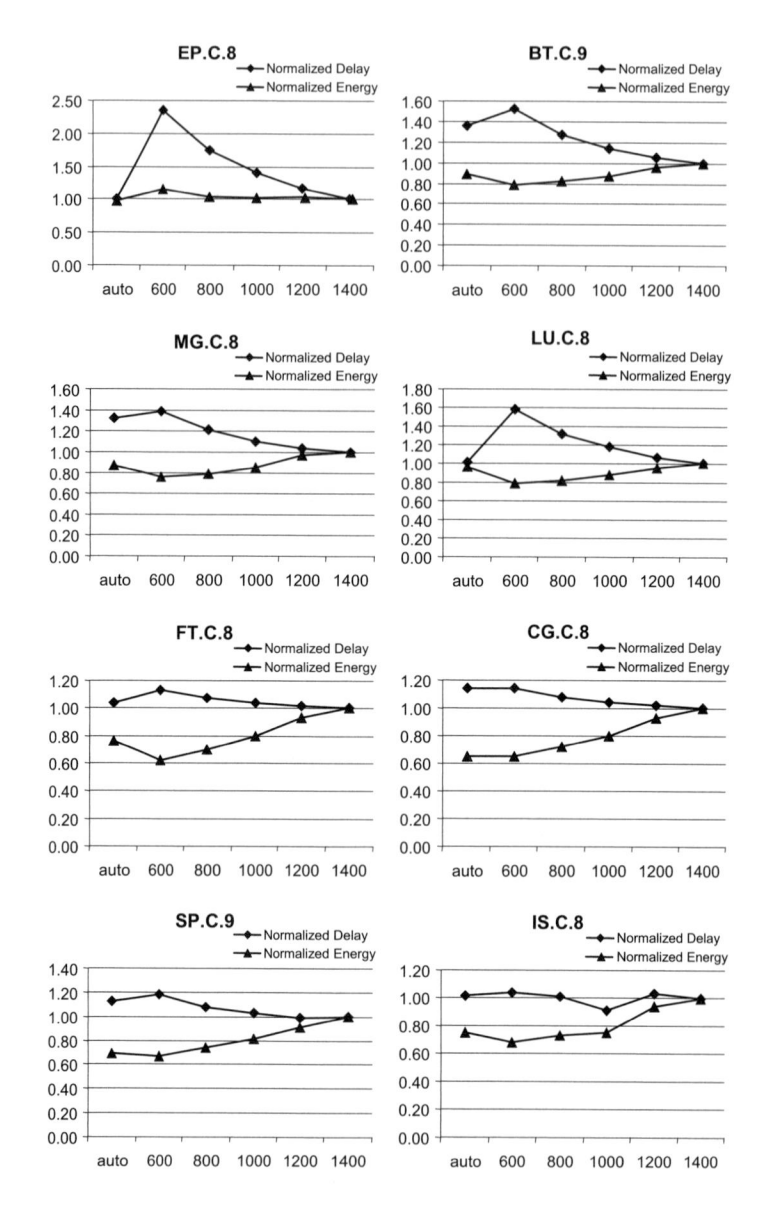

FIG. 20. Energy-delay crescendos for the NPB benchmarks. For all diagrams, x-axis is CPU speed, y-axis is the normalized value (delay and energy). The effects of DVS on delay and energy vary greatly.

Type III (FT, CG and SP): near linear energy reduction and linear delay increase, where the rate of delay increase is smaller than the rate of energy reduction. These codes can use DVS to conserve energy effectively. Communication or memory to computation ratio is quite high in many phases of these codes. However, the EXTERNAL control course granularity means parts of the code suffer performance loss. In some cases, the performance is minimal, in others it is not.

Type IV (IS): near zero performance decrease, linear energy saving when scaling down CPU speed. This code is almost completely communication bound (integer parallel sort). Thus frequency of the processor has little effect on performance and running at low frequency will save energy. Codes in this category can be run at low frequency and meet HPC users' needs.

This classification reveals the inherent limitations to external control. First, the energy-performance impact is a function of an application's performance phases. Yet, the granularity of EXTERNAL control is to try a best-fit operating point for the entire application. This causes additional performance delay and does not meet the dynamic criteria we described as characteristic of a good DVS scheduler for HPC applications. Second, the homogeneity of setting all processors to the same frequency limits effectiveness to homogeneous applications. Workload imbalance, common to scientific application such as adaptive mesh refinement, is not exploited using external control.

6.4.3 Internal Scheduling

We use FT.C.8 and CG.C.8 as examples to illustrate how to implement internal scheduling for different workloads. Each example begins with performance profiling followed by a description of the DVS scheduling strategy derived by analyzing the profiles.

6.4.3.1 FT Performance.
Figure 21 shows the performance profile of FT generated with the MPICH trace utility by compiling the code with "–mpilog" option. The following observations are drawn from this profile:

- FT is communication-bound and its communication to computation ratio is about 2:1.
- Most execution time is consumed by all-to-all type communications.
- The execution time per communication phase is large enough to compensate for the CPU speed transition overhead (20–30 μs observed).
- The workload is almost homogeneous and balanced across all nodes.

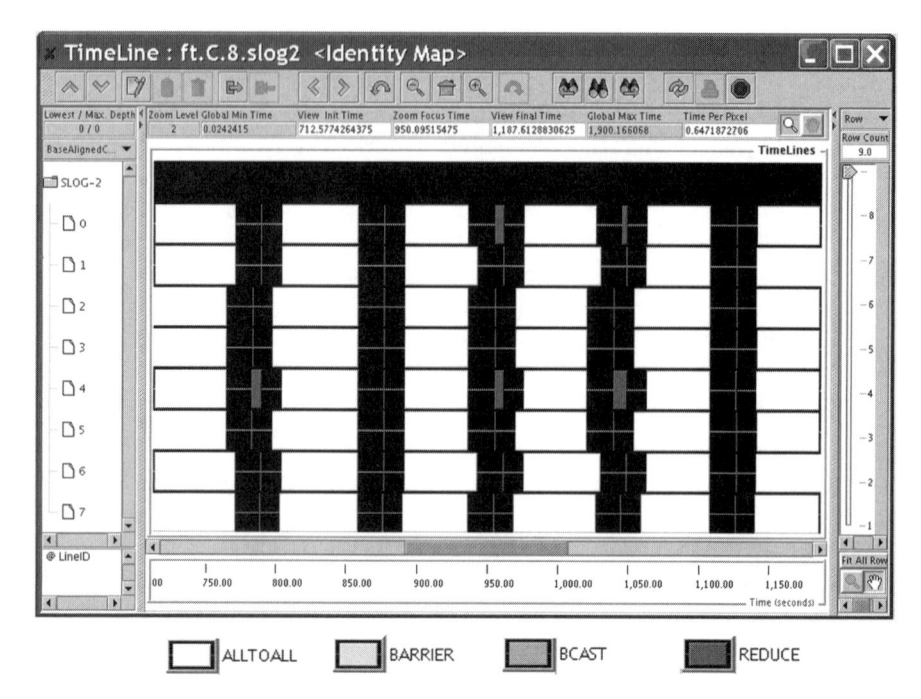

FIG. 21. A performance trace of FT.C.8 using the MPI profiling tool (MPE) in MPICH. Traces are visualized with Jumpshot. x-axis is execution time, y-axis is processor number involved in computation; graph shows work by processor.

```
...
call set_cpuspeed( low_speed)
call mpi_alltoall( ... )
call set_cpuspeed( high_speed)
...
```

FIG. 22. INTERNAL control for FT.

6.4.3.2 An Internal DVS Schedule for FT.
Based on these observations, we divide time into all-to-all communication phases and other phases. We will schedule the CPU for low speed during all-to-all communication phases and high speed elsewhere. Figure 22 shows how we use our PowerPack API to control DVS from within the source code of the FT application.

6.4.3.3 Energy Savings for FT.
Figure 23 shows the energy and delay using internal scheduling. We are not limited to using only the highest and lowest

Normalized Energy and Delay of INTERNAL Control for FT.C.8

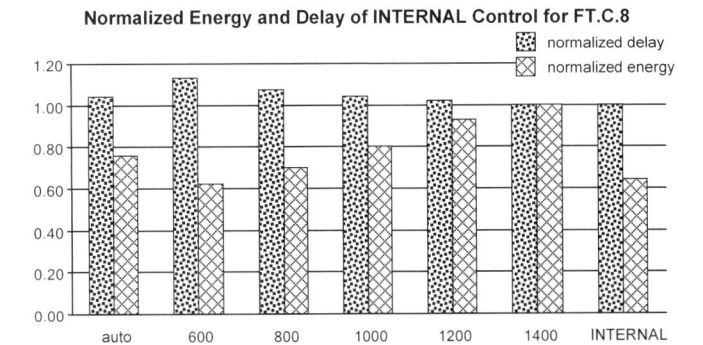

FIG. 23. Normalized energy and delay of INTERNAL, EXTERNAL and CPUSPEED scheduling. In INTERNAL control, high speed and low speed are set as 1400 and 600 MHz respectively. All EXTERNAL control's decisions (600–1400 MHz) are given on the x-axis. CPUSPEED is shown as auto in this figure. Normalized delay is total application execution time with DVS divided by total application execution time without DVS. Values <1 indicate performance loss. Normalized energy is total system energy with DVS divided by total system energy without DVS. Values <1 indicate energy savings.

processor frequencies. However, using the highest and lowest frequency settings between the phases of FT provided better results than all other combinations. Hence, in INTERNAL results for FT we use 600 MHz for the all-to-all communication phase and 1400 MHz for all other phases. The best overall result for FT is 36% energy without noticeable delay increase (<1%). This is a significant improvement over both external control and CPUSPEED. External control at 600 MHz saves 38% energy but at a cost of 13% delay increase. CPUSPEED saves 24% energy with 4% delay increase. This shows internal scheduling is preferred when the application contains obvious CPU-bound phases and non-CPU bounded phases and each phase lasts long enough to compensate for the CPU speed transition overhead.

6.4.3.4 CG Performance:

Figure 24 shows the performance profile of CG generated with the MPICH trace utility by compiling the code with "–mpilog" option. The following observations are drawn from this profile:

- CG is communication intensive and synchronizes all nodes between phases.
- Wait and Send are major communication events that dominate execution time.
- The execution time of each phase is relatively small, the message communications are frequent and CPU speed transition may impact delay significantly.
- Nodes exhibit heterogeneous behavior. Nodes 4–7 have larger communication-to-computation ratio than nodes 0–3.

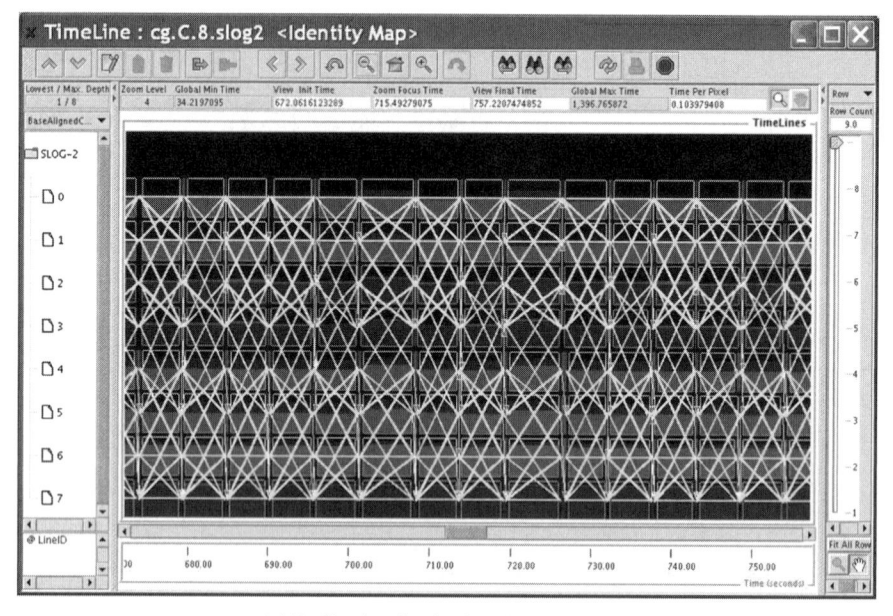

(a) Profile visualized at iteration granularity

FIG. 24. Performance trace of CG.C.8 using MPE tool provided with MPICH. The traces are visualized with Jumpshot. x-axis is execution time, y-axis is processor number involved in computation; graphs show work by processor; arrows indicate message source and destination. (a) Iteration granularity shows the application is regular and can be partitioned into phases. (b) Message granularity reveals different communication types and workloads on different processors.

6.4.3.5 An Internal Schedule for CG.
Based on the performance observations, we found it challenging to improve power-performance efficiency in CG. Thus, we implemented two distinct phase-based dynamic scheduling policies within CG. The first policy (applied to nodes 4–7) scales down the CPU speed during any communication. The second policy (applied to nodes 0–3) scales down CPU speed only during the MPI_Wait phases. Both policies increase energy and delay (1∼3%). Since the performance behavior on each node is asymmetric, we can set different speeds for each execution node. The DVS controls are applied to CG as shown in Fig. 25.

6.4.3.6 Energy Savings for CG.
Figure 26 shows the energy and delay using internal scheduling. We provide results for two configurations: internal I which uses 1200 MHz as high speed and 800 MHz as low speed and internal II which uses 1000 MHz as high speed and 800 MHz as low speed. Experiments show that internal

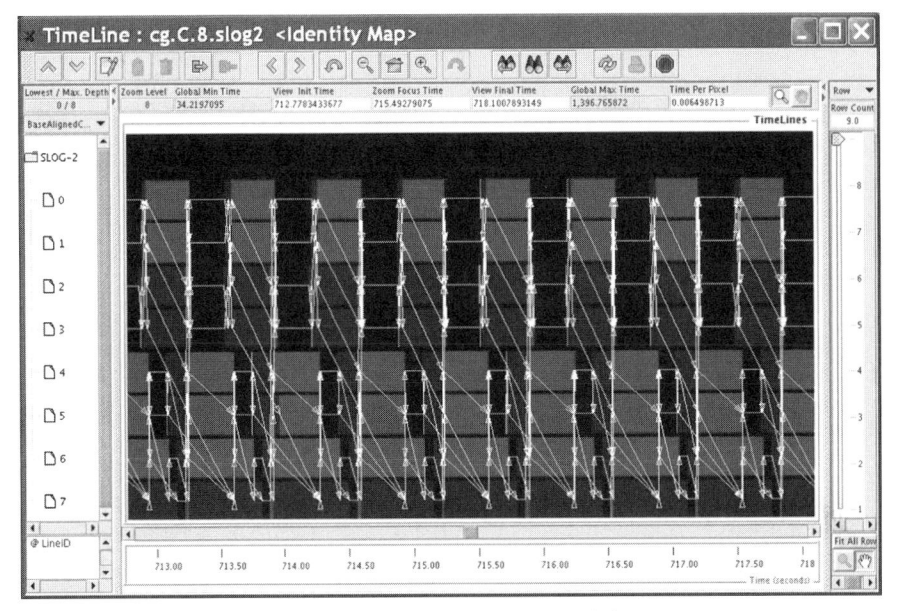

(b) Profile visualized at message granularity

FIG. 24. (*continued*)

```
...
if ( myrank .ge. 0 .and. myrank .le. 3)
call set_cpuspeed( high_speed)
else
call set_cpuspeed( low_speed)
endif
...
```

FIG. 25. INTERNAL control for CG.

I saves 23% energy with 8% delay increase and internal II saves 16% energy with 8% delay increase. Both internal I and II scheduling for CG do not provide significant advantages over external scheduling at 800 MHz. The frequency of communication phases in CG requires more transitions per unit time than FT. The overhead for frequency transition is more costly in CG. Thus, while energy savings are possible, the additional overhead adds to the observable delay for CG. Since external schedul-

Normalized Energy and Delay of INTERNAL Control for CG

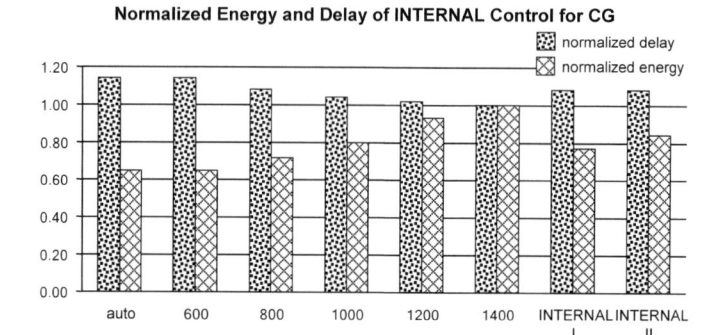

FIG. 26. Normalized energy and delay of INTERNAL scheduling, EXTERNAL control and CPUSPEED scheduling for CG. For INTERNAL I, high speed is 1200, and low speed is 800; for INTERNAL II, high speed is 1000 and low speed is 800.

ing does not incur overhead after the initial transition, the performance it is able to perform as well as the internal scheduling.

6.4.3.7 *Overall.*

Internal scheduling provides DVS control with finer granularity than external scheduling. Internal scheduling achieves better (or at least as good) energy-performance efficiency. FT shows the benefit of phased-based internal scheduling; CG shows the benefit of heterogeneous internal scheduling.

6.5 Lessons from Power-Aware Cluster Design

High-performance power-aware distributed computing is viable. DVS scheduling policies are critical to automating middleware that alleviates users from thinking about power and energy consumption. Our results indicate given user-defined energy-performance efficiency metrics, our schedulers can reduce energy and guarantee performance. Our experiments all indicate that no single scheduling strategy fits all scientific codes.

Our contributions to power-aware HPC were the first of their kind [16]. One of the big hurdles early on was convincing the HPC community that power was indeed a problem and not something the microarchitecture community would solve singlehandedly. Early work by Rutgers [17] and IBM [26] highlighted the power issues in commercial servers. However, while the problems were similar, the techniques used to conserve power and energy in commercial server farms would simply not work in the 24/7 all-performance-all-the-time systems commonplace in HPC. We've now shown conclusively that the power issue is critical to HPC and power-aware techniques can be adapted to address power without killing performance.

Since the first appearance of our work, others have joined the fray. Our initial techniques were entirely manual. Our colleagues at the University of Georgia and North Carolina State University showed how to automate DVS transitions by filtering the MPI communication library functions [35]. Others at Los Alamos National Laboratory use performance prediction to identify slack in parallel codes and set DVS transitions accordingly [42].

Of course, there is still work to be done. The CPU is but one of many devices in the system. Depending on the workload, other system components may dominate the power usage. Disks in particular can consume and enormous amount of power for applications with extremely large data sets. In the codes we observed, memory was a significant consumer of power. Since scientific codes often use as much memory as available, so for large-memories (or fat nodes) power-aware memory could save significant amounts of power in clusters. Lastly, there has been little work on holistic approaches to energy conservation. Power-aware techniques are mostly localized and independent. Little is known about the effects of multiple power-aware components on total system power.

7. Conclusions

Power is now a critical design constraint in clusters built for high-performance computing. Profiling techniques pinpoint exactly where power and energy are consumed in clusters. Low-power approaches use hardware design to reduce the power profiles of cluster systems and applications. Power-aware techniques provide dynamic control to reduce power and energy consumption in clusters. For benchmarks applications, energy savings of 30% are possible with less than 1% performance impact.

REFERENCES

[1] Adiga N., Almasi G., Barik R., et al., "An overview of the BlueGene/L supercomputer", in: *Proc. of IEEE/ACM SC 2002, Baltimore, MD*, 2003.
[2] Allen G., Dramlitsch T., Foster I., et al., "Supporting efficient execution in heterogeneous distributed computing environments with cactus and globus", in: *Proc. of SC 2001, Denver, CO*, 2001.
[3] AMD, "Mobile AMD Duron Processor Model 7 Data Sheet", http://www.amd.com/usen/assets/content_type/white_papers_and_tech_docs/24068.pdf, 2001 (last accessed).
[4] Bailey A.M., "Accelerated Strategic Computing Initiative (ASCI): Driving the need for the Terascale Simulation Facility (TSF)", in: *Proc. of Energy 2002 Workshop and Exposition, Palm Springs, CA*, 2002.

[5] Bailey D., Harris T., Saphir W., et al., "The NAS Parallel Benchmarks 2.0", NASA Ames Research Center Technical Report #NAS-95-020, December 1995.
[6] Bailey D.H., Barszcz E., Barton J.T., et al., "The NAS Parallel Benchmarks", *Internat. J. Supercomputer Applications and High Performance Computing* **5** (3) (1991) 63–73.
[7] Bellosa F., "The benefits of event-driven energy accounting in power-sensitive systems", in: *Proc. of 9th ACM SIGOPS European Workshop, Kolding, Denmark*, 2000.
[8] BlueGene/LTeam, "An overview of the BlueGene/L supercomputer", in: *Supercomputing 2002 Technical Papers*, 2002.
[9] Bohrer P., Elnozahy E.N., Keller T., et al., "The case for power management in Web servers", in: Graybill R., Melhem R. (Eds.), *Power Aware Computing*, Kluwer Academic, IBM Research, Austin TX 78758, USA, 2002.
[10] Borkar S., "Low power design challenges for the decade", in: *Proc. of the 2001 Conf. on Asia South Pacific Design Automation, Yokohama, Japan*, 2001.
[11] Brooks D., Tiwari V., Martonosi M., "Wattch: A framework for architectural-level power analysis and optimizations", in: *Proc. of 27th International Symposium on Computer Architecture, Vancouver, BC*, 2000.
[12] Brooks D.M., Bose P., Schuster S.E., et al., "Power-aware microarchitecture: Design and modeling challenges for next-generation microprocessors", *IEEE Micro* **20** (6) (2000) 26–44.
[13] Burger D.C., Austin T.M., "The SimpleScalar Toolset, Version 2.0", *Computer Architecture News* **25** (3) (1997) 13–25.
[14] Cai G., Lim C., "Architectural level power/performance optimization and dynamic power optimization", in: *Proc. of Cool Chips Tutorial at 32nd ISCA*, 1999.
[15] Cameron K.W., Ge R., Feng X., Varner D., Jones C., "POSTER: High-performance, power-aware distributed computing framework", in: *Proc. of 2004 ACM/IEEE Conference on Supercomputing, SC 2004*, 2004.
[16] Cameron K.W., Ge R., Feng X., Varner D., Jones C., "[Poster] High-performance, power-aware distributed computing framework", in: *Proc. of IEEE/ACM SC 2004, Pittsburgh, PA*, 2004.
[17] Carrera E.V., Pinheiro E., Bianchini R., "Conserving disk energy in network servers", in: *Proc. of 17th International Conference on Supercomputing*, 2003.
[18] Chandra S., "Wireless network interface energy consumption implications of popular streaming formats", in: *Multimedia Computing and Networking, MMCN'02*, in: *Proc. SPIE*, vol. 4673, The International Society of Optical Engineering, San Jose, CA, 2002.
[19] Chaparro P., Gonzalez J., Gonzalez A., "Thermal-effective clustered microarchitectures", in: *Proc. of First Workshop on Temperature-Aware Computer Systems, Munich, Germany*, 2004.
[20] Company I., "IXIA Product Catalog".
[21] Culler D.E., Singh J.P., Gupta A., *Parallel Computer Architecture: A Hardware/Software Approach*, Morgan Kaufmann Publishers, San Francisco, CA, 1999.
[22] Dhodapkar A., Lim C.H., Cai G., Daasch W.R., "TEM2P2EST: A thermal enabled multi-model power/performance ESTimator", in: *Proc. of the First International Workshop on Power-Aware Computer Systems*, 2000.

[23] Dongarra J., "An overview of high performance computing", http://www.netlib.org/utk/people/JackDongarra/SLIDES/hpcasia-1105.pdf, 2005 (last accessed).
[24] Dongarra J., "Present and future supercomputer architectures", http://www.netlib.org/utk/people/JackDongarra/SLIDES/HK-2004.pdf, 2004 (last accessed).
[25] Dongarra J.J., Bunch J.R., Moller C.B., Stewart G.W., *LINPACK User's Guide*, SIAM, Philadelphia, PA, 1979.
[26] Elnozahy M., Kistler M., Rajamony R., "Energy conservation policies for Web servers", in: *Proc. of 4th USENIX Symposium on Internet Technologies and Systems, Seattle, WA*, 2003.
[27] Fan X., Ellis C.S., Lebeck A.R., "Memory controller policies for DRAM power management", in: *Proc. of International Symposium on Low Power Electronics and Design, ISLPED*, 2001.
[28] Fan X., Ellis C.S., Lebeck A.R., "The synergy between power-aware memory systems and processor voltage scaling", Department of Computer Science, Duke University, Durham, TR CS-2002-12, 2002.
[29] Feng W., "Making a case for efficient supercomputing", *ACM Queue* 1 (7) (2003) 54–64.
[30] Feng W., Warren M., Weigle E., "The bladed Beowulf: A cost-effective alternative to traditional Beowulfs", in: *Proc. of IEEE International Conference on Cluster Computing, CLUSTER'02, Chicago, IL*, 2002.
[31] Feng W., Warren M., Weigle E., "Honey, I shrunk the Beowulf!", in: *Proc. of 2002 International Conference on Parallel Processing, ICPP'02, Vancouver, BC, Canada*, 2002.
[32] Feng X., Ge R., Kirk C., "ARGUS: Supercomputing in 1/10 cubic meter", in: *Parallel and Distributed Computing and Networks, PDCN 2005*, 2005.
[33] Feng X., Ge R., Cameron K., "Power and energy profiling of scientific applications on distributed systems", in: *Proc. of 19th International Parallel and Distributed Processing Symposium, IPDPS 05, Denver, CO*, 2005.
[34] Flinn J., Satyanarayanan M., "Energy-aware adaptation for mobile applications", in: *Proc. of 17th ACM Symposium on Operating Systems Principles, Kiawah Island Resort, SC*, 1999.
[35] Freeh V.W., Lowenthal D.K., Springer R., Pan F., Kappiah N., "Exploring the energy-time tradeoff in MPI programs", in: *Proc. of 19th IEEE/ACM International Parallel and Distributed Processing Symposium, IPDPS, Denver, CO*, 2005.
[36] Ge R., Feng X., Cameron K.W., "Improvement of power-performance efficiency for high-end computing", in: *Proc. of 1st Workshop on High-Performance, Power-Aware Computing, HPPAC 2005, in conjunction with IPDPS'2005, Denver, CO*, 2005.
[37] Gropp W., Lusk E., "Reproducible measurements of MPI performance", *Proc. of PVM/MPI '99 User's Group Meeting*, 1999.
[38] Grunwald D., Levis P., Farkas K.I., "Policies for dynamic clock scheduling", in: *Proc. of 4th Symposium on Operating System Design & Implementation, San Diego, CA*, 2000.
[39] Gurumurthi S., Sivasubramaniam A., Irwin M.J., Vijaykrishnan N., Kandemir M., "Using complete machine simulation for software power estimation: The SoftWatt approach", in: *Proc. of Eighth International Symposium on High-Performance Computer Architecture, HPCA'02, Boston, MA*, 2002.

[40] HECRTF, "Federal plan for high-end computing: Report of the high-end computing revitalization task force", 2004.
[41] J.L. Hennessy, D.A. Patterson, *Computer Architecture: A Quantitative Approach*, third ed., Morgan Kaufmann Publishers, San Francisco, CA, 2003.
[42] Hsu C.-H., Feng W.-C., "A power-aware run-time system for high-performance computing", in: *Proc. of IEEE/ACM Supercomputing, SC\05, Seattle, WA*, 2005.
[43] Hsu C.-H., Kremer U., "The design, implementation, and evaluation of a compiler algorithm for CPU energy reduction", in: *Proc. of ACM SIGPLAN Conference on Programming Languages, Design, and Implementation, PLDI'03, San Diego, CA*, 2003.
[44] Standard Performance Evaluation Corporation, "The SPEC benchmark suite", http://www.spec.org, 2002.
[45] IBM, *PowerPC 604e User's Manual*, IBM, 1998.
[46] Intel, "Developer's manual: Intel 80200 Processor Based on Intel XScale Microarchitecture", http://developer.intel.com/design/iio/manuals/273411.htm, 1989 (last accessed).
[47] Intel, "Intel Pentium M Processor datasheet", 2004.
[48] Isci C., Martonosi M., "Runtime power monitoring in high-end processors: Methodology and empirical data", in: *Proc. of the 36th annual IEEE/ACM International Symposium on Microarchitecture*, 2003.
[49] J G., "A high-level language benchmark", *BYTE* **6** (9) (1981) 180–198.
[50] Joseph R., Brooks D., Martonosi M., "Live, runtime power measurements as a foundation for evaluating power/performance tradeoffs", in: *Proc. of Workshop on Complexity-Effective Design, Goteborg, Sweden*, 2001.
[51] Kurita T., Takemoto M., "Design of low power-consumption LSI", *Oki Technical Review* **68** (4) (2001).
[52] Laird D., "Crusoe processor products and technology", http://www.transmeta.com/press/download/pdf/laird.pdf, 2000 (last accessed).
[53] LBNL, *Data Center Energy Benchmarking Case Study*, LBNL, 2003.
[54] Lorch J.R., Smith A.J., "PACE: A new approach to dynamic voltage scaling", *IEEE Trans. Comput.* **53** (7) (2004) 856–869.
[55] Lorch J.R., Smith A.J., "Software strategies for portable computer energy management", *IEEE Personal Communications Magazine* **5** (1998) 60–73.
[56] McMahon F.H., "The Livermore Fortran Kernels: A computer test of numerical performance range", Lawrence Livermore National Laboratory, UCRL-53745, December 1986.
[57] McVoy L., Staelin C., "lmbench: Portable tools for performance analysis", in: *Proc. of USENIX 1996 Annual Technical Conference, San Diego, CA*, 1996.
[58] Mudge T., "Power: A first class design constraint for future architectures", *Computer* **34** (4) (2001) 52–57.
[59] Phillips J.C., Zheng G., Kumar S.. Kale L.V., "NAMD: Biomolecular simulation on thousands of processors", in: *Proc. of 14th International Conference on High Performance Computing and Communications, SC 2002, Baltimore, MA*, 2002.
[60] Rosenblum M., Herrod S.A., Witchel E., Gupta A., "Complete computer simulation: The SimOS approach", in: *IEEE Parallel and Distributed Technology, Fall 1995*, 1995.
[61] Sakagami H., Murai H., Seo Y., Yokokawa M., "TFLOPS three-dimensional fluid simulation for fusion science with HPF on the Earth Simulator", in: *Proc. of SC2002*, 2002.

[62] Smith J.E., "Characterizing computer performance with a single number", *Comm. ACM* **32** (10) (1988) 1202–1206.
[63] Tennessee U., Manheim U., NERSC, "Top 500 Supercomputer list", SC|05, http://www.top500.org/, 2005 (last accessed (1/6) 2006).
[64] Tiwari V., Singh D., Rajgopal S., et al., "Reducing power in high-performance microprocessors", in: *Proc. 35th Conference on Design Automation, San Francisco, CA*, 1998.
[65] Top500, "27th edition of TOP500 list of world's fastest supercomputers released: DOE/LLNL BlueGene/L and IBM gain top positions", 2006.
[66] Vachharajani M., Vachharajani N., Penry D.A., Blome J.A., August D.I., "Microarchitectural exploration with liberty", in *Proc. of 35th International Symposium on Microarchitecture, Micro-35*, 2002.
[67] Vargas E., "High availability fundamentals", http://www.sun.com/blueprints/1100/HAFund.pdf, 2000 (last accessed).
[68] Vijaykrishnan N., Kandemir M., Irwin M., Kim, H., Ye W., "Energy-driven integrated hardware-software optimizations using SimplePower", in: *Proc. of 27th International Symposium on Computer Architecture, Vancouver, BC*, 2000.
[69] Wang H.-S., Zhu X., Peh L.-S., Malik S., "Orion: A power-performance simulator for interconnection networks", in *Proc. of 35th Annual IEEE/ACM International Symposium on Microarchitecture, MICRO-35, Istanbul, Turkey*, 2002.
[70] Warren M.S., Weigle E.H., Feng W.-C., "High-density computing: A 240-processor Beowulf in one cubic meter", in: *Proc. of IEEE/ACM SC 2002, Baltimore, MA*, 2002.
[71] Weissel A., Bellosa F., "Process cruise control-event-driven clock scaling for dynamic power management", in: *Proc. of International Conference on Compilers, Architecture and Synthesis for Embedded Systems, CASES 2002, Grenoble, France*, 2002.
[72] Zhu Q., Chen Z., Tan L., et al., "Hibernator: Helping disk array sleep through the winter", in: *Proc. of the 20th ACM Symposium on Operating Systems Principles, SOSP'05*, 2005.
[73] Zhu Q., Zhou Y., "Power aware storage cache management", *IEEE Transactions on Computers (IEEE-TC)* **54** (5) (2005) 587–602.

Compiler-Assisted Leakage Energy Reduction for Cache Memories

WEI ZHANG

Department of Electrical and Computer Engineering
Southern Illinois University Carbondale
Carbondale, IL 62901
USA
zhang@engr.siu.edu

Abstract

With the scaling of technology, leakage energy reduction has become increasingly important for microprocessor design. Being the major consumer of the on-chip transistor budget, it is particularly critical to mitigate cache leakage energy. In contrast to many recent studies that attempt to minimize cache leakage by exploiting architectural-level information, this chapter introduces two compiler-assisted approaches to manage the cache leakage dissipation without significant impact on either performance or the dynamic energy consumption. More specifically, the first approach exploits static and profiling information to detect the sub-bank transitions at the compilation time, which can improve the energy efficiency of the drowsy instruction caches. The second approach exploits the fact that only a small portion of the data caches will be accessed during the loop execution, the compiler can provide hints to place other non-active cache blocks into the low power mode during the loop execution to save the data cache leakage energy. Our experiments on a state-of-the-art VLIW processor indicate that the proposed compiler-based approaches can improve the energy-efficiency of both instruction and data caches effectively.

1. Introduction 156
2. Related Work 159
3. Static Next Sub-Bank Prediction for Drowsy Instruction Caches 161
    3.1. Overview 161
    3.2. The Memory Sub-Bank Prediction Buffer and Its Dynamic Energy Overhead . 162
    3.3. Transitional Instructions 165
    3.4. Instruction Cache Addressing Schemes 167
    3.5. Identify Transitional Instructions 168

3.6. ISA and Compiler Support 169
4. Compiler-Assisted Loop-Based Data Cache Leakage Reduction 170
5. Evaluation Methodology 172
    5.1. Experimental Results for Static Next Sub-Bank Prediction 174
    5.2. Experimental Results for Loop-based Data Cache Leakage Reduction 182
6. Conclusion 186
    References 187

1. Introduction

As transistor counts and clock frequencies increase, power and energy consumption has become an important design constraint for modern microprocessors [1]. While energy-aware design is obviously critical for battery-driven mobile and embedded systems where the battery lifetime is a primary constraint, it has also become crucial for plugged computing devices such as desktops and servers due to the packaging and cooling requirements where power consumption has grown from a few watts per chip to over 100 watts [23]. Therefore, in these systems, performance may be limited by the inability to mitigate the heat dissipation produced by power-hungry circuits operating at high speeds. Moreover, when circuits work in high temperature, the reliability of the system decreases.

Power consumed in microprocessors can be classified into dynamic and static (leakage) power. While dynamic power arises due to signal transitions (i.e., the switching activities of repeated capacitance charge and discharge on the output of gates), leakage power is consumed constantly (independent of any activity), which is mainly due to subthreshold and gate leakage [7]. Dynamic power consumption is proportional to the square of the supply voltage, thus it can be reduced by scaling down the supply voltage. However, to maintain high switching speed under reduced voltage levels, the threshold voltage must also be scaled, making it easier for current to leak through the transistors. As a result, the leakage power dissipation will be increased significantly [12]. Moreover, the increases in device speed and chip density aggravate the leakage consumption problem. In addition, new techniques for reducing dynamic power consumption and for improving performance, such as low threshold voltage [29] and gate-oxide scaling [30], further increase the relative importance of leakage power [28]. Therefore, while dynamic power has been a major source of power dissipation for current microprocessors, leakage power is expected to grow exponentially in upcoming generations [1,2]. It is projected that as processor technology moves below 0.1 micron, static (leakage) power consumption is set on the path to dominate the total power used by the CPU [8] (see Fig. 1).

While all the on-chip components consume leakage energy, it is particularly important for optimizing the leakage dissipation of the cache memory subsystem.

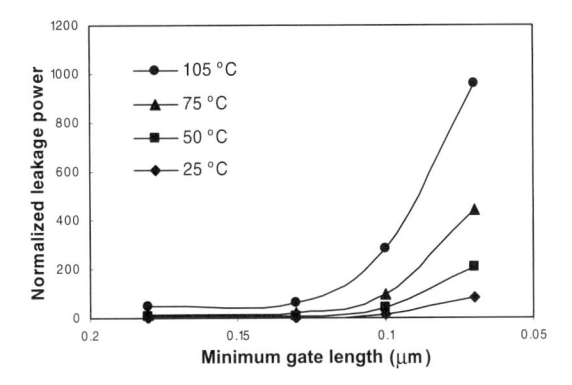

FIG. 1. Normalized leakage power through an inverter [8].

Current microprocessors typically use two levels of on-chip caches (including an L1 instruction cache, an L1 data cache, and a unified L2 cache) to mitigate the performance gap between the processor and the memory, which consume a large and growing fraction of on-chip real estate. Since leakage energy will be consumed no matter the transistors are switching or not, cache memories are a good target to optimize the leakage energy dissipation. Actually, it has been estimated that leakage will amount to more than 70% of energy consumed in caches if left unchecked for 0.07 micron process [7]. Therefore, it is critical to minimize the cache leakage energy dissipation without significantly impacting performance, dynamic energy or cost.

Various techniques have been proposed to reduce the cache leakage energy. Circuit level techniques include adaptive substrate biasing, dynamic supply scaling and supply gating [19]. Many of these circuit mechanisms can be exploited at the architectural level to control leakage at the cache line and cache bank granularities [19]. Currently, many leakage optimization approaches rely on employing hardware counters to monitor and predict the access patterns of the cache lines. In those approaches, cache lines that are not accessed for a fixed time period are predicted to be dead and are placed into the leakage control mode for reducing leakage energy consumption. When the cache lines in the low power mode need to be used, it will be activated [8,7] or re-fetched from the L2 [5], resulting in performance and dynamic energy penalties. While the hardware-based approach is simple and reasonably effective to manage the cache leakage dissipation, it needs to spend both time and energy in monitoring the cache line access patterns, and to make simple predictions based on the history cache access information. Unfortunately, the counter-based prediction may not be very accurate due to the limited runtime information available. Moreover, the counter-based prediction approach is not adaptive to program behavior, since the time interval to monitor the cache accesses is typically fixed (otherwise different time intervals must be used for different programs or even different phases of the same program, which will complicate the counter design). The mis-predictions of cache access patterns can cause the excessive and late activation of cache lines, which can impact both performance and energy consumption. On the other hand, more complex and expensive hardware-based predictors themselves can incur additional energy and area cost, which may compromise the leakage energy savings of cache memories.

Due to the deficiencies of hardware-centric approaches, this chapter attempts to exploit useful compiler information to manage cache leakage more cost-effectively. It should be noted that while the energy is directly consumed by hardware, software (and specifically the compiler) can determine how the hardware will be exercised, and thus can impact the overall energy consumption. Traditionally, compiler optimizations have been studied extensively for optimizing performance [25]; however, recently compiler-based optimizations have also been used to improve the energy efficiency [35–39]. In contrast to previous compiler-directed approaches targeting reducing the dynamic energy dissipation, this chapter studies compiler-assisted strategies to optimize the leakage energy consumption. Compared with the hardware-centric approaches [8,7,5], the compiler-assisted leakage reduction has the following advantages, including: (1) the compiler-assisted strategies can detect leakage reduction opportunities without incurring hardware overhead; (2) since compiler can analyze a large scope of program and transform the code, it can potentially identify or create opportunities for conducting leakage control mechanisms profitably. In this chapter, we propose two compiler-assisted approaches to minimizing the leakage energy dissipation of the instruction cache and data cache respectively, which are the two major leakage consumers among on-chip components. The basic ideas of these two approaches are described below:

- The first approach aims at minimizing the leakage energy of the instruction cache without significant dynamic energy overhead [10]. Recent research in drowsy instruction cache shows that the leakage energy of the instruction cache can be significantly reduced with little performance degradation by exploiting the instruction spatial locality at the cache sub-bank level [7]. The performance penalty due to the sub-bank wake-up latency can be dramatically reduced by using a prediction buffer to pre-activate the next sub-bank at runtime. However, consulting the prediction buffer at every cache access consumes non-trivial dynamical energy, which can compromise the overall energy savings. This chapter introduces a compiler-assisted approach to capturing the sub-bank transitional behavior at the compilation time and pre-activating the instruction cache sub-bank that will be accessed at runtime according to the compiler-directed hints. We also investigate a hybrid approach to exploiting both the static and dynamic information for reducing the performance penalty further with little dynamic energy overhead.

- The second approach targets data cache leakage reduction, which is based on the observation that typically only a relatively small portion of the data are actively used during certain program phases and the program often spends a large fraction of execution time in loops [9]. Consequently, a large portion of data cache lines that are not accessed by the loops can be placed into the leakage control mode during the loop execution to save the data cache leakage energy consumption. Since loops can be nested, we propose an optimistic approach to placing the non-active cache lines into the low leakage mode during the execution of the innermost loop (note that non-nested loop is regarded as innermost loop in our work). We also investigate the impact of loop transformations such as loop tiling and loop distribution [25] on the reduction of the active data set accessed by innermost loops, which can determine the effectiveness of the proposed leakage optimization strategy. We find that making use of loop transformations can reduce the active data set successfully and thus the proposed approach can be applied or adapted to a wider variety of applications.

Overall, our experimental results indicate that compiler-assisted approaches are very successful in reducing the leakage of both the instruction cache and the data cache. Also, the compiler-assisted leakage optimization approaches have minimal impact on the dynamic energy consumption and the performance.

The rest of the chapter is organized as follows. We discuss related work in Section 2. The compiler-assisted approaches for instruction caches and data caches are introduced in Sections 3 and 4, respectively. The evaluation methodology and experimental results are presented in Section 5. Finally, we conclude this chapter in Section 6.

2. Related Work

There have been a great deal of research efforts recently for reducing the cache leakage energy [3–8,19,11,31,32,27] at different levels, ranging from circuit-level to architectural and compiler levels. The circuit level leakage control mechanism can be broadly divided into two categories: the state-destroying mechanisms and the state-preserving mechanisms. Powell et al. [4] developed the gated V_{dd} technique to switch off a cache line for reducing leakage energy. Since the data in the cache line is lost, the decision to turn off a cache line must be made cautiously. Recently Kaxiras et al. proposed the use of time-based strategies to turn off cache lines [5], which strike a balance between leakage energy saved and dynamic energy induced (i.e., due to extra cache misses). In contrast, Flautner et al. proposed to use dynamic voltage scaling (DVS) for reducing the leakage power of cache cells [8]. By scaling the voltage of the cell to approximately 1.5 times V_t, the state of the memory cell can be retained and the leakage energy dissipation is reduced dramatically [8]. While voltage scaling does not reduce leakage as much as gated V_{dd}, it has an important advantage of being able to preserve the data in the drowsy mode, and thus extra cache misses can be avoided. The only performance penalty of the state-preserving mechanism is the wake-up latency, since it takes time to reinstate the power supply lines of the drowsy cache lines to the normal voltage level [7]. Without appropriate management of cache lines power status, however, it is shown that the performance penalty of a 32 KB direct mapped drowsy instruction cache by using a simple policy—to periodically put all cache lines into the drowsy mode and to wake up a cache line only when it is accessed again—can be as high as 5.7% [7]. Recently Kim et al. proposed an instruction cache leakage energy reduction strategy at the cache sub-bank granularity by exploiting the spatial locality of instructions [7]. To hide the wake-up delay and thus to reduce the performance penalty, Kim et al. utilized a sub-bank prediction buffer to store and predict the transition points and to pre-activate the next sub-bank at runtime [7].

At the architectural level, Powell et al. proposed the DRI-cache, which uses the gated-V_{dd} technique to dynamically adjust the size of the active portion of the cache by turning off a bank of cache lines according to the miss rates [4]. Zhou et al. proposed the AMC (Adaptive Mode Control) cache [6] to adjust the cache turn-off intervals while controlling the performance overhead by keeping the tag array alive and tracking the miss rate with respect to the ideal miss rate. Velusamy et al. applied formal feedback-control theory to adjust the cache-decay intervals adaptively [31]. Li et al. studied several architectural techniques that exploit the data duplication across the different levels of cache hierarchy and found that the best strategy in terms of energy and energy-delay product is to place the L2 sub-block into a state-preserving mode as soon as its contents are moved to L1 and to reactivate it only when it is accessed [32]. Compared with all the above work that exploits circuit-level or architecture-level techniques for energy reduction, the approaches studied in this chapter utilize compiler information intelligently to minimize cache leakage energy dissipation in a cost-effective fashion.

Recently, there are also a number of research efforts in exploiting compiler information for leakage energy optimizations. Zhang et al. [33] investigated a dataflow analysis to identify large slacks for functional units of statically multi-issued processors, which can be exploited to mitigate the leakage consumption of functional units by inserting turn_on and turn_off instructions during the compilation. Rele et al. [34] studied a compiler-based approach to reducing leakage power dissipation by functional units for superscalar processors. To optimize the cache leakage, a compiler-directed strategy was proposed for instruction caches [19], in which the information of the last usage of instructions obtained by the compiler is encoded and utilized at runtime to control the instruction cache lines dynamically. For the data cache, Zhang [27] et al. presented code restructuring techniques for array-based and pointer-intensive applications to reduce leakage energy. While this approach [27] can reduce the data cache energy consumption significantly, it may increase the code size and impact the performance by inserting activate/deactivate instructions at the cache line granularity. In contrast, in this chapter, we propose a compiler-directed leakage optimization strategy at the loop granularity (i.e., innermost loops) for L1 data caches. Also, we investigate a compiler-assisted approach to reducing the performance and dynamic energy overheads for drowsy instruction caches.

3. Static Next Sub-Bank Prediction for Drowsy Instruction Caches

In this section, we introduce a compiler-assisted approach to predicting the next sub-bank statically and accurately to reduce the performance and dynamic energy overheads for drowsy instruction caches. The overview of this approach is given in Section 3.1. In Section 3.2, we provide the background information about the sub-bank prediction buffer [7] and quantify its dynamic energy overhead. We explain the transitional instructions and the instruction cache addressing schemes in Sections 3.3 and 3.4, respectively. In Section 3.5, we present the approach to identifying transitional instructions based on static and profiling information. The ISA and compiler support of the proposed approach are described in Section 3.6.

3.1 Overview

As mentioned in Section 2, the hardware-based next sub-bank prediction [7] is the state-of-the-art approach to mitigating the run-time overhead of drowsy instruction caches; however, the downside of this approach is its dynamic energy overhead, which can compromise the overall energy savings (i.e., the leakage energy savings minus the dynamic and leakage energy overheads) of the drowsy instruction cache substantially. Note that in [7], Kim et al. also proposed a less-expensive tag-based next sub-bank prediction scheme, which extends the tag array to contain the block address, the next sub-bank address, and a valid bit. Nevertheless, such a scheme needs to write to the instruction cache for saving the next sub-bank information, which may raise severe security concerns since I-cache is typically read-only and protected by the operating system. Moreover, writing to the instruction cache will make it much harder to protect I-cache from transient errors, since instruction caches are often protected by parity bits and the error correction is simply implemented by re-fetching from L2 under the assumption that the L1 instruction cache is read-only. Thus, we will not consider the tag-based prediction strategy in this chapter. In order to reduce the overall energy consumption without significant impact on performance for drowsy instruction caches, we must develop more energy-efficient sub-bank prediction schemes. In this section, we propose a compiler-assisted sub-bank prediction and pre-activation strategy to mitigate the dynamic energy overhead while still reducing the leakage energy substantially. *The idea of this approach is to use compilers to find the transitional points statically and to insert the pre-activation instructions appropriately in the program to pre-activate the next sub-bank promptly and accurately at runtime.* Compared to the hardware-centric next sub-bank prediction approach (i.e., the next cache sub-bank prediction buffer [7]), which requires additional hardware resources and needs to consult the prediction buffer for every cache access, the compiler-assisted approach just needs a simple ISA extension (i.e., the hint bits to annotate the sub-bank prediction information). Our experiments reveal that the compiler-assisted approach is very successful in capturing the sub-bank transitional behavior to reduce the performance penalty and the dynamic energy overhead of drowsy instruction caches. In addition, we propose a hybrid approach to exploiting both the static and dynamic information efficiently for pre-activating the next cache sub-bank, and our results show that the hybrid approach is the best strategy for the drowsy instruction cache to optimally balance leakage energy reduction and high performance.

3.2 The Memory Sub-Bank Prediction Buffer and Its Dynamic Energy Overhead

Since the drowsy caches can preserve the data in the drowsy mode, the only cost of being wrong is an additional delay and energy overhead to wake up a drowsy cache line. Therefore, simple policies can be used to reduce leakage energy without much impact on performance. For instance, Flautner et al. proposed a simple policy to periodically put all cache lines into the drowsy mode and a line is woken up only when it is accessed again, which was shown to be effective at reducing data cache leakage energy without affecting performance by more than 1% [8]. However, Kim et al. also found that such a policy is not as effective for instruction caches because data caches tend to have better temporal locality while instruction caches normally exhibit better spatial locality [7]. More precisely, Kim et al. found that such a simply policy may have a run-time impact of as much as 5.7% and the percentage of drowsy lines can be as low as 68.5% [7]. Kim et al. then proposed an approach to managing the drowsy instruction cache at the sub-bank granularity (instead of

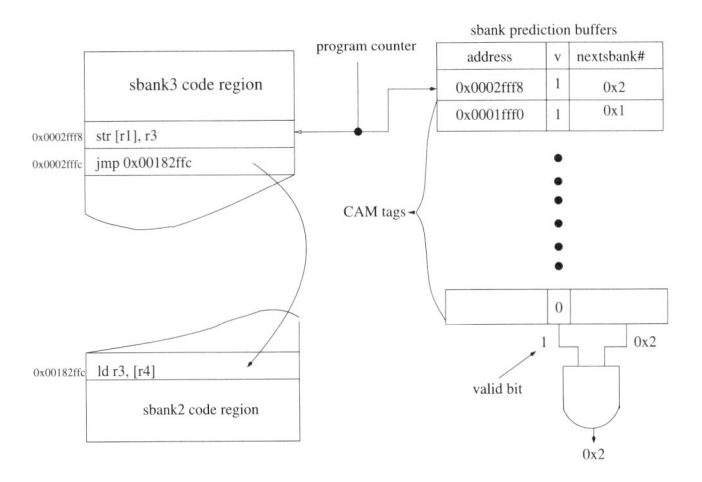

FIG. 2. The next sub-bank prediction buffer [7].

cache line granularity) to exploit the spatial locality [7]. In this scheme, only one sub-bank is active at a time and all other sub-banks are placed into the drowsy mode. Whenever the processor accesses a cache line in a non-active sub-bank, the pre-decoder activates the next target sub-bank, and puts the currently active sub-bank into the drowsy mode [7].

Since there is a wake-up latency to activate the next sub-bank, Kim et al. proposed to use a memory sub-bank prediction buffer to predict and pre-activate the next sub-bank, and it is shown to be effective at reducing the performance penalty significantly [7]. Figure 2 illustrates the next sub-bank prediction buffer scheme presented in [7]. Each entry of the prediction buffer contains an instruction address of the instruction one before the instruction causing the transition to another sub-bank (assuming a single cycle wake-up latency). The buffer entry also contains a valid bit and the index of the next target sub-bank. The buffer is consulted at each cache access to determine whether to awake a new sub-bank or not. In case of misprediction or no prediction, the old entry is updated or a new entry is allocated [7].

The hardware-based sub-bank prediction is not energy-efficient since the prediction buffer needs to be consulted at each cache access, resulting in significant extra dynamic energy consumption. While Kim et al. presented the area overhead of the hardware prediction buffer, its dynamic energy overhead is not quantified [7]. In this work, we use cacti 3.2 [18] to calculate the dynamic energy consumption per access to the prediction buffer and compare it with the leakage energy savings of the drowsy instruction cache for various applications. We model a next sub-bank prediction buffer with 128 entries, which is shown to be effective at reducing performance

TABLE I
LEAKAGE ENERGY DISSIPATION FOR CACHE SUB-BANKS WITH DIFFERENT SIZES

Leakage per bit	Leakage for cache sub-banks			
	1K	2K	4K	8K
1.63E–15 J	1.19E–11 J	2.39E–11 J	4.78E–11 J	9.56E–11 J

penalty in [7]. The dynamic energy per access reported by cacti is 2.39E–10 J. To estimate the leakage energy savings for an instruction cache (note that we actually calculate the total leakage energy consumption of the instruction cache, which is the upper bound of the leakage energy savings), we refer to the data presented in [8] under 0.07 um technology. Table I gives the leakage energy consumption results for an instruction cache sub-bank with different sizes. If we compare the dynamic energy per access to the prediction buffer with the leakage energy of a cache sub-bank, we can see that the dynamic energy per access is much larger. For a 1 KB or 2 KB sub-bank, the dynamic energy overhead is even an order of magnitude larger than the leakage energy consumption of the cache sub-bank.

Since the prediction buffer is accessed per cache access (not per cycle), we run experiments to collect the total number of accesses to a 16 KB direct-mapped L1 instruction cache (see Section 5 for the detailed experimental framework and configuration), and the results are presented in Table II. The dynamic energy overhead in Table II is the extra dynamic energy consumption due to the accesses to the sub-bank prediction buffer at runtime. The calculation of the I-cache leakage energy is based on the leakage per bit number listed in Table I. The last column of Table II gives the ratio of the dynamic energy overhead of the prediction buffer to the total leakage energy consumption of the instruction cache. As can be seen, the dynamic energy overhead is in the same order of magnitude as the total I-cache leakage energy consumption. Specifically, the dynamic energy overhead is equal to 48.6% of the total I-cache leakage on average. Note that the data listed in Table II is the total leakage energy consumption of the instruction cache, the total leakage savings of the instruction cache should be less than it regardless of the drowsy management policy, since cache lines in the drowsy mode still consume non-zero leakage energy. In other words, the total leakage energy consumption is the upper bound of the possible leakage savings for the drowsy instruction cache. Therefore, the overall energy savings of the drowsy instruction cache (i.e., the total leakage savings minus the dynamic and leakage energy overheads) can be substantially compromised by taking the dynamic energy overhead into account.

It should be noted that the relative ratio of dynamic energy per access and sub-bank leakage consumption per cycle can vary widely with design style and fabrication technology [5]. The comparison we made in this chapter is only used to demonstrate

TABLE II
COMPARING THE DYNAMIC ENERGY OVERHEAD OF THE NEXT SUB-BANK PREDICTION BUFFER
AND THE TOTAL LEAKAGE ENERGY CONSUMPTION OF A 16 KB INSTRUCTION CACHE

Benchmark	Execution cycles	I-cache accesses	I-cache leakage (J)	Dynamic energy overhead (J)	Ratio
164.GZIP	1292173435	408374714	0.247063561	0.097601557	0.395046345
181.MCF	21663288437	12124968426	4.142020749	2.897867454	0.699626494
256.BZIP2	13942496443	8449023200	2.66580532	2.019316545	0.757488377
MPEG2ENC	598165	49160	0.000114369	1.17492E−05	0.102730985
RAWDAUDIO	21886688	8420938	0.004184735	0.002012604	0.480939458
RAWCAUDIO	40543714	10261086	0.007751958	0.0024524	0.316358726
POLYPHASE	933323	532981	0.000178451	0.000127382	0.713823173
PARAFFINS	362685	177828	6.93E−05	4.25009E−05	0.612886969
DJPEG	11780602	5700216	0.002252451	0.001362352	0.604830748
DES	32820484	2769360	0.006275277	0.000661877	0.105473757
CORDIC	139599	39060	2.67E−05	9.33534E−06	0.349752166
CJPEG	31530629	17505754	0.006028656	0.004183875	0.693998

Notes. The last column is the ratio of the dynamic energy consumption of the prediction buffer to the total leakage energy consumption of the instruction cache. Note that the total leakage savings of the instruction cache should be less than its total leakage energy consumption, since cache lines in drowsy mode still consume non-zero leakage.

quantitatively that one should be very cautious to apply leakage control techniques that can incur extra dynamic energy, since the dynamic energy overhead can easily nullify or significantly compromise the leakage energy savings. This motivates us to develop a compiler-assisted approach for reducing the leakage energy and the performance penalty without incurring substantial dynamic energy overhead.

3.3 Transitional Instructions

By analyzing the program, we find that most of the instructions triggering transitions to other sub-banks (called transitional instructions in this chapter) can be identified statically. Therefore, the compiler can detect those transitional points and provide hints to the processor to pre-activate the next sub-bank at runtime, thus eliminating the hardware overhead of recording and re-discovering this knowledge dynamically. The transitional instructions that can be identified by the compiler are divided into three categories as follows.

1. Frontier instructions: These are normal instructions (excluding branches) that cause the transitions from an active sub-bank to another sub-bank. When the program is executed in the streaming mode, for instance,

running the instructions within a basic block, the PC (Program Counter) address is updated by PC+4 sequentially. If the current instruction address is mapped to the last cache block of a sub-bank (note that such instructions are called border instructions in this chapter), the next sequential instruction will be mapped to the first cache block of another sub-bank if the previous instruction (i.e., the border instruction) is not a branch instruction, as shown in Fig. 3. To hide the performance penalty, the compiler can annotate information in the border instruction to pre-activate the next sub-bank at runtime (note that although frontier instructions can trigger the activation of drowsy cache lines automatically, it is important to pre-activate the corresponding sub-banks when executing the border instructions, so that the performance penalty due to the activation delay can be hidden).

2. Unconditional branches with static target addresses: Since the target addresses can be identified at the link time, the compiler can annotate the previous instruction(s) to pre-activate the target sub-bank.

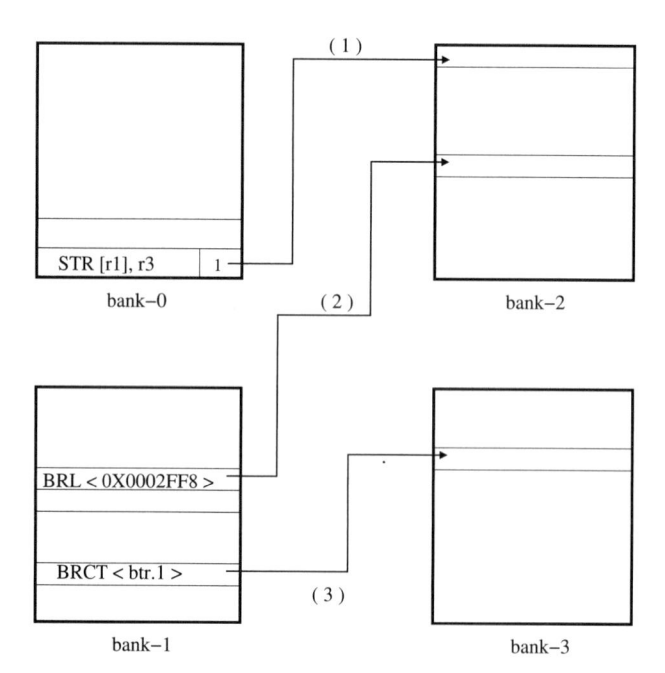

FIG. 3. Example of three types of transitional instructions that can be identified by compilers, including: (1) frontier/border instructions, (2) unconditional branches with static target addresses, and (3) conditional branches with static target addresses.

3. Conditional branches with static target addresses:
   While the target addresses of this type of instructions can be computed at link time, the transitions to other sub-banks are determined by whether the branches are taken or not (which is not perfectly known at the compilation time). If the conditional branch is not taken, the next subsequential address may reside in the same sub-bank and hence does not need the pre-activation. Consequently, the compiler needs to predict the directions of these conditional branches in addition to generating the target addresses statically. While there are some effective approaches to estimating the branch behavior at the compilation time [21,22], we use a profiling based approach in this chapter. More specifically, the compiler only selects the dynamic branches with a taken probability larger than a given threshold (80% in this chapter) as the possible transitional instructions, and previous instructions of those branches are annotated to pre-activate the target sub-banks.

3.4 Instruction Cache Addressing Schemes

The L1 instruction cache (also called L1-Icache or iL1 in this chapter) can be addressed by using different schemes [20], as discussed below.

1. Physically-indexed, physically-tagged iL1: In this scheme, the physical address must be obtained before the instruction cache is indexed. Therefore, the TLB (Translation Look-aside Buffer) must be consulted before retrieving the instruction cache, which can lengthen the critical path of the processor. In terms of energy consumption, the TLB needs to be consulted for each instruction access no matter hitting in iL1 or not, thus making it not energy efficient. Currently, this configuration is not very popular for the L1 instruction cache, due to its significant impact on the overall performance and energy consumption. However, the advantage of this scheme is that it has no aliasing problems.

$$\left| \lfloor ((PC \bmod S_{cache}) \div S_{sub\text{-}bank}) \rfloor - \lfloor (((PC - 4) \bmod S_{cache}) \div S_{sub\text{-}bank}) \rfloor \right| \geqslant 1, \qquad (1)$$

$$\left| \lfloor ((Target \bmod S_{cache}) \div S_{sub\text{-}bank}) \rfloor - \lfloor ((PC \bmod S_{cache}) \div S_{sub\text{-}bank}) \rfloor \right| \geqslant 1, \qquad (2)$$

2. Virtually-indexed, physically-tagged iL1: In this configuration, both the TLB and the iL1 can be accessed simultaneously by using the virtual address of the instruction. After the physical address is obtained from the TLB, it can be used for comparison with the tag bits in iL1 to determine

cache hit or miss. Therefore, the TLB is not on the critical path, but it is still not energy efficient since the TLB still needs to be accessed for each instruction fetch. Another disadvantage of this scheme is the aliasing problem, where two or more virtual addresses are mapped to the same physical address, and thus multiple copies of the same data can be present in the cache simultaneously. The solution of this problem is to either limit the size of iL1 (within one page) or to add a few bits to differentiate between different address spaces. Currently, many processors use this addressing scheme, such as AMD K6, MIPS R10K and PowerPC.

3. `Virtually-indexed, virtually-tagged iL1`: In this configuration, the L1 instruction cache is both indexed and tagged with the virtual address. Consequently, TLB accesses are not needed at all until an iL1 miss occurs (depending on the addressing scheme for the L2 cache), which can benefit both the performance and energy consumption. However, the downside is the aliasing problem.

In this chapter, we assume a virtually-indexed instruction cache (including virtually-indexed virtually-tagged iL1 and virtually-indexed physically-tagged iL1), which has been used in some embedded processors, such as MC68030 [16], StrongARM and its descendant the XScale [17]. Built upon this assumption, for each instruction in the program, the compiler can determine its location in iL1 according to its virtual address, making it possible to identify transitional instructions statically.

3.5 Identify Transitional Instructions

To determine whether the current instruction will cause a transition to another sub-bank or not, we use formula (1) for frontier instructions and formula (2) for branch instructions. In both formulas, the *PC* is the address of the current instruction, the S_{cache} is the size of the instruction cache and the $S_{sub-bank}$ is the size of the sub-bank. The *Target* in formula (2) represents the target address of a branch, which can be calculated at compilation time. In formula (1), if the previous instruction and the current instruction belong to different sub-banks, the current instruction is predicted to be the frontier instruction. Similarly, formula (2) calculates the sub-bank index of the target address and the sub-bank index of the current instruction address. If they differ, the current branch is predicted to cause the transition to another sub-bank (note that for conditional branches, we also need to take the profiling information into account as aforementioned).

3.6 ISA and Compiler Support

To activate the sub-bank at runtime, this chapter assumes the existence of an ISA extension to annotate the sub-bank pre-activation information at the compilation time. We propose the addition of 4 hint bits to each instruction, as shown in Fig. 4. The first bit indicates whether the next sub-bank should be pre-activated or not, and the last three bits represent the next sub-bank index (note that we use 3 bits because we focus on experiments with an instruction cache of 8 sub-banks. For instruction caches with more than 8 sub-banks, more hint bits are needed). The extra leakage energy due to the hint bits are accounted for estimating the overall leakage savings compared to the default instruction cache. Note that since only one cache sub-bank is active at runtime, the leakage overheads due to the extra hint bits are much smaller than the dynamic energy overhead to access a prediction buffer (see the leakage per bit data from Table I for an estimation). It should also be noted that for processors that already provide extra hint bits to encode useful information at the compilation time, these hint bits can be exploited to eliminate or mitigate the space overhead.

After identifying the transitional instructions, the compiler then inserts hint bits one instruction ahead of the transitional instructions in the program. At runtime, the processor then decodes the hint bits and pre-activates the corresponding sub-banks for reducing the performance penalty.

It should be noted that the goal of this chapter is to *reduce the leakage energy consumed in the instruction cache without compromising performance*. The proposed compiler-directed scheme, however, inevitably has several energy (and performance) overheads. For instance, the extended hint bits will increase the dynamic energy per instruction access and also consume leakage energy. Also, there is a dynamic energy overhead to decode the hint bits and to turn on/off sub-banks at runtime. In this chapter, where significant, we quantify these overheads to examine the energy behaviors of various schemes. In the rest of this chapter, when we mention *energy*, we mean the leakage energy consumed by the instruction cache plus any energy overheads (dynamic or static) associated with the energy optimization scheme.

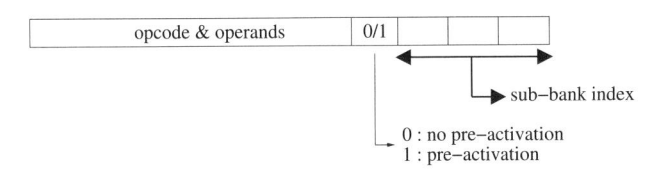

FIG. 4. Adding 4 hint bits to each instruction. The first bit indicates whether the next sub-bank should be pre-activated or not. The last three bits comprise the next sub-bank index for an instruction cache of 8 sub-banks.

4. Compiler-Assisted Loop-Based Data Cache Leakage Reduction

Besides the instruction cache, the data cache is another major consumer of the on-chip transistors, which should also be a main target for on-chip leakage energy control. While a number of circuit and architectural level techniques have been studied to mitigate the data cache leakage energy, this section investigates a simple yet effective compiler-based approach to minimizing the leakage dissipation for data caches. This approach is based on the observation that only a small portion of the data are active at certain program phases runtime and the program typically spends a significant time in loops. Consequently, a large portion of data cache lines, which are not accessed by the loops, can be placed into the leakage control mode to reduce the data cache leakage energy. Since loops can be nested, we propose an optimistic approach to placing the cache lines into low leakage mode during the execution of the innermost loop (note that non-nested loops are treated as innermost loops without outermost loops in this work).

It is well known that in general, a program spends a large percentage of time in executing only a small portion of the program, i.e., loops, which has been the focus of traditional performance-oriented compiler optimizations [25]. Similarly, compiler can also exploit the loop behavior to significantly benefit the energy dissipation. More specifically, if the size of the data accessed by the loop is much less than the size of the data cache, the rest of the cache lines that are not accessed by the loops can be placed into the low leakage mode during the loop execution. Since loops typically take a long period of time to execute, this strategy can potentially lead to substantial leakage energy savings for data caches. On the other hand, the set of data accessed by the loops must stay in the active mode in order to not impact performance, since these cache lines may be accessed repeatedly and activating a cache line from the low leakage mode incurs performance penalty.

Based on the above observation, we can formalize the loop-based data cache leakage problem as follows:

> *For a given code region R, suppose it takes T cycles to execute and the data it accesses contains S cache lines. Assume the data cache (L1-Dcache) consists of D cache lines. In order to save leakage energy, we can place $D - S$ (if $S < D$) cache lines into the low leakage mode for T cycles. The leakage energy savings is proportional to the product of $(D - S) * T$ (we ignore the dynamic energy penalties due to turn on/off cache lines at this analytic stage, and our simulation results show that the leakage energy savings dominate).*

According to this abstraction, it is obvious that the larger the T and the smaller the S, the more leakage energy can be potentially reduced (note that D is fixed for a given cache configuration). The execution time T and the set of active data cache lines S, however, are conflicting with each other. If a large region of code is chosen, it may take more time to execute (i.e., T is increased); but also access more data (i.e., S is increased). Obviously, if S is equal or larger than D, there will be no leakage savings no matter how long T is (unless the leakage energy is controlled at a finer granularity). On the other hand, if a very small region of code is selected, even though it only accesses a small set of data (i.e., S is small), the execution time is short too (i.e., T is small); then the leakage energy saving will be insignificant. Consequently, there is a tradeoff to choose the region of code by considering both the execution time and the data size so that the leakage energy of data caches can be minimized.

While theoretically, the compiler can partition the program into regions with arbitrary sizes to apply the proposed strategy for reducing the data cache leakage energy, the exploration of the large design space is beyond the scope of this chapter. In this work, the proposed strategy works on the innermost loop granularity to manage leakage energy, since the innermost loops provide a good tradeoff between large execution time and small data footprint. Specifically, the compiler divides the program into two different leakage mode: active mode (also called normal mode) and non-active mode, based on the loop analysis. The program begins at the active mode. Just before the execution of the loop, the mode is transitioned into non-active, where all the cache lines will be placed into the low leakage mode. At the end of the innermost loop, the mode is changed back into active again. To reduce the performance penalty of activating each data cache line that is accessed sequentially, we apply the just-in-time activation technique proposed in [26] for the drowsy data cache lines. More specifically, when the current cache line is accessed in the active mode, the next cache line will be activated to avoid the performance degradation by exploiting the spatial locality of data accesses (note that for a set-associative data cache, the next set will be activated).

We assume the existence of an instruction to set the cache lines into different leakage modes: active mode and non-active mode. With this abstraction, the compiler's job is to insert the instruction statically to set the program into different leakage modes based on the loop analysis. Precisely, the compiler needs to insert the instruction to set the program into the non-active mode at the beginning of the innermost loops, and insert the instruction to set the program into the active mode at the end of the innermost loops. When the program is executed at runtime, all the data cache lines will be placed into low leakage mode (deactivate) at the non-active mode. In contrast, at the active (normal) mode, the just-in-time activation will be employed to pre-activate the next data cache line.

The idea behind the proposed strategy is illustrated in Fig. 5. D1, D2 and D3 represent the cache lines that are accessed by block B1, B2 and B3, respectively. B2 contains a loop, which may take much longer time to execute than both B1 and B3.

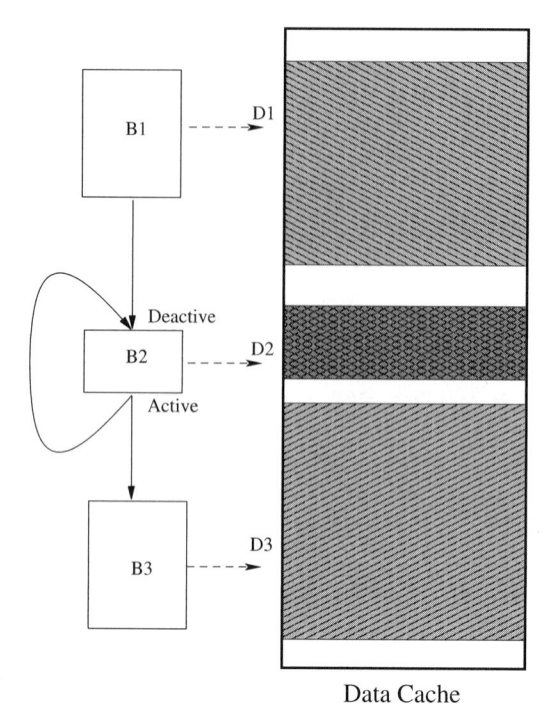

FIG. 5. A code fragment with three blocks, including a loop (i.e., B2), and the sets of data cache lines accessed by these blocks.

Compiler inserts the instructions to set the deactive mode at the beginning of loop blocks (i.e. B2 in this example) and to set the active mode at the end of the loop (see Fig. 5). At runtime, D1 and D3 will be placed into the low leakage mode when the deactivate instruction is executed (just before the execution of loop B2); and these cache lines will stay in the low leakage mode until they are accessed again after B2 is finished. If B2 takes a large fraction of the total execution time, cache lines in D1 and D3 can be placed into the low leakage mode for long time, leading to substantial leakage energy savings.

5. Evaluation Methodology

We use simulation to evaluate the proposed compiler-assisted leakage reduction schemes. We target the instruction and data caches for a state-of-the-art VLIW

TABLE III
DEFAULT PARAMETERS USED IN THE PERFORMANCE AND ENERGY SIMULATIONS

Parameter	Value
Feature size	0.07 micron
Supply voltage	1.0 V
L1 instruction cache	16 KB direct-mapped, 8 sub-banks
L1 instruction cache latency	1 cycle
L1 data cache	32 KB 2-way cache
L1 data cache latency	1 cycle
Unified L2 cache	512 KB 4-way cache
L2 cache latency	10 cycles
L1 cache line size	32 B
L2 cache line size	64 B
L1 cache line leakage energy	0.33 pJ/cycle
L1 drowsy cache line leakage energy	0.01 pJ/cycle
L1 state-transition (dynamic) energy	2.4 pJ/transition
L1 state-transition latency	1 cycle
L1 dynamic energy per access	0.11 nJ
L2 dynamic energy per access	0.58 nJ

processor, since VLIW architectures[1] are not only used in high-performance microprocessors (e.g., Intel IA-64), but also increasingly used in DSP and embedded systems [24]. The Trimaran v3.7 [13] was used for the compiler implementation and architecture simulation. Trimaran is comprised of a front-end compiler IMPACT, a back-end compiler Elcor, an extensible intermediate representation (IR) Rebel, and a cycle-level VLIW/EPIC simulator that is configurable by modifying the machine description file [13]. The virtual/real register allocation algorithm was implemented as the last optimization phase in Elcor. The machine description file of Trimaran was configured to simulate VLIW processors with various number of real and virtual registers. By default, the simulated VLIW processor consists of two IALUs (integer ALUs), two FPALUs (floating-point ALUs), one LD/ST (load/store) unit and one branch unit. The compiler-assisted sub-bank prediction and the loop-based data cache leakage reduction are implemented in Elcor after the instruction scheduling. The cycle-level simulator was augmented to recognize the ISA extensions. Other system parameters used for our default setting are provided in Table III. The energy values reported are based on circuit simulation [19]. We select a diverse set of benchmarks from the SPEC 2000, SPEC 95 [14], and Mediabench [15] for the energy and performance evaluation.

---

[1] The EPIC architecture is regarded as an extension of VLIW architectures, which also employ architectural features of dynamic-issued processors, such as superscalars.

5.1 Experimental Results for Static Next Sub-Bank Prediction

5.1.1 Performance Overhead Reduction

Accessing an instruction in the drowsy mode has performance penalty, since it takes time to reinstate the power supply lines of the drowsy cache lines to the normal voltage level [7]. We assume that it takes one cycle to activate the drowsy cache sub-bank as in [7]. Figure 6 compares the performance penalty reduction of the hardware-based approach [7] and the compiler-based approach. We find that on average, the compiler-based approach reduces 6% more performance penalty than the prediction buffer approach, illustrating that the compiler-assisted sub-bank prediction is very effective at detecting the transitional instructions and to instruct the processor for pre-activating the next sub-banks accurately.

Since the compiler-assisted approach can find and distinguish three types of transitional instructions: the frontier instructions, unconditional branches and conditional branches without indirect addressing, Table IV lists the number of each type of transitional instructions captured by the compiler. The compiler can identify more than 300 transitional instructions statically for 2/3 applications, and for 164.gzip, mpeg2enc, djpeg, des and cjpeg, more than 2000 instructions are labeled as transitional instructions. Such a large number of static transitional instructions make it hard to save and predict the sub-bank transition behavior by using a prediction buffer, unless the buffer contains many entries or is fully associative for reducing the number of conflict misses. The hint bits are used by the processor to trigger the pre-activations of the corresponding sub-banks at runtime, which can effectively reduce the perfor-

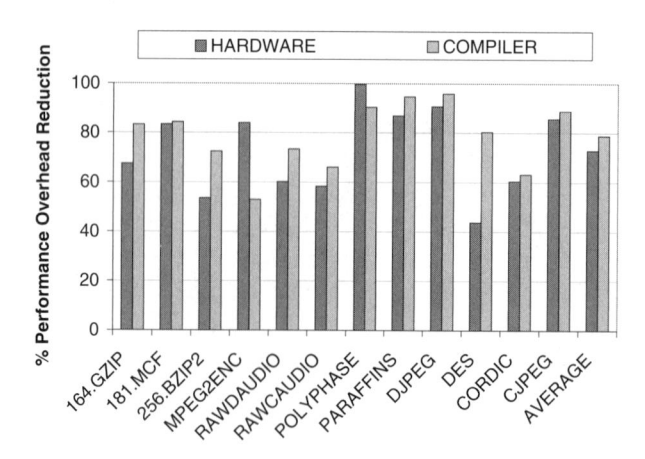

FIG. 6. Comparison of performance penalty reduction of the next sub-bank prediction buffer and the compiler-assisted sub-bank prediction approach.

TABLE IV
THE NUMBER OF THREE TYPES OF TRANSITIONAL INSTRUCTIONS IDENTIFIED BY THE COMPILER

Benchmark	Frontier instructions	Unconditional branches	Conditional branches	Total instructions
164.GZIP	524	818	858	2200
181.MCF	246	193	275	714
256.BZIP2	192	794	678	1664
MPEG2ENC	800	1371	1173	3344
RAWDAUDIO	23	37	40	100
RAWCAUDIO	22	40	43	105
POLYPHASE	100	141	60	301
PARAFFINS	53	52	77	182
DJPEG	1997	2591	2618	7206
DES	1053	627	414	2094
CORDIC	56	69	48	173
CJPEG	2015	2665	2688	7368

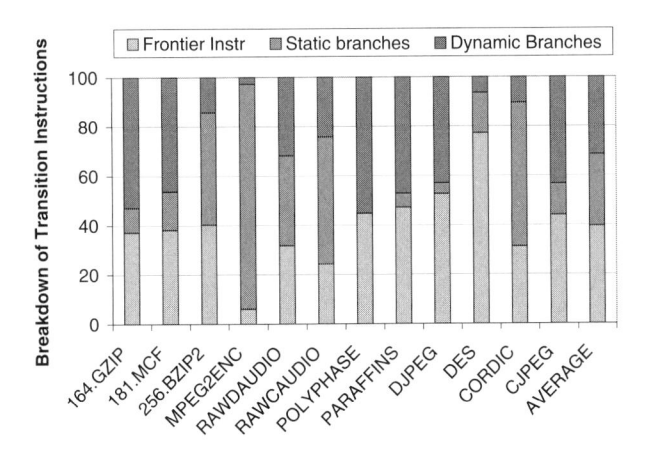

FIG. 7. The breakdown of frontier instructions, unconditional branches and conditional branches with static target addresses that are pre-activated at runtime to minimize the performance overhead.

mance penalty. Figure 7 provides the breakdown of the three types of transitional instructions that are pre-activated at runtime. As can be seen, different types of transitional instructions make various contributions to the overall performance overhead reduction. An interesting result is that the frontier instructions cause a fairly large percentage of transitions to other sub-banks, which is quite stable for all the benchmarks. On average, 39.6% wake-up latencies are reduced by the pre-activation of frontier instructions. Thus, besides the sub-bank transitions caused by the disruptions of control flows due to branches, the sequential instruction flow can also induce a fairly large number of sub-bank transitions, which are appropriate for the compiler to discover and exploit in a low-cost manner.

5.1.2 Leakage Energy Savings

Both the hardware-centric approach and the compiler-assisted approach can place the non-active sub-banks into the drowsy mode for saving leakage energy. Figure 8 compares the percentage of time that the cache lines are placed into the drowsy mode for both approaches, which determines the leakage energy savings that can be achieved. There is no notable difference between these two approaches since they only differ in the way to pre-activate the next sub-banks, which implies that the instruction cache leakage savings (without considering the dynamic energy overhead) achieved by both approaches are comparable. However, since the compiler-based approach eliminates the extra dynamic energy for consulting the prediction buffer, the overall savings by the compiler-based approach can be larger if we take both the dynamic and leakage energy overhead into consideration (note that for the compiler-assisted approach, the dynamic and leakage energy overheads come from the extra hint bits). Figure 9 compares the overall energy savings for the hardware-based approach and the compiler-based approach by considering the energy overheads, with respect to the total L1 I-cache leakage energy consumption as listed in Table II. We can observe that the compiler-based approach can achieve much more energy savings than the hardware-based approach, except for mpeg2enc and des, in which the number of dynamic accesses to the I-cache is significantly low, resulting in small

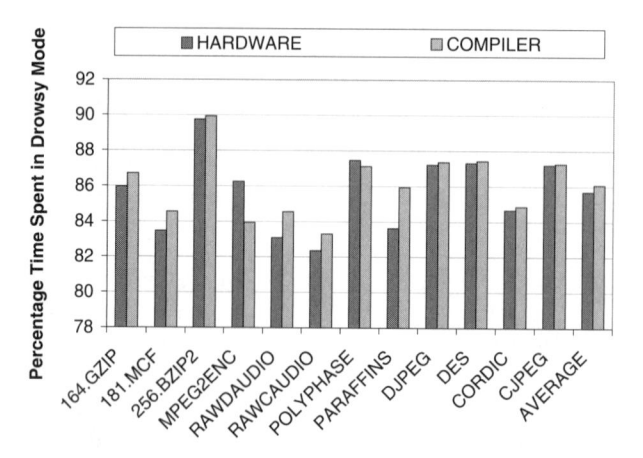

FIG. 8. Comparison of the percentage of time that the cache lines are placed into the drowsy mode for the hardware approach and the compiler-assisted approach.

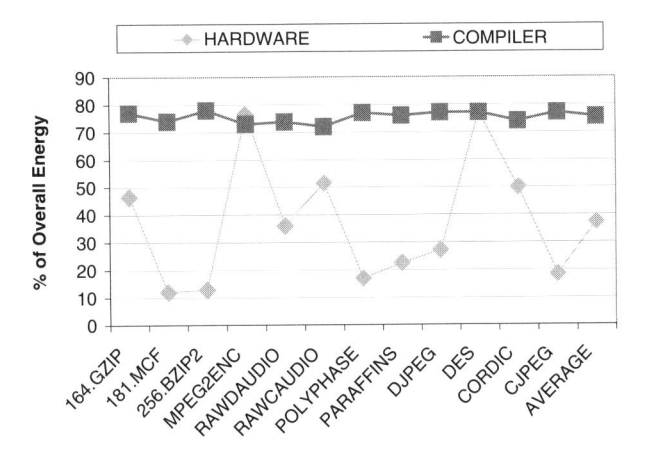

FIG. 9. Comparison of the leakage energy savings for the hardware-based approach and the compiler-assisted approach. The results are normalized with respect to the total L1 I-cache leakage energy consumption as listed in Table II.

dynamic energy overhead (see Table II). On average, we find that the compiler-based approach can save 38.2% more overall energy than the hardware-based approach for the instruction cache.

5.1.3 Sensitivity Analysis

In this section, we examine how the effectiveness of the compiler-assisted approach is affected by the configuration of the L1 instruction cache. We mainly study two factors—the size of the cache and the number of sub-banks. For simplicity, we select one benchmark (i.e., 164.gzip) from SPEC 2000 and another benchmark (i.e. des) from Mediabench. We run experiments on these two benchmarks by varying the L1 instruction cache configuration parameters.

Figure 10 illustrates how the performance overhead reduction of the compiler-based approach is impacted by the size of the instruction cache and the number of sub-banks respectively. The number of sub-banks is fixed to be 8 in (a) and the size of the instruction cache is 16 KB in (b). As can be seen, the performance penalty reduction decreases as the instruction cache size increases. The reason is that each cache sub-bank can contain more instructions with the increase of the instruction cache size, so there are less number of transitional instructions to access other sub-banks. As a result, fewer number of transitional instructions can be identified by the compiler. In contrast, when the number of sub-banks is increased, the performance penalty reduction tends to be increased (except for des) when the number of sub-banks is increased from 4 to 8, as illustrated in Fig. 10(b). With the increase of

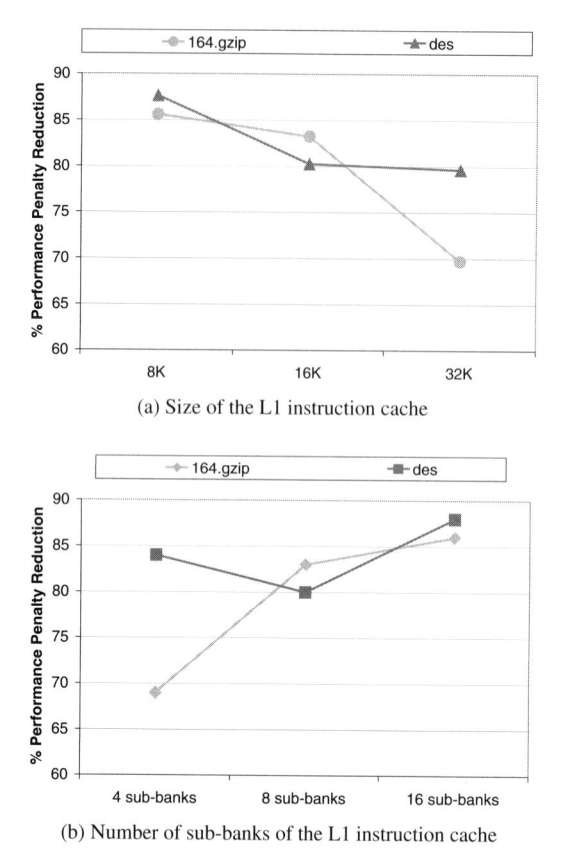

FIG. 10. The impact of the L1-Icache size (a) and the number of sub-banks on the performance penalty reduction for the compiler-assisted approach. The number of sub-banks is fixed to be 8 in (a) and the size of the instruction cache is 16 KB in (b).

the number of sub-banks, each sub-bank contains less number of instructions, given a fixed instruction cache size. Therefore more instructions can become transitional instructions and the performance penalty tends to increase without pre-activating (however, it should be noted that the performance penalty is also determined by the program behavior, i.e., the runtime branch behavior and the address mappings). In general, there is a clear trade-off between performance and leakage reduction in choosing the size of sub-banks (or the number of the sub-banks given a fixed instruction cache). The performance penalty is less severe with a larger sub-bank. For instance, in the extreme case, there will be no performance penalty if the instruction cache only has one bank. However, the smaller the sub-bank is, the more leakage energy reduction can be achieved (without considering the energy cost due to sub-bank transitions). Since the performance penalty can be reduced by employing the proposed compiler-assisted approach without significant dynamic energy overhead, an attractive strategy is to select a small sub-bank size to reduce the instruction cache leakage energy substantially at a finer granularity.

5.1.4 Hybrid Strategy

Although the compiler-assisted sub-bank prediction has the advantage of reducing the performance degradation of drowsy instruction caches without significant dynamic energy overhead, it can only annotate those transitional instructions whose target addresses can be determined statically. In contrast, the hardware-based approach can exploit the runtime information to capture the behavior of the transitional instructions, some of which cannot be discovered by the compiler. For instance, the target addresses of branch instructions with indirect addressing mode or return instructions typically cannot be determined statically by the compiler. On the other hand, the hardware-based approach needs to compare the current instruction address with the transitional instructions stored in the prediction buffer for every instruction cache access, which can result in tremendous dynamic energy overhead (compared to the leakage savings) as aforementioned. To combine the advantages of both approaches, we propose a hybrid strategy to pre-activate the next sub-bank of drowsy instruction caches.

The hybrid strategy makes intelligent use of the static information as the compiler-assisted approach. In addition, the compiler annotates those branch instructions whose target addresses cannot be calculated at the link time, which will be handled by the hardware-based approach. The proposed hybrid strategy also employs a next sub-bank prediction buffer similar to the hardware-centric approach, however, this buffer is not consulted for each cache access, but only for those instructions that cannot be identified statically by the compiler. Therefore, the transitions caused by these types of instructions can still be captured by the hybrid approach, while the dynamic energy overhead associated with the hardware prediction buffer is reduced significantly by accessing the prediction buffer selectively.

Figure 11 compares the percentage of performance overhead reduction for all three approaches, i.e., the hardware-based approach, the compiler-based approach and the hybrid approach. As can be seen, the hybrid approach out-performances both the hardware and compiler-based approaches in terms of performance. On average, the hybrid approach reduces the performance overhead by 14.2% and 8.2% more than the hardware-based approach and the compiler-based approach, respectively.

Besides the performance improvement, the hybrid approach is also capable of reducing the total number of accesses to the prediction buffer, compared to the

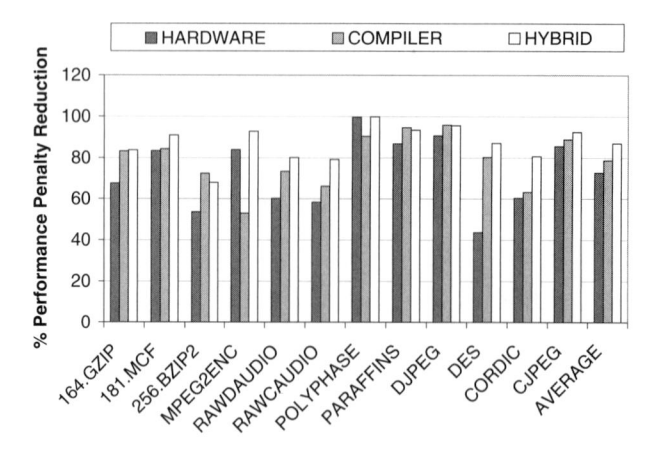

FIG. 11. The comparison of the performance penalty reduction of the drowsy instruction cache for the hardware-based approach, the compiler-based approach and the hybrid approach.

TABLE V

COMPARISON OF THE TOTAL NUMBER OF ACCESSES TO THE PREDICTION BUFFER FOR THE HARDWARE-CENTRIC APPROACH AND THE HYBRID APPROACH THAT ALSO EXPLOITS COMPILER INFORMATION

BENCHMARK	HARDWARE	HYBRID	Ratio
164.GZIP	408374714	48362123	0.118425851
181.MCF	12124968426	1591420167	0.13125149
256.BZIP2	8449023200	910433789	0.1077561
MPEG2ENC	49160	7989	0.162510171
RAWDAUDIO	8420938	1475648	0.175235585
RAWCAUDIO	10261086	1770081	0.172504255
POLYPHASE	532981	49484	0.092843835
PARAFFINS	177828	13133	0.073852262
DJPEG	5700216	186057	0.032640342
DES	2769360	8894	0.003211572
CORDIC	39060	1703	0.04359959
CJPEG	17505754	2027866	0.115839969

hardware-based approach only. Table V lists the total number of accesses to the prediction buffer for the hardware-based approach and the hybrid approach. The last column gives the ratio of the number of accesses to the prediction buffer of the hybrid approach to that of the hardware-based approach. Due to the selective accesses to the prediction buffer, the hybrid approach reduces the number of accesses to the prediction buffer by 89.2%, in comparison to the hardware-based approach. As a re-

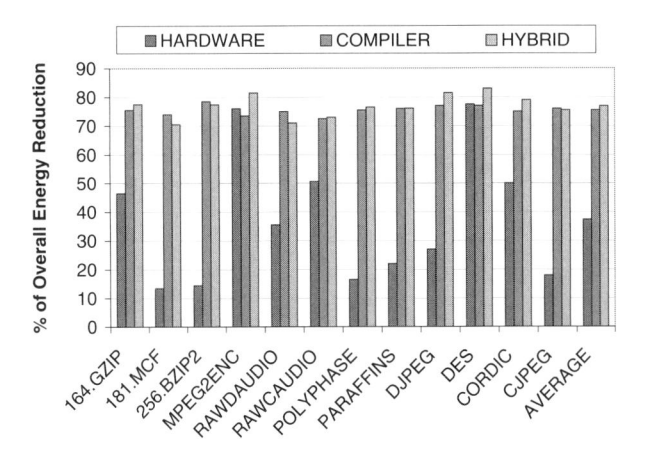

FIG. 12. The comparison of the overall energy savings for the hardware-based approach, the compiler-assisted approach and the hybrid approach.

sult, the dynamic energy overhead of the prediction buffer for the hybrid approach is much smaller.

Figure 12 presents the overall energy savings for the hardware-based approach, the compiler-based approach and the hybrid approach. The hybrid approach saves more energy than the hardware-based approach across all applications, because it reduces the number of accesses to the prediction buffer without compromising the performance. The hybrid approach is very comparable to the compiler-based approach in the overall energy savings. For 8 out of the 12 benchmarks, the hybrid approach actually saves more overall energy than the compiler-based approach. On average, the hybrid approach reduces 1.4% more overall energy than the compiler-based approach, because it improves the performance by predicting the behavior of those transitional instructions that are not identifiable by the compiler. In general, the hybrid approach shows superior results in both the performance and energy. Therefore, it is the best strategy for the drowsy instruction cache to balance leakage reduction and performance.

5.1.5 Energy-Delay Product (EDP) Results

Since both the performance and energy consumption are important design goals, we use energy-delay product to compare the schemes proposed in this chapter. Figure 13 gives the energy-delay product for the hardware-based approach, the compiler-based approach and the hybrid approach, which are normalized with respect to the original scheme without applying any energy control techniques. As

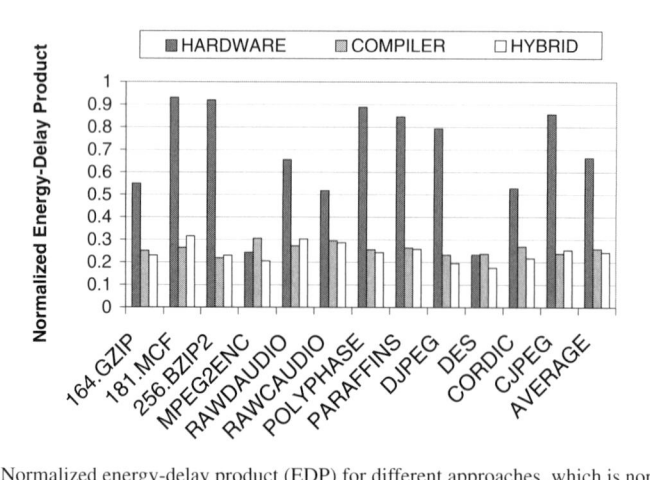

FIG. 13. Normalized energy-delay product (EDP) for different approaches, which is normalized with the original scheme without applying any energy reduction techniques.

can be seen, both the compiler-based approach and the hybrid approach can reduce the EDP significantly, compared with the hardware-based approach. The hybrid approach is the best among all three schemes, which is better than the hardware-based approach across all the benchmarks. On average, the hybrid approach can reduce the EDP by 42.1% more than the hardware-based approach, mainly because it can has much less dynamic energy overheads than the hardware-based approach by controlling the accesses to the prediction buffer.

5.2 Experimental Results for Loop-based Data Cache Leakage Reduction

We implement the proposed loop-based strategy to reduce the data cache leakage energy, and compare the results with the pure hardware based approach. We fix the window size to be 2000 clock cycles for the hardware counter based approach, since such a window size is shown to be able to maximize the leakage energy savings effectively [7]. Figure 14 gives the percentage of time that the cache lines can be placed into the low leakage mode for both the pure hardware based approach and the loop-based approach. For all the benchmarks except 129.compress, we find that the proposed loop-based strategy can place cache lines into the low leakage mode longer than the pure hardware based approach. The reason is that for the loop-based approach, all the cache lines except those storing the hot data (i.e., the data accessed by the loops) are placed into the low leakage mode immediately when encounting loops; while for the pure hardware based approach, it has to wait for a fixed time

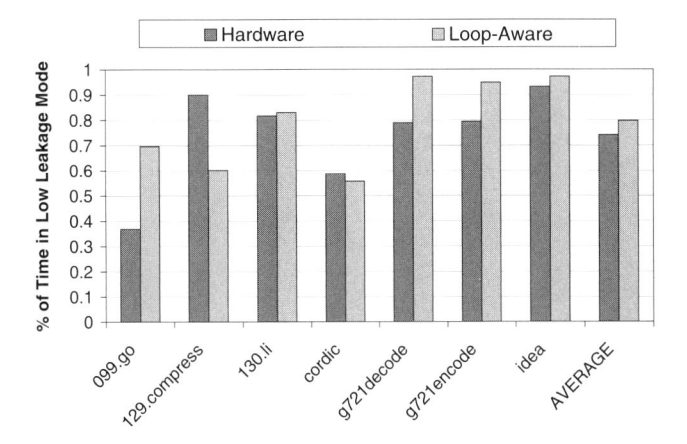

FIG. 14. Percentage of times that cache lines are in low leakage mode.

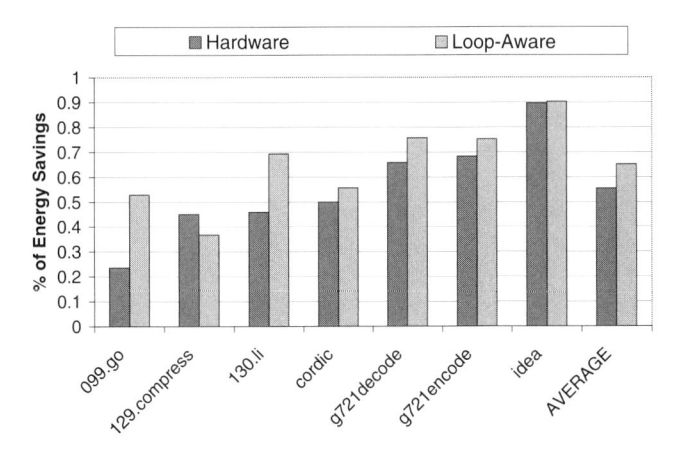

FIG. 15. Percentage of leakage energy savings.

interval (i.e., 2000 cycles in our experiments) before it can put those non-active cache lines into the leakage-saving mode.

Figure 15 presents the percentage of leakage energy savings for the loop-based approach and the pure hardware based approach. Except 129.compress, we find the proposed strategy leads to more leakage energy reduction. For cordic, although the loop-based approach has a less percentage of time for placing cache lines into the low leakage mode than the pure hardware based approach; the just-in-time activa-

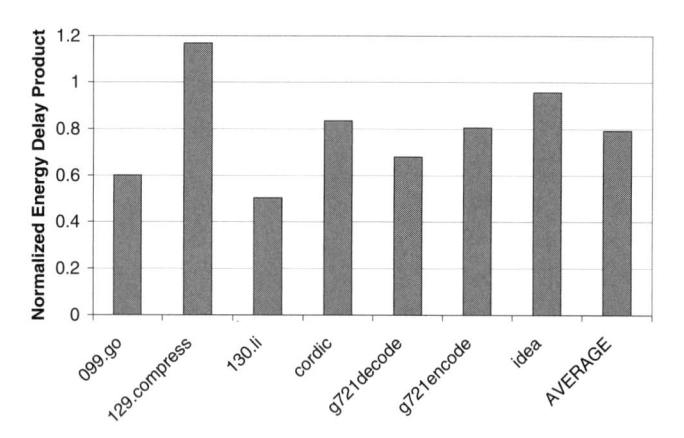

FIG. 16. Normalized energy delay product for the loop-based approach. The results are normalized with the energy delay product of the pure hardware-based approach.

tion can reduce the performance penalties by pre-activating the drowsy cache lines, which also results in leakage energy savings. In average, the loop-based approach can save 9.7% more leakage energy for L1 data cache than the pure hardware based approach.

While energy efficiency is important, in most systems, performance is also an important goal that can not be compromised. The energy-delay product provides a metric to evaluate both the energy consumption and the performance behavior. We present the energy delay product results of the proposed loop-based approach in Fig. 16. To compare our approach with the pure hardware based approach, the results are normalized with the energy delay product of pure hardware based approach. We find that except 129.compress, the loop-based approach has a lower energy delay product than the pure hardware based approach (the less the better). These results indicate that the loop-based approach is an effective and comparable approach to reducing the leakage energy for data caches, by considering both the energy consumption and performance.

We have also studied the effectiveness of the loop-based approach by varying the size of the L1 data cache, since the set of data cache lines that can be placed into the low leakage mode is sensitive to the size of the cache. Figure 17 presents the percentage of time spent in the low leakage mode by varying the size of the L1 data cache from 16 KB to 32 KB and 64 KB, while fixing other system parameters. In all these three difference configurations, cache lines can be put into the low leakage mode for more than 80% of the total execution time. For 130.li, we notice that this percentage increases slightly by increasing the size of the cache. The reason is that

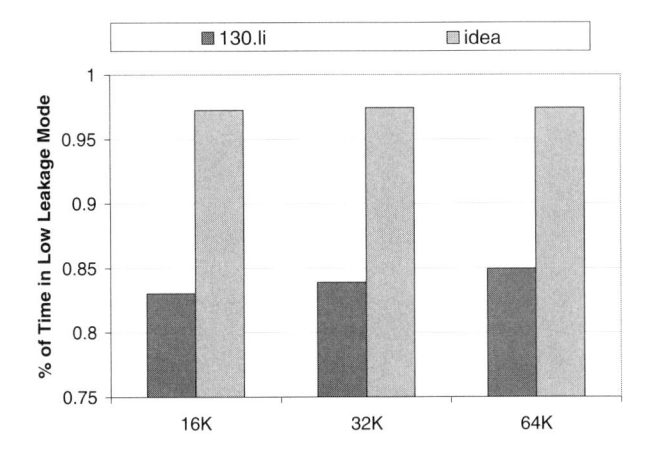

FIG. 17. Percentage of time spent in the low leakage mode by varying the data cache size.

the hot cache lines accessed by the loop are fixed. Therefore, the larger the data cache is, the more cache lines can be placed into the low leakage mode.

5.2.1 Impact of Compiler Optimizations

As aforementioned, the leakage energy savings of the loop-based approach is proportional to the product of $(D - S) * T$, where D is the number of cache lines of the L1 data cache, S is the active cache lines accessed by the loop and T is the total execution time of the given code region (it is loop for our approach). If S is a very large number, for instance, assuming S is close to or even larger than D, there will be very few or no energy savings. In other words, if the number of cache lines accessed by the loop is close to the size of the cache, our approach will result in negligible energy savings. For some array intensive applications, however, we find that the size of the data accessed by the innermost loop can be very large, sometimes larger than the size of a typical L1 data cache. Therefore, to apply the loop-based approach to such programs (we focus on the loop intensive benchmarks in this section), we propose to utilize loop transformations [25] to create opportunities for leakage energy reduction of data cache lines.

There are various loop transformation techniques in the literature [25] to enhance the data locality or extract higher instruction level parallelism (ILP). In this chapter, we concentrate on applying loop tiling and loop distribution to reduce the data memory footprints accessed by the innermost loops. The less the number of cache lines that are accessed by the loop, the more the cache lines that can be placed into the low leakage mode during the loop execution.

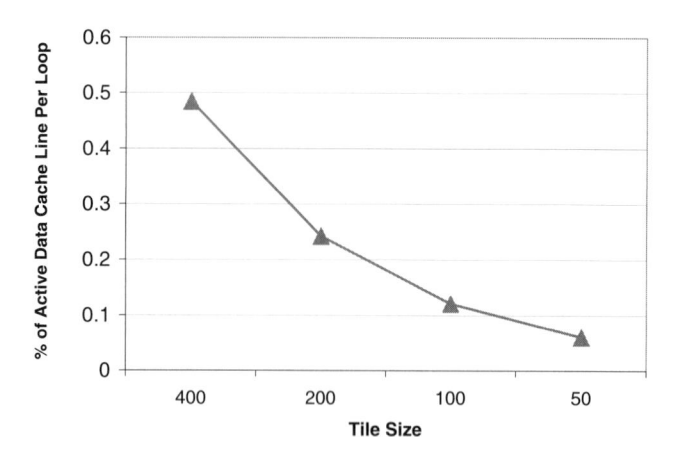

FIG. 18. Average data footprint accessed by innermost loops in terms of the percentage of data cache lines. In this experiment, we fix the size of the data cache to be 32 KB. We focus on applying loop tiling and loop distribution. The loop tile size is varied from 400 to 200, 100 and 50.

Figure 18 shows the impact of the combination of loop tiling and loop distribution on the average data footprint of innermost loops in terms of the percentage of data cache lines. The loop tile size is varied from 400 to 200, 100 and 50. We select one loop intensive benchmark btrix from SPEC FPT 92 [14] for this evaluation. From Fig. 18 we can see that loop tiling and loop distribution are very successful in reducing the data footprint of innermost loops. For loop tile size 200 and below, the average data footprint of innermost loops is less than 24.3% of the data cache lines, implying a large fraction of cache lines can be placed into low the leakage mode during the execution of innermost loops. As a result, leakage energy consumption can be potentially reduced significantly. Thereby applying loop transformations to array-intensive applications containing innermost loops that access lots of data can enhance the proposed loop-based approach for mitigating the data cache leakage dissipation.

6. Conclusion

This chapter presents two novel and cost-effective compiler-assisted approaches to reducing the leakage energy of the instruction cache and data cache without significant performance or dynamic energy overheads. The static next sub-bank prediction for drowsy instruction caches exploits static and profiling information to identify as many transitional instructions as possible at the compilation time, and provides useful hints for the processor to pre-activate the next sub-bank at runtime for avoiding performance degradation and dynamic energy overhead. The loop-based data cache leakage optimization approach exploits the fact that the program hotspots often only access a limited number of data during the program phases (i.e., innermost loops in this work), thus the rest of the "cold" data cache lines can be placed into the low leakage mode during the loop execution by using compiler-directed loop information. Compared to the hardware-based leakage reduction approaches, such as [7,5, 8], the compiler-assisted approaches can exploit application behavior intelligently with low hardware cost, thereby leading to comparable or even more leakage energy reduction with less performance and dynamic energy overhead.

It should be noted that the hardware-centric leakage energy reduction techniques have the advantage of being transparent to the software, and thus re-compilation is not needed. By comparison, in the compiler-based approaches (for both the instruction cache and the data cache), the programs need to be re-compiled in order to extract and annotate the useful application-specific hints for the leakage energy management at runtime. While this disadvantage may limit the use of the proposed approaches for legacy code where recompilation is difficult or impossible, we believe the compiler-assisted approaches can be effective for embedded systems where the entire systems (including the code) typically need to be rebuilt for new products. Moreover, since embedded processors are often constrained by cost and energy, the lightweighted compiler-based approaches studied in this chapter can be particularly suitable for these systems to minimize the cache leakage energy dissipation.

REFERENCES

[1] Ronen R., Mendelson A., Lai K., Lu S.-L., Pollack F., Shen J., "Coming challenges in microarchitecture and architecture", *Proc. IEEE* **89** (3) (March 2001).

[2] Semiconductor Industry Association, "The International Technology Roadmap for Semiconductors", http://www.semichips.org, 2005.

[3] Ye Y., Borkar S., De V., "A new technique for standby leakage reduction in high-performance circuits", in: *Proc. of the Symposium on VLSI Circuits*, 1998, pp. 40–41.

[4] Powell M.D., Yang S., Falsafi B., Roy K., Vijaykumar T.N., "Reducing leakage in a high-performance deep-submicron instruction cache", *IEEE Trans. VLSI* **9** (1) (February 2001).

[5] Kaxiras S., Hu Z., Martonosi M., "Cache decay: Exploiting generational behavior to reduce cache leakage power", in: *Proc. of ISCA*, 2001.

[6] Zhou H., Toburen M.C., Rotenberg E., Conte T.M., "Adaptive mode control: A static power-efficient cache design", in: *Proc. of PACT*, 2001.

[7] Kim N.S., Flautner K., Blaauw D., Mudge T., "Drowsy instruction caches", in: *Proc. of the 35th Annual ACM/IEEE International Symposium on Microarchitecture*, 2002.

[8] Flautner K., Kim N.S., Martin S., Blaauw D., Mudge T., "Drowsy caches: Simple techniques for reducing leakage power", in: *Proc. of ISCA*, 2002.
[9] Zhang W., "Compiler-directed data cache leakage reduction", in: *Proc. of the IEEE Computer Society Symposium on VLSI (ISVLSI04)*, February 2004.
[10] Allu B., Zhang W., "Static next sub-bank prediction for Drowsy instruction cache", in: *Proc. of the International Conference on Compilers, Architecture, and Synthesis for Embedded Systems (CASES'04), Washington DC*, September 2004.
[11] Heo S., Barr K., Hampton M., Asanovic K., "Dynamic fine-grain leakage reduction using leakage-biased bitlines", in: *Proc. of ISCA*, 2002.
[12] Butts J.A., Sohi G., "A static power model for architects", in: *Proc. of the International Symposium on Microarchitecture*, December 2000.
[13] http://www.trimaran.org.
[14] http://www.spec.org.
[15] Lee C., Potkonjak M., Mangione-Smith W.H., "MediaBench: A tool for evaluating and synthesizing multimedia and communications systems", in: *Proc. of the International Symposium on Microarchitecture*, 1997, pp. 330–335.
[16] Motorola, *Motorola MC68030 Enhanced 32-bit Microprocessor User's Manual*, third ed., Motorola, 1992.
[17] Intel, *Intel StrongARM SA-1100 Microprocessor Developer's Manual*, Intel, August 1999.
[18] Shivakumar P., Jouppi N., "CACTI 3.0: An integrated cache timing, power and area model", WRL Research Report 2001.
[19] Zhang W., Hu J.S., Degalahal V., Kandemir M., Vijaykrishnan N., Irwin M.J., "Compiler-directed instruction cache leakage optimization", in: *Proc. of the 35th Annual ACM/IEEE International Symposium on Microarchitecture*, 2002.
[20] M. Cekleov, M. Dubois, "Virtually-address caches. Part 1: problems and solutions in uniprocessors", *IEEE Micro* **17** (5) (1997) 64–71.
[21] Ball T., Larus J.R., "Branch prediction for free", in: *Proc. of SIGPLAN Conference on Programming Language Design and Implementation*, 1993.
[22] Hwu W.W., Conte T.M., Chang P.P., "Comparing software and hardware schemes for reducing the cost of branches", in: *Proc. of ISCA*, 1999.
[23] M. Kandemir, N. Vijaykrishnan, M.J. Irwin, "Compiler optimizations for low power systems", in: *Power Aware Computing*, Kluwer Academic, 2002 (Chapter 10).
[24] J.A. Fisher, P. Faraboschi, C. Young, *Embedded Computing: A VLIW Approach to Architecture, Compilers, and Tools*, Morgan Kaufmann Publishers, 2005.
[25] S. Muchnick, *Advanced Compiler Design Implementation*, Morgan Kaufmann Publishers, San Francisco, CA, 1997.
[26] Hu J.S., Nadgir A., Vijaykrishnan N., Irwin M.J., Kandemir M., "Exploiting program hotspots and code sequentiality for instruction cache leakage management", in: *Proc. of the International Symposium on Low Power Electronics and Design, Seoul, Korea*, August 25–27, 2003.
[27] Zhang W., Karakoy M., Kandemir M., Chen G., "A compiler approach for reducing data cache energy", in: *Proc. of ICS*, 2003.

[28] Meng Y., Sherwood T., Kastner R., "Exploring the limits of leakage power reduction in caches", *ACM Transaction on Architecture and Code Optimization* **2** (3) (September 2005) 221–246.
[29] Liu D., Svensson C., "Trading speed for low power by choice of supply and threshold voltages", *IEEE Journal of Solid State Circuits* **18** (1) (January 1993).
[30] Lee D., Blaauw D., Sylvester D., "Gate oxide leakage current analysis and reduction for VLSI circuits", in: *IEEE Transaction on Very Large Scale Integration*.
[31] Velusamy S., Sankaranarayanan K., Parikh D., Abdelzaher T., Skadron K., "Adaptive cache decay using formal feedback control", in: *Proc. of 2002 Workshop on Memory Performance Issues in Conjunction with ISCA-29*, 2002.
[32] Li L., Kadayif I., Tsai Y., Vijaykrishnan N., Irwin M.J., Sivasubramaniam A., "Managing leakage energy in cache hierarchies", *J. Instruction-Level Parallelism* (2003).
[33] Zhang W., Kandemir M., Vijaykrishnan N., Irwin M.J., De V., "Compiler support for reducing leakage energy consumption", in: *Proc. of the 6th Design Automation and Test in Europe Conference (DATE-03)*, March 2003.
[34] Rele S., Pande S., Onder S., Gupta R., "Optimization of static power dissipation by functional units in superscalar processors", in: *Proc. of the International Conference on Compiler Construction*, in: *Lecture Notes in Comput. Sci.*, vol. 2304, Springer-Verlag, April 2002, pp. 261–275.
[35] Tiwari V., Malik S., Wolfe A., "Compilation techniques for low energy: An overview", in: *Proc. of the International Symposium on Low Power Electronics*, October 1994, pp. 38–39.
[36] Su C.L., Tsui C.-Y., Despain A.M., "Low power architecture design and compilation techniques for high performance processor", in: *Proc. of IEEE Compcon.*, 1994, pp. 489–498.
[37] Hsu C.H., Kremer U., "The design, implementation, and evaluation of a compiler algorithm for CPU energy reduction", in: *Proc. of the ACM SIGPLAN Conference on Programming Language Design and Implementation (PLDI'03), San Diego, CA*, June 2003.
[38] U. Kremer, "Low power/energy compiler optimizations", in: Piguet C. (Ed.), *Low-Power Electronics Design*, CRC Press, 2005.
[39] Valluri M., John L., "Is compiling for performance = compiling for power?", in: Lee G., Yew P.-C. (Eds.), *Interaction Between Compilers and Computer Architectures*, Kluwer Academic Publishers, 2001 (Chapter 6).

Mobile Games: Challenges and Opportunities

PAUL COULTON, WILL BAMFORD, FADI CHEHIMI, REUBEN EDWARDS, PAUL GILBERTSON, AND OMER RASHID

Lancaster University
Bailrigg, Lancaster,
UK LA1 4YW

Abstract

Mobile games are expected to play significant role in future mobile services by evolving beyond the largely single player titles that currently dominate the market to ones that take advantage expanding mobile phone functionality and the wide demographic of the mobile phone user. However, because of the fragmented nature of the mobile software development market, and the restrictions imposed by the mobile phone hardware, the skills required for games development are more akin to embedded software development and those used in the game development of the early 1980s, rather than those currently practiced amongst console and PC game developers. In the first half of this chapter we discuss the hardware restrictions and the different software environments encountered together with methodologies so that mobile game developers can produce effective designs.

    Having discussed the challenges in developing mobile games in the second half of the chapter we discuss the opportunities for innovation provided by the mobile phones through both their anywhere connectivity and the ever enhancing feature set. To this end we present examples using Cameras, RFID, Bluetooth, and GPS that illustrate how through careful design mobile games do not simply have to be cut down versions of console games but can provide uniquely mobile gaming experiences.

1. Introduction 192
2. Challenges 193
   2.1. Who Plays Mobile Games? 193
   2.2. Physical Constraints 195
   2.3. Software Development Environments 201

2.4. 3-D Mobile Games 214
3. Opportunities 217
   3.1. Text Games 217
   3.2. Camera Games 224
   3.3. Location Based Games 225
   3.4. Proximity Games 231
   3.5. IP Multimedia System (IMS) 237
4. Conclusions 239
   Acknowledgements 239
   References 240

1. Introduction

As the term mobile has a different meaning when used with respect to communications rather than computing it is worthwhile providing a clear definition at the start of this chapter to place the discussions in the correct context. In this chapter we base our discussions with respect to communications and thus conform to the International Telecommunications Union (ITU) definition, which states that

> "the term mobile can be distinguished as applying to those systems designed to support terminals that are in motion when being used".

In other words, unless a game on a cellular device uses a connection as part of the game, there is a strong argument that it is essentially a hand-held, portable, or nomadic game. This also means that although some innovative work on game play has been achieved using Personal Digital Assistants (PDAs) with WiFi access they have greater limitations in terms of truly mobile connectivity and lack the user penetration of mobile phones with global mobile subscriptions having reached two billion users [1].

Mobile games are seen as an important service for many of these consumers by both the network operators and the games industry. Operators are currently trying to drive up the Average Revenue Per User (ARPU) by encouraging greater use of data services and games are seen as a means to achieve this. The games industry also sees this as an enormous opportunity for increasing sales and customer base, indeed, mobile games already represent 14% of $43 billion total world gaming revenue [2] and many current predictions would suggest that the mobile platform will become the dominant force in games. However, the mobile games market is currently dominated by single player or quasi Peer to Peer (P2P) games [3] (using short range communications such as Bluetooth) which do little to increase data traffic. Whilst there are some successful massively multiplayer games which do generate data traffic on the Internet, the nature of mobile users and their environment means that new design approaches need to be considered. In this chapter we explore the challenges and opportunities faced by developers and researchers in this field. The challenges manifest in a number of aspects from the nature of the mobile gamer to the various software and hardware capabilities of the mobile phones themselves. In terms of opportunities the unique nature of the mobile environment and the many new features that are continually appearing on the mobile phone such as cameras or Radio Frequency Identification (RFID) readers.

2. Challenges

2.1 Who Plays Mobile Games?

In relation to the demographic of mobile users playing games, Glu Mobile's 2005 UK survey showed that the classic image of the PC or console gamer as an 18–35 year old male is incorrect for mobile. The main demographic is much more varied as 16% of all phone users regularly play games on their phone, although this rises to 29% in the 16–24 age group, and just as many women as men play games on their mobile phones. In terms of ethnicity, where gamers have traditionally been predominantly white, a report from 2005 by NPD group on US mobile gamers indicate that they are twice as likely to be African-American, Hispanic or Asian. Although the demographic of mobile gamers is different to that of console gamers, they are often simply divided into the two traditional broad categories used by the games industry, that of casual gamer and hardcore gamer [4]. Whilst many in the console games industry have fairly well established techniques for targeting the hardcore gamer, the casual gamer has proved more elusive and success stories in this area, such as 'The Sims' by EA Games, have often proved surprising. In fact there is growing criticism, both inside and outside the games industry, that the very large budgets now required to produce a game result in creativity being stifled with too few risks taken in relation to the game play design and that the dominant features are related to improved graphics and sounds.

Before we consider mobile gamers specifically, it is worthwhile considering the popular profiles of hardcore and casual gamers [5], as they are undoubtedly influencing industry expectations about the mobile gamer.

Hardcore Gamers:

- purchase and play many games;
- enjoy longer play sessions and regularly play games for long periods;
- are excited by the challenge presented in the game;

- will tolerate high levels of functionality in the user interface and often enjoy mastering the complexity;
- often play games as a lifestyle preference or priority.

Casual Gamers:

- buy fewer games, buy popular games, or play games recommended to them;
- enjoy shorter play sessions—play in short bursts;
- prefer having fun, or immersing themselves in an atmospheric experience;
- generally require a simple user interface (e.g. puzzle games);
- consider game playing as another time-passing entertainment like TV or films.

It is often assumed within the games industry that casual gamers form the majority of mobile gamers [6]. We believe this is an over simplification, as the nature of the mobile environment is a major contributor to the formation of gaming habits. There is a strong argument that the game industry must establish new definitions specifically for mobile. In an interview for the Game Daily Biz in February 2006, Jason Ford, General Manager for Sprint Entertainment, defined two specific types of hardcore mobile gamer:

> "First there are the 'cardcore' mobile gamers. These are people who play casual games in a hardcore fashion. The type that might spend hours and hours trying to get a Bejewelled high score but don't own a gaming console."

> "Second is the 'hard-offs'. These are your more typical hardcore gamers, who are playing off their normal platform. They're the type more likely to check out the mobile version of a hit console title, because they know and like the brand."

Additionally at the 2006 CES panel discussion 'Future of Mobile Games', IDG's CEO, Dan Orum, stated:

> "We're seeing an emergence of the 'social gamer', they're like the typical 'hardcore gamer', but with social lives."

Whilst these definitions form part of the profile of mobile gamers, it further highlights that the industry perceives the player demographic for mobile from its experience of the PC and console market and it is still fixated with the hardcore gamer. We would advocate the games developers and researchers should seek out newer ways of both understanding and then attracting mobile gamers rather than trying to pigeon hole them within the existing system and in the opportunities section of this chapter we present games specifically aimed at mobile gamers who may not be attracted to console gaming.

2.2 Physical Constraints

Programming mobile phones requires a considerably different approach for developing applications than those developed for the Personal Computer (PC) market. These days even the most basic PC has large amounts of memory and huge disk space relieving the PC developer from the need to minimise the usage of their particular applications resources. Whereas, mobile phones have limited memory and currently no disk drive storage meaning that whatever space your application requires is taken away from your overall data space [7]. Even the most advanced mobile phone has a significantly more limited screen size compared to that of a PC, which when coupled with the different way users interact with handheld devices means that developers must get their ideas across in a very small footprint [7]. In essence this is the skill set of the real time embedded systems programmer and in the following sections we will highlight some of the issues, that are independent of the software, that need to be considered.

2.2.1 Memory

In terms of memory constraints of mobile phones there are two basic constraints: the first covers the overall size of the application and the second relates to the application memory space which is the memory required for the application to run.

2.2.1.1 Application Size. The memory available for a mobile game is often governed by the distribution mechanism that is to be used to deliver the game to the end user. For, instance many of the Nokia N-Gage games come on a memory card that is inserted into the phone once purchased. This means that the size of the application can be in many megabytes although it limits the ability for wide scale sales as the user has to either purchase the game from an outlet or purchase online and wait for delivery. The most popular means of delivery is installation Over-The-Air (OTA), where users simply select the application they want to purchase and download it to their phone. Although most mobile phones, and in particular smart phone, will allow games to be downloaded and run that are in the multiple of megabytes there are often limits (64 K or 128 K are typical) defined by the operators. There is also generally an associated data cost, in this type of delivery, in addition to the game purchase which can be off putting to the user if the game size is excessive. Overall, there is still a distinct advantage to utilising efficient programming techniques for games, such as fixed point arithmetic and tiled graphics, and developers should bear in mind the distribution mechanism from the outset [8].

2.2.1.2 Application Memory Space. When running an application more memory is required than the actual application file which is often termed the heap and is used, for example, to store graphics, objects created at run time, etc. [9]. This memory varies with device but normally comes with some limit that is available to the developer, for example, with Nokia series 60 phones this is currently 2 Mb. Although, for most environments this usually is not a too big problem for game developers, in general, developers should ensure that their application can run in the available heap space for the devices and environments they wish to support.

2.2.2 Processors

Games have always been among the most processor-intensive applications and typically the majority of processor cycles are spent performing the calculations necessary to modify the game view and update the display. Mobile device manufacturers rarely state the processing power of the chips their devices contain but they are always going to be much lower than those of a PC. For example many series 60 phones typically use a 204 MHz ARM processor which is sufficient to support real-time rendered 3-D graphics, though not with the number of polygons or special effects typical of high-end console and PC games. However, it is here that the smaller screen size works in the developers' favour as fewer pixels on the screen mean fewer pixels to process [9].

2.2.3 Networking

There are a number of issues relating to networking that are particularly relevant to mobile game design:

- Mobile games usage tends to be spontaneous rather than planned and is often used to fill spare moments resulting in short game sessions.
- Users dropping out of games due to either accidental loss of connectivity (such as loss of signal while going through a tunnel) or deliberate disconnection to perform another task (such as answering a call).
- The network latency can be as high as a few seconds in mobile, which is too long for the majority of fast action multiplayer games.

Latency is the amount of time it takes a system to respond and is of particular importance when trying to develop multiplayer games operating over a mobile network. With a stand-alone PC, latency is normally measured in microseconds while latency over the conventional Internet is usually in the order of 100 to 200 milliseconds [10]. Although the advent of Quality of Service (QoS) in 3G systems will address some aspects of latency it will be variable dependent on customer subscription and unavailable in regions covered only by the General Packet Radio System (GPRS).

The other significant possibility for networking multiplayer games is to utilise Bluetooth although that does require the players to be in close proximity and the number of devices that can be supported is typically limited to around 4 or 5 [11]. Over a Bluetooth device the latency is typically on the order of 20 to 50 milliseconds what is generally sufficient for fast action multiplayer games [11].

2.2.4 Screen Size and Aspect Ratio

Obviously, mobile phones have screens that are tiny compared to those on PCs and games must therefore be carefully designed to fit in the device's screen. Generally if mobile games are to be commercially successful they must sell in very large numbers which means that they should be capable of running on as many devices, with various screen sizes and colour capabilities, as possible. For example, an application could have to cope with a monochrome 128×128 pixel screen on an older Nokia series 40 phone and a 12 bit colour 176×208 pixel screen Nokia series 60 phone [9]. Further while PC screens are usually wider than they are high, mobile phone screens are often either square or higher than they are wide, which can impact on the applications such as side scrolling games. For instance, a player may not be able to see an enemy closing on his position until they are right on top of him. Other aspects to good design are that any information that the users requires regularly, such as life force, should be clearly visible; generally high contrast colours work best because of variable lighting conditions.

2.2.5 Phone User Interface

Except for a couple of special examples (e.g. The N-Gage) mobile phone controls are not optimised for playing games and they are first and foremost a device for voice telephony. Although keypad layouts vary in general it consists of a standard ITU-T keypad, two soft keys, and buttons for starting and ending phone calls (although the latter two cannot be utilised by applications) and developers must limit the number of 'actions' in a game to this limited range of keys [9]. Symbian UIQ mobile phones have pointing devices although this is likely to be an advantage in only certain types of games. Other games have utilised more novel approaches using the phone camera or RFID but we shall save the discussion of these to the forthcoming section on opportunities. The main element to bear in mind is that the game interface should be simple, consistent (i.e. left key positive and right key negative), and intuitive otherwise players will quickly become frustrated and lose interest in the game.

2.2.6 Sound Support

Sounds in games can be a tricky area as they are a good way of providing feedback especially when working with a small screen but can consume a lot of memory. There is also the fact that mobile phones have variable capabilities and sound formats differ from model to model which increases porting issues.

Other considerations relate to the fact that mobile games are often played in surroundings that include non-players and disturbing these individuals can be a real possibility. It is generally advised [12] that sounds default to 'off' at the start of the game and that sound volume is adjustable within the settings for the game.

2.2.7 Coding Optimisation

Writing efficient code is an important technique for programmers of mobile phone games and in this section we will present a number of optimizations which have been split into three categories: (1) memory optimization, (2) object oriented optimization and (3) coding style optimization. For each category we will describe practices that can be used to improve execution performance of mobile games [13].

2.2.7.1 Memory.

There are certain hidden memory overheads caused by inefficient variable usage and allocation. Such practices can often inflate code size causing slower performance and sub-optimal execution. The following list highlights some techniques that can be used to combat these effects [13]:

- Use array notation for arrays rather than pointer notation.
- Pass array function parameters as pointers not with array notation ([]).
- Rearrange structure or class members.

2.2.7.2 Object Orientated Optimisation.

Object-orientation has introduced to software engineering a collection of practices and methodologies that make the software production more portable, the code more reusable, and the whole process less time consuming. The majority of mobile operating systems utilise object orientated methods and on mobile phones programmers are required to deal with objects more cautiously than they would normally do in relation to PC programming. There are certain hidden memory overheads that may inflate code caused by inefficient use of objects and in the following points are practices that can alleviate the situation [13]:

- Initialize objects at declaration time.
- Use constructor initialization lists.
- Declare objects when needed.
- Pass function parameters by reference.

2.2.7.3 Coding Style Optimisation.
C and C++ are very lean and efficient programming languages as they allow the programmer to get too close to the hardware as no other programming language can [14]. However, to make best use of these languages on mobile phones, programmers have to write clean, safe and well-designed programs. Here we list several coding techniques that can be used to leverage performance with C/C++ code (although some could equally be applied to other programming languages) [13]:

- Always decrement the counter in a loop.
- Unroll small fixed-count loops.
- Use registers for loop counters and frequently used variables.
- Use dependency chains rather than single data dependencies in loops.
- A "switch" is generally more efficient than an "if–else".
- Arrange if-else statements or switch cases with the most common appearing on top.
- Try to Inline small functions.
- Declare functions as static.
- Declare local variables as static.

2.2.8 Testing

In this respect developing a mobile phone game is the same as any other software engineering exercise and we do not believe it is necessary to document these within this chapter. However, there are certain elements in the design of your game that you should pay attention to in order to ensure that the application is of a sufficient standard to be released to the general public. The following list has been compiled from various manufacturers of both phones and software environments and represent generic recommendations that you should address in your game design, these are:

- The game speed does not compromise the use and purpose of the game.
- The game does not consume the device's processor power and memory excessively.
- The game does not affect the use of the system features or other applications.
- The game does not cause any harm to the user, other applications, or data.
- Only vital data is saved locally to the phone's internal memory.
- Communication to and from the device/application is kept within reasonable limits, i.e. your application does not unnecessarily cost the user extra through excessive network usage.

- Occasional tasks, exceptional tasks (for example, for emergency conditions), and tasks that cope with errors (for example, caused by the interruption of a network connection during the application's use) must be considered and treated appropriately.
- The game must be able to handle exceptional, illegal, or erroneous actions.
- The game can be installed and uninstalled completely i.e. all files, icons, images, etc.
- The game must be capable of being installed in either the phone memory or on the memory card if one is available on the particular phone model.

2.2.9 Distribution

For any mobile games developer one challenge is how to sell any game that you produce and the most common method is through portals on which there appear to be three basic operational models.

In the first model, content site providers and content developers (we are considering games but the same applies to other content e.g. videos, ring tones, etc.) make use of the operator's subscriber base and billing capabilities to sell their content. In practice the content site providers promote and price their content/services on their own portals and liaise with the operators for charging and collection of payments. The operators can charge for the use of their billing infrastructure and also receive traffic revenue from delivering the content. In the case of a content site provider the onus is on them to source and maintain a large range of content. An example of this type of operation is used very successfully in the companies such as Handango (www.handango.com), MonsterMob (www.mob.tv), SymbianGear (www.symbiangear.com).

In the second model, the operators act as aggregators, effectively promoting the content of third party developers/providers on their own portal site. In this model the content revenue is split between the developers/providers and the operators. Obviously in this model it is the operators who must source and maintain a large range of content if their site is to continue to be attractive to their customers. An example of this type would be Vodafone live (vodafone-i.co.uk/live/) and Planet three (www.three.co.uk/planetthree/).

In the final model, the operator brands all the content as their own and must then pay the content developers either for the content itself or for licenses from them. In this case there is no revenue sharing and the operator is solely responsible for marketing, delivery, charging and billing. Although some operators are operating this model for certain applications, none seem to be operating it exclusively and in most case it is combined with the second model.

At present the downloading business seems to be based on a volume rather than on a margin. Users are encouraged to try the content even if it is to only play the game once. Thus the price at which your application is sold should not be seen to deter this practice. In effect the portals aim for users to download content frequently and erase it when they get bored so that they can personalise their phone with new content.

2.3 Software Development Environments

Mobile game development spans a wide range of both hardware and software options and is one of the major challenges for the mobile games developer. Unlike consoles where there is a clear life cycle between versions of devices, in the mobile world there are over 250 device platforms plus around 80 different mobile operators. Just collecting the information and guidelines on these devices and markets can prove to be overwhelming and very costly task. For example, a game publisher has 20 games in its portfolio and wants these available globally in 5 languages. It would have to create close to 5000 different builds (20 games × 5 languages × 50 top devices).

Whilst we are unable to offer a solution to the porting problem in this section we discuss the software possibilities for game development and try to highlight the various advantages and disadvantages of each so that a developer can way these factors of against the requirements of a particular game.

2.3.1 Symbian

Symbian was founded originally by Psion, Nokia, and Ericsson in July 1998 with Motorola joining later that year, Matsushita in 1999, and Siemens in 2002. Motorola sold its stake in the company to Psion and Nokia in September 2003, although they continue to produce some phones based on Symbian OS. Psion's stake was then bought by Nokia, Panasonic, Siemens AG and Sony Ericsson in July 2004 [15].

Symbian OS has its roots in the Psion 'EPOC' software and the early versions of the operating system still carried that name. The first open Symbian OS phone was based on Version 6.0 of the operating system. 'Open' meaning that users of Symbian OS-based phones were able to install their own software which was an important step forward for developers. This version appeared in 2001 and was shipped on the Nokia 9210 Communicator. Symbian OS continued to evolve with improved API functionality and market-leading features, with the next big change occurring in early 2005 when Symbian OS Version 9.0 was announced. This version is designed principally to be more secure which has meant that it would not support software compiled for earlier releases [15].

The major advantage of Symbian OS is that, unlike some of its competitors, it was specifically designed for mobile phones that have relatively limited resources. There is a strong emphasis on conserving memory through specific programming idioms and other techniques such that memory usage is low, which reduce the likelihood of memory leaks. There are similar techniques for conserving disk space. Furthermore, all Symbian OS programming is event-based, and the CPU is powered down when applications are not directly dealing with an event. This is achieved with the aid of a programming idiom called active objects. Correct use of these techniques helps ensure longer battery life [15].

The different user interfaces available for Symbian OS are designed to allow manufacturers to produce a range of phones in different styles, addressing many different market segments. They essentially fall into three categories: S60 (formerly known as Series 60), Series 80, and UIQ. There was a fourth, Series 90, but this has now been amalgamated into S60.

S60 consists of a suite of libraries and standard applications and is intended to power fully-featured modern mobile phones with large colour screens, which are most often referred to as smartphones. S60 is most often associated with Nokia, which is by far the largest manufacturer of these mobile phones, although it is also licensed to other major mobile phone manufacturers such as BenQ-Siemens, Samsung and Panasonic. S60 is most often associated with the standard ITU keyboard layout for one-handed entry, although recent versions also offer support for stylus entry [15].

S60 supports applications developed in J2ME, native Symbian C++, Python and Flash Lite. The most important feature of S60 phones from an application developer's perspective is that they allow new applications to be installed after purchase.

With the emergence of Symbian OS as the industry standard open operating system, there is not only the opportunity to create applications that enhance the phone's functionality, but also a risk of poor-quality or malicious applications being developed that cause the phone to function incorrectly. Symbian Signed was created in response to this potential for bad practice and effectively uses a verifiable signature to establish a formal link between an application and its origin. Many operators and manufacturers (e.g. Nokia, Sony Ericsson, Orange, and T-Mobile) have already decided that their portals will only feature Symbian Signed applications. There are changes in the Symbian Signed process for pre and post version 9.0 so it is worth checking the criteria at the Symbian Signed site (www.symbiansigned.com).

There are a number of major advantages to developing games in Symbian apart from the fact that it has a 70% share of the smartphone market. As it is a native OS developers are able to maximise performance particularly for graphically intensive games. Symbian allows developers complete access to incorporate new features appearing on phones immediately appear and enables greater innovation. The incorporation of OpenGL ES means that powerful 3-D games are now possible using a known standard.

The only disadvantage we often here expressed is the alleged 'steep learning curve' associated with Symbian. However, this should not prove significant to any experienced C++ programmer as Symbian now has a large developer community and support network. Overall there could be better support for the games developer, in terms of better support for sprites and tile graphics for instance, Symbian always provides the earliest access to new phone features on the devices which it installed and the most comprehensive access to existing phone features, which normally means some of the most innovative games often appear first on Symbian.

2.3.2 Brew

The Binary Runtime Environment for Wireless (BREW) is an application development and execution platform created by Qualcomm specifically for mobile phones (brew.qualcomm.com). A wide range of content is available for BREW enabled handsets, ranging from games and multimedia content through to business and enterprise applications. Currently, BREW is only implemented on CDMA-based handsets, although it is technically feasible to port BREW to other chipsets (Qualcomm have previously demonstrated a limited version running on GSM handsets).

The largest market for BREW is in North America and Japan, although even in these areas, J2ME represents a more popular platform. Availability in Europe is very limited due to stronghold of GSM and limited support for CDMA. Qualcomm has always maintained that J2ME is not a direct competitor to BREW, because it is possible to create a Java virtual machine (JVM) on top of BREW's architecture. However, some manufacturers are beginning to ship devices with a pre-loaded J2ME extension and it is also possible to integrate the J2ME extension after manufacture using BREW's over the air delivery mechanism (known as the BREW Delivery System—BDS). Other languages such as Flash and XML can also now be supported through additional BREW extensions.

Qualcomm markets BREW as a complete end-to-end solution. This means that, unlike most other development platforms such as J2ME and Symbian, the BREW Solution, as Qualcomm terms it, manages the testing, distribution and billing as well as development of wireless applications.

In order to develop for BREW handsets, Qualcomm provide a free software developer kit (the BREW SDK), which includes sample applications, the simulator (similar to the J2ME and Symbian emulators) and extensive online help. They also supply a free tools package (BREW SDK Tools) which is a set of utilities which includes the MIF Editor for editing the MIF files which define a BREW application's capabilities, privileges, dependencies as well as the application's icon, the Resource Editor and Compiler for editing the resources used in the application (such as text, images, GUI elements and other objects/binaries), and a Visual Studio add-in, for compiling BREW applications from within Microsoft's Visual Studio IDE. Other development tools are only available to registered BREW developers and these include the AppSigner Tool, for signing your BREW application with a digital certificate (to verify your developer credentials) and tools for testing your application prior to it getting sent away for independent tests (such as the Shaker and Grinder). Different SDKs are provided for targeting each significant revision of the BREW application execution environment (AEE), of which there are currently five versions (1.0, 1.1, 2.0, 2.1 and 3.0). If you have a compatible IDE, such as Microsoft Visual Studio, Qualcomm provide all the necessary tools in order to test your application for free through the BREW Simulator in Windows. However, unlike J2ME and Symbian, the process for compiling your application for the simulator is different to compiling your code for a real handset (the simulator cannot load BREW native binaries and vice versa). The complete process involved in becoming a commercial BREW developer is fairly convoluted and, unlike J2ME, where applications can essentially be developed for free, the following cost incurring steps need to be taken in order to develop a commercial BREW application:

- A developer must pay a yearly subscription for a digital signature from VeriSign.
- A developer must purchase a C/C++ compiler which can compile to a binary format compatible with the ARM chipsets used by BREW devices—the RealView ARM compiler for BREW is currently one of the few options although some developers have had limited success using the free GNU ARM compiler.
- In order to verify that your application is suitable for deployment, it must undergo a rigorous independent third-party test procedure known as TRUE BREW Testing. These tests will try to determine if there are any bugs in the application and whether the application responds properly to suspend/resume methods (which may be invoked if the handset receives an external event such as a phone call or SMS). Additionally, they will also consider usability issues (look and feel) and whether the application fulfils its functional requirements. If the application fails the test, it must be modified and resubmitted for testing (at a further cost).
- Further costs may be incurred if you chose to seek official help/services from Qualcomm. For instance, BREW labs.

In order to help pay for the cost of creating and deploying commercial BREW applications, some developers have sought publishers to help subsidize some of these costs (who in turn will take a share of the revenue). It is also important to mention that as Qualcomm manages the delivery of your application, and they, together with the network operator, will also take a percentage of the revenue generated by the application.

Most modern BREW phones have good support for game operations such as buffered graphics, sprite handling and transformations, tile-mapping and extensive sound playback capabilities. There is also good support for multiplayer game features such as HTTP or low-level socket communications. Some of the newer high-end BREW handsets also have 3-D accelerated graphics capabilities which can be harnessed using a cut down version of OpenGL for limited devices (OpenGL ES).

2.3.3 Windows Mobile

Microsoft's early forays into the mobile market were largely unsuccessful, but with the release of Windows Mobile 5.0 they have a platform that developers are much happier with. Windows Mobile 5 has gained most ground with arguably the second tier of mobile phone manufacturers like E-ten, HPC and Gigabyte, although it is beginning to capture some interest from major manufacturers like Samsung and Motorola.

The major advantage of Windows Mobile 5 is its compatibility with PDAs. Most applications developed for Windows Mobile 5 will run on previous version of the OS (Windows Mobile 2003 Phone Edition) and with earlier Microsoft based PDAs (Pocket PC 2003 and 2003 SE). Windows Mobile devices largely fit into the gap between PDAs and traditional mobile phones and are typically aimed at business users. Facilities such as push email which made the Blackberry devices such a success worldwide are being integrated into Windows Mobile 5, and other third party developers are providing integration between the mobile device and business critical server applications. This makes the target market for games developed on this platform different from those developed for more mass market mobile phones.

Compatibility between Windows Mobile 5 and Windows Mobile 2003 PDAs is due to the fact that both are built upon the same underlying operating system (www.microsoft.com/windowsmobile/developers/). Windows CE has been used for a variety of devices including the failing Gizmondo handheld gaming platform.

Based on a subset of the Win32 API core developed for Microsoft's desktop operating systems, these OSs share some limited compatibility with desktop operating systems. Each subsequent version of the OS, however, offers improved functionality and new APIs. Games written that use these features will not be backward compatible. Games written for Windows Mobile 2003 will however run under Windows Mobile 5.0.

Windows Mobile 5 devices support a minimum level of hardware functionality. These devices have a limited number of supported form factors. Each form factor has a specific screen resolution and colour depth which can be easily determined programmatically. Telephony, Bluetooth, storage and camera support and API accessibility are all standardized.

Input methods are less standardized across the range of devices. PDA and PDA-like smartphones rely on stylus input with a virtual keyboard or writing recognition system. Some will have mini-keyboards and more traditional phones layouts will only support standard mobile phone keypad. By supporting only stylus input to simulate mouse clicks will limit titles to the majority of devices that support this input method. Stylus input is most intuitive for tactical and strategy titles.

One important feature of Windows Mobile 5 is the inclusion of DirectX Mobile Edition. DirectX is the de facto standard for 3D games development for the Desktop PC and XBox. Its introduction into the mobile arena enables developers to leverage their desktop experience onto a new platform.

Programming for Windows Mobile 5 comes in two forms. Native coding uses the same methods as writing for Win32 systems. Typically done in C++, WM5 presents a cut down version of the standard Windows APIs. This environment is called eMbedded C++ by Microsoft.

A .NET virtual environment is available for Windows Mobile 5 called .NET Compact Framework. This framework provides a rapid development environment using either C# or VB.Net, and is an excellent method for people approaching mobile devices for the first time. Java Virtual Machines are also available for Windows Mobile 5, but library compatibility is reported to be low.

2.3.4 Linux

There is a growing operator and manufacturer interest in Linux as a mobile phone operating system and in some quarters it is being viewed as a serious alternative to market dominant Symbian, Windows Mobile and other proprietary operating systems. With mobile phones increasing in complexity there is a clear trend towards off the shelf third party operating systems based upon industry standards and Linux seems to be a natural choice due to its royalty free nature. The only drawback is the fact that these attempts so far have not been standardized and each vendor has its own implementation. Linux is already a popular operating system for mobile phones in China and is rapidly gaining momentum in Europe. Several device manufacturers are marketing smart phones running Linux including Motorola, Samsung, Philips, NEC, Panasonic and Siemens etc. [16]. Motorola is leading this market and the initiative for a standardized Linux based mobile operating system. Recently a group of top mobile manufacturers and operators are launching a foundation to create an open Linux-based software platform for mobile devices. The group includes industry giants such as Motorola, Vodafone, NTT DoCoMo, Samsung, NEC and Panasonic. Their objective is to create and market an application programming interface (API) specification, architecture and source reference to fulfil the lack of an open and common approach that has so far denied Linux handsets their deserved share of the market. This will be the third group to launch such an initiative following Linux Phone Standards Forum (LiPS) (www.lipsforum.org) and the Mobile Linux Initiative (MLI) [17].

Further comprehensive attempts to consolidate Linux as a mobile platform have been from A La Mobile Inc. (www.a-la-mobile.com) and Trolltech (www.trolltech.com). A La Mobile has announced an open Linux mobile smart phone platform. Their convergent Linux Platform is not just a kernel, middleware or applications, it is a complete software stack for wireless handsets integrated, tested, certified and supported. The architecture is open, configurable and scalable. It is designed to give handset manufacturers the autonomous selection of software components and functionality. On the other hand it enables network providers to customize the user interface and experience to suit different requirements.

Trolltech's Qtopia [18] is a comprehensive application platform and user interface for Linux-based mobile phones. Qtopia Phone Edition not only allows manufacturers and designers to build feature packed phones but also provides developers with necessary tools and API's to create innovative applications. Trolltech have already got some big names of the industry as their customers. Trolltech provide full access to a modifiable source code and complete degree of freedom to customize the user experience and functionality behind it. Qtopia Phone edition offers a fully customizable interface with improved input methods and comes bundled with integral media features. Supported by an open development environment it is very easy to create and test for a target device.

The application development in Linux is mainly a C++ framework but so far handsets available in the market only allow downloadable Java applications. Qtopia on the other hand offers a free SDK to develop application based upon Qt [19] which is a C++ application development framework. Qtopia has different versions available to suit the hardware requirements of different handsets. Developers can use several other tools available for Qt to create and design their applications. One of the key advantages being the portability from one platform to another. It is possible to port a game developed with Qt from e.g. KDE to handheld running Qtopia [20].

Mobile Linux market is likely to consolidate, there are clear indications that handset makers and operators will converge around a single mobile Linux platform. This means that there are loads of opportunities for game developers to create new and exciting games that harness the power of Linux and OpenGL.

2.3.5 J2ME

The Java programming language may be syntactically similar to C++ but differs from it fundamentally. C++ requires programmers to allocate and reclaim memory for their applications whilst Java utilises an automated garbage collector to free up memory. Furthermore Java runs on a virtual machine which means that the Java virtual machine (JVM) can be implemented to run on top of a range of operating systems using different hardware configurations. This virtual machine architecture gives a strict control of binary code being executed which ensures safe execution or only trusted code to be executed. A comprehensive set of API provides developers with necessary means to create their applications. Putting it all together a Java platform consists of the language itself, the virtual machine and the set of API's. In 1999 Sun (www.sun.com) decided to move away from their paradigm of "compile once and run anywhere" and divided Java into three distinct parts:

- Java 2 Standard Edition (J2SE) to be targeted towards desktop systems.
- Java 2 Enterprise Edition (J2EE) targeted towards multi-user enterprise environment as a platform.
- Java 2 Micro Edition (J2ME) a set of specifications targeted for handheld devices like mobile phones, PDA's and set top boxes. Essentially J2ME is a cut down version of J2SE and uses a small foot print virtual machine known as the kilobyte virtual machine (KVM) to suit the needs of a some what resource constrained device it runs on.

Unlike J2SE, J2ME is not a software or single specification. It can be considered as a platform, or more specifically as a collection of technologies and specifications that are specifically designed for different functionalities of a mobile device. The specifications for all Java platforms are developed under the Java Community Process (JCP) (jcp.org). An expert group consisting of interested organizations work together to create the specification which is referred to as the Java Specification Request (JSR). Each JSR is given a unique numeric identifier. J2ME is a collection of specific JSRs and are often referred to by their JSR number. Mobile devices are diverse in nature and it is very difficult to create a platform that will suit the capabilities of each device. It is this need to support a diverse range of devices that J2ME is divided into configurations, profiles and optional packages.

2.3.5.1 Configurations.
Configurations provide details about a virtual machine and a basic set of APIs that can be used with a particular range of devices. These devices are grouped together by similarities in terms of total memory and processing power available. A configuration supported by a device means that it has certain virtual machine features and certain minimum libraries that a developer or

content provider can expect to be available on the device. So far Sun has introduced two configurations, namely:

- Connected Device Configuration (CDC): This configuration is designed for devices that have more processing power, memory and better network connectivity.
- Connected Limited Device Configuration (CLDC): As the name suggests this configuration is designed for devices with limited network connectivity, relatively slow processors and less memory as compared to the devices falling under the CDC. Devices likely to fall in this configuration are pagers, mobile phones and entry level PDAs.

2.3.5.2 Profiles. Whilst a configuration provides the lowest common denominator for a set of devices, profiles, provide an additional layer on top of configurations. This layer provides the APIs for a specific class of device. This combination of configuration and profiles insures that whilst some devices may have similar functionalities, they do indeed have different requirements in terms of the available APIs and interfaces to their hardware. There are currently four java profiles:

- Mobile Information Device Profile (MIDP).
- Information Module Profile (IMP).
- Foundation Profile.
- Personal Profile.

Doja is another profile defined by NTT DoCoMo which operates under CLDC in i-mode devices. In our discussions we shall only focus on MIDP which is targeted towards mobile phones and PDAs.

2.3.5.3 MIDP. MIDP provides the basic functionality required by a mobile application such as network connectivity, persistent data storage on device, application control, application management and user interface etc. When MIDP is combined with CLDC it provides a more resolute platform for mobile phones and PDAs. Two version of MIDP have been released so far, MIDP 1.0 and the current MIDP 2.0. Current structure of J2ME comprises of CLDC with MIDP 2.0 on top with a few optional packages running on top which provide functionality and access to unique features of the phone such as access to Bluetooth radio or access to RFID etc.

In terms of game development MIDP provides both a set of high level API's and low level API's. The high level API's allow the creation of user interfaces by keeping the development time to a minimum whilst low level API's, as the name suggests, provide low level access at pixel level and facilitate the creation of new graphics objects on the screen. 'Canvas' is the base class for creating low level graphics and is the most commonly used class in J2ME game development. Graphics can be animated and there is support provided for to capture the user input, however, a specific Game API is also available which provides access to specific game related functions such as animation control, frame and sprite management, collision detection and layers. The Game API was added as part of MIDP 2.0 and is not available in MIDP 1.0 and is generally considered to be the most useful addition to MIDP 2.0. Sound support is provided by Mobile Multimedia API (MMAPI) which is an optional MIDP package which provides support for audio and video handling on the phone. The availability of this API varies from device to device, but all new phones being introduced provide support for MMAPI. There is also a cut down version of MMAPI often referred to as Media API which provides support for basic tone generation for devices that do not provide support for MMAPI.

Before we leave game development in J2ME we thought it was worthwhile to briefly consider multiplayer mobile gaming using the recently introduced Scalable Network Application Package (SNAP) from Nokia (snapmobile.nokia.com). SNAP is end to end solution for creating online multiplayer mobile games build with J2ME and it not only provides support for multiplayer gaming but also the online gaming community features.

Developers have broad range of tools at their disposal to create and add multiplayer/community features into their games. These include emulators, development tools and optional access to live development server. Sun have also included support for SNAP development in their Wireless Toolkit 2.3 Beta which is one of the most widely used tools for mobile application development.

2.3.6 Python

Python is an object-oriented programming language that is used in a wide variety of application domains. It was devised and created by Guido van Rossum early in the 1990s and although already well established on PCs, Python was only introduced as a programming platform for mobile phones in December 2004 by Nokia for their Series 60 devices. Early in 2006 they declared the project, known as PyS60, open-source (sourceforge.net/projects/pys60), allowing the mobile development community to implement new library modules, fix implementation bugs and even port the interpreter to new mobile platforms—it is now also being developed for the Nokia Series 80 platform and a version for UIQ has been reported on the web.

Python is a developer-friendly platform due to its easy-to-use syntax, highly dynamic capabilities and its extensive standard library. Further, PyS60 allows for immediate and easy access to powerful phone functionality through the modules that are provided as standard with the run-time. Often, several lines of complex native code can be replaced with a single line of Python code. For these reasons, many developers use Python to develop proof-of-concept or rapid prototype applications.

You may think that abstracting functionality a layer up from the native layer will necessarily take away some of the low-level choices that you, as a developer, can make, but this is not necessarily the case—it is possible to extend the Python run-time by developing your own dynamically loadable native extensions. These are essentially wrapper classes for compiled C++ code, which provide a special interface with which your Python code can communicate.

Python scripts are written in plain text, so it is possible to develop applications using any standard text editor and there are many excellent free Integrated Development Environments that provide features to aid the developer. Technically, it is even possible to edit the scripts on your phone, using a simple phone-based editor.

Nokia provide an emulator that can be used for testing Python scripts before deploying them to the phone—although there may be some implementation differences between the real and emulated environment. Although, testing features such as camera access, Bluetooth communications and messaging is best realized by trying your application on a real handset. It is possible to compile your code into a stand-alone Symbian installation file (.sis) using the py2sis application that is provided with the PyS60 SDK, although you'll still need the Python run-time installed on the phone.

In regards to limitations, as Python is a scripted language it means that a phone user has to install the interpreter in order to execute any Python code. This interpreter presents an overhead, in terms of processor load and memory consumption, before any application is even loaded, and in the constrained world of mobile devices, every last KB/CPU cycle counts. Further, as Python code is not compiled, as is the case with C++/J2ME, so it is relatively easy for anybody to open up your source code and see exactly how your application was implemented or share the application with others. Finally, unlike J2ME, the interpreter is only available for a small proportion of consumer handsets and because its implementation is closely tied to the underlying Symbian OS, it is unlikely that we will see Python introduced for other mobile platforms anytime soon (and if it were, the functionality and performance could vary considerably).

If we consider the use of Python for game development then we are able to develop games utilising lots of novel features as the latest version of the device installation package includes modules supporting real-time full screen 2D graphics, networking over Bluetooth and GPRS, Cell ID access, SMS sending, high and low-level GUI widgets, file IO, local database support, access to phone's camera, sound playback and recording, and access to detailed system information such as the phone's IMEI number, battery and signal levels, disk space and the free and total memory available on the device. Further, native threading and exception handling are also supported in line with the standard language specification on which PyS60 is currently based (version 2.2.2+). These modules are written at native level, which means that they are very fast (J2ME will always pay a performance penalty because the source code is compiled to an intermediate bytecode format, which then has to run on the phone's virtual machine).

Unlike the latest version of J2ME (MIDP 2.0), PyS60 does not ship with any specific game library modules. However, the key to Python's power is with its object-orientation and this, together with the high-level modules specified above, allow you to develop your own set of classes and functions, which support your game operations, relatively quickly.

The current generation of mobile phones which support Python generally have a high specification. Typically, they will feature a 220 MHz processor and a 16 or 18-bit colour screen with a resolution of 176×208 or 354×416 pixels, at least 2 MB of RAM and 10 MB of shared memory. However, if you plan on creating a real-time game, with multiple graphics and/or images moving on screen at a time (a platform game is an example of one such game), you will need to construct a very careful design. Despite the fast processor, graphics processing and rendering can still slow your game to a crawl if too many objects—characters, tiles, HUD elements and other sprites—are drawn to the screen at a time. As graphics presents the main bottleneck in the system, the key is to only draw items that are within the screen's viewport, and to limit the number of rendering operations which take a long time to process (transparent images are very problematic). Also, all resources should be compressed as much as possible, whilst retaining an acceptable level of quality.

2.3.7 Flash Lite

Flash Lite is a version of Adobe's highly successful Flash technology developed for mobile phones. The main attraction of Flash is that it allows the significant developer base that already exists for Flash in the PC community the ability to quickly develop applications for the mobile environment without having to learn new skills. Flash Lite players are available for both Symbian and Brew phones and in essence builds upon the success that Flash gave to NTT DoCoMo's imode platform, were it has become the de-facto standard. At the end of 2005 more than 20 million subscribers, which is 45 per cent of imode users, had Flash Lite enabled phones and of the 4600 official imodes sites 2000 supported Flash Lite [21]. In March 2006 Adobe estimated that there were already 60 million devices equipped with Flash Lite in the market and with manufacturers, such as Nokia and Sony Ericsson, shipping their new devices with Flash Lite pre-installed the market is expanding rapidly.

Flash Lite is integrated into Flash Professional 8 which allows the development of Flash applications for the two current versions of the Flash mobile players, namely 1.1 and 2.0, which are present on mobile phones.

Flash Lite 1.1 is based on Flash version 4, with similar scripting functionality and media support to the old desk top version. Whilst many developers bemoan the limitations imposed compared to the newer versions of Flash there are ways of working around such limitations although with a small increase in development effort.

Flash Lite 2.0 is effectively a subset of Flash 7 incorporating approximately 80 per cent of the functions and operations of Flash version 7 [21]. Thus for many developers Flash Lite 2.0 would be the referred option although few phones have been shipped thus far with this version installed so 1.1 currently allows greater penetration into the market.

One noticeable difference in Flash Lite compared with Flash is that some mobile phone functionality has been built in. Although this is limited compared to the other software platforms previously discussed it does include the ability to dial a phone number and send a Short Message Service (SMS) or Multimedia Message Service (MMS) from within the Flash application.

Flash Lite obviously has great potential for bringing information services onto phones and producing richer interface, however, in terms of games Flash Lite is predominantly seen as being most applicable for the casual games market as it will suffer in terms of operational speed for very complex graphical games. However, as we have already discussed the fact that casual games are a large proportion of the mobile games market there is an obvious potential for rapid development of simple games. The main limitation would come in terms of being able to innovate within the games by incorporating some of the functionality present on many mobile phones such as Bluetooth or the camera and if these could be accessed from within Flash Lite it could make significant in roads within the games sector.

2.3.8 Choosing Your Software Environment

In this section we have discussed many of the software environments available to the games developer and in essence the final decision is based on the requirements for the individual game. Some factors that should be considered are:

- target audience for game and what devices are likely to be owned by that group;
- does game use any specific hardware (i.e. Bluetooth) and how well is that supported;
- ease of development.

Whereas in the console games world we generally design for the latest or next generation device in mobile it is often desirable to design for the most prevalent and often then simplest device commonly available in the market as current pricing strategies are volume based.

2.4 3-D Mobile Games

Three dimensional (3D) computer graphics has been an important feature in games development since it was first introduced in the early 1980s. There is no doubt that 3D-based content is often viewed as more attractive in games than the more abstract 2D graphics [22]. This is because it enables players to become immersed in the game, as it more closely resembles their vision within the world around them, and when used in conjunction with large screens and surround sound systems it is particularly powerful. Immersion is often seen as a key component to the commercial success of a console game as immersive play encourages longer and often more repeated game play which is likely to mean the game attracts attention from the game playing community and often leads to brand loyalty [22]. In terms of the console gamer demographic there is an obvious need for 3D game titles; however, as we discussed at the start of this chapter the mobile gamer does conform to these existing models and it remains to be seen if 3D becomes the dominant style in mobile games. In this section we will avoid trying to answer this question but rather look at the current situation in terms of both software and hardware for the development of 3D graphics on mobile phones.

2.4.1 3-D Mobile Games—Software Support

Until recently, interfaces that enabled 3D graphics on mobile phones were limited as no standards were available, no common engines were used and developments made were based on personal implementations of graphics APIs. This limitation has been rectified with the introduction of OpenGL ES and M3G.

OpenGL ES is a lightweight well-defined subset of desktop OpenGL which is optimized by removing some classes and APIs that are expensive for the mobile platform and introducing smaller data types and support for fixed-point arithmetic [22]. There are three versions of OpenGL ES available currently. The first is v1.0 which was introduced in July 2003 by the Khronos Group and which emphasizes software rendering and basic hardware acceleration for mobile 3D graphics. The Khronos Group is the premier global association that strives to stimulate the growth of OpenGL ES and other graphics technologies for mobile devices.

In mid 2004, Khronos released its newer version of the API, OpenGL ES v1.1, which emphasizes hardware acceleration. Later in 2005 Khronos introduced the latest version, OpenGL ES version 2.0, which provides the ability to create shaders and write to vertex using OpenGL ES Shader Language, OpenSL ES [23]. OpenGL ES v1.0 is currently the only version supported by manufacturers. Early 2004, Nokia was the first to release a mobile phone implementing OpenGL ES, Nokia 6630. Now, all Symbian- (v8.0 and above) and BREW- (v1.1 and above) based phones implement this API, in addition to some Mobile Linux phones. However, there is an OpenGL ES implementation for Windows CE-based Pocket PCs and smartphones. As previously discussed Microsoft has stripped off its DirectX 3D API (MDX, DirectX Mobile) to suit devices running its newest mobile operating system: Windows Mobile 5.0.

M3G is the first Java-specific standard for 3D graphics on mobile phones. This API is an optional package to be used with profiles like the Mobile Information Device Profile (MIDP). It is based on OpenGL ES in low-level and crafted for J2ME development platform. This implies that any device embedding J2ME interpreter, given that its profile is aligned with OpenGL ES, will be able to present and render 3D graphics [24]. Although M3G is based on OpenGL ES, the API does not implement fixed-point arithmetic. It uses floating-point instead which causes loss in programming efficiency and resources consumption. The first device to support M3G in software was Nokia 6630 and the first to accelerate it in hardware is the W900 Walkman phone by Sony Ericsson released in October 2005 [25].

2.4.2 3-D Mobile Games—Hardware Support

Unlike software APIs and operating systems, hardware architecture for mobile phones is almost similar. The majority of mobile Central Processing Units (CPU), and hence processor Instructions Set, are based on the ARM technology. All Symbian OS, Microsoft Windows Mobile, Mobile Linux and Savaje support the ARM architecture [26].

There is currently a plethora of companies introducing optimized CPUs for the mobile platforms. In the next subsection we shall mention only those processors that are dedicated for graphics acceleration in general and 3D graphics in particular.

With the explosion in the demand for high-quality graphics presentation on cell phones, and the continuous appeal for sophisticated multimedia services, graphics hardware technology providers have leveraged their extensive experience and knowledge to indulge new solutions to meet users' and companies' expectations and requirements. PowerVR MBX graphics processor by Imagination Technologies, a leader in system-on-chip intellectual property (SoC IP), is developed to meet the growing needs of multimedia applications on power-conscious mobile devices like smartphones. It implements OpenGL ES 1.0/1.1. PowerVR MBX introduces benefits of low memory bandwidth, high image quality, low power demands, fill rate of 300 million pixels/sec, throughput of 3.7 million triangles/s [26] and refresh rate up to 37 frames/s [26].

NVIDIA® GoForce® 3D 4800 handheld graphics processing units on the other hand delivers dazzling 3D graphics, multi-megapixel digital still images, video capture and playback, and extended battery life to advanced handheld devices. The processor is capable of rendering 250 million pixels per second and 5 million triangles per second while a frame rate of 30 frames/s [26]. The Sonny Ericsson W900 Walkman phone mentioned earlier is the first phone to ship with NVIDIA GoForce® 3D 4800 on board.

Other industry leaders like Bitboys, ATI and Qualcomm are emphasizing on developing hardware solutions to accelerate 3D graphics on mobile handsets in hardware rather than purely in software.

2.4.3 3-D Mobile Games—Optimisation

To display a 3-D game object on a mobile phone screen requires [27]:

1. Rotation and translation of the vertices of the game object's polygons to reflect its orientation and position relative to the camera.
2. The vertices must then be projected onto the camera's X–Y plane.
3. Each polygon is clipped to remove any regions that lie outside the camera view, any polygons that lie entirely outside this view are discarded.
4. The resulting polygons are mapped onto the screen.

This sequence of tasks is known as the rendering pipeline and in all four stages of the rendering pipeline, we run up against variations of the central problem which is the need to maintain numerical precision without sacrificing speed. Precision and speed are important factors in making a game world seem real. If we lose precision in any of the steps of the pipeline, objects will seem to change shape at random from one display frame to the next, undermining the illusion of solidity. On the other hand, if we take too long over any of the steps of the pipeline, our frame rate will fall off, undermining the illusion of smooth movement. Maintaining precision without sacrificing speed is difficult on mobile phones, because the data types and arithmetic operations that make it easy to maintain precision are also the ones that tend to produce the slowest-executing code [27].

To get the fastest possible 3-D game code we should only use, in order of preference:

- integer shift operations (« and »);
- integer add, subtract and Boolean operations (&, | and ˆ);
- integer multiplication (∗).

In other words, in speed-critical code we must represent all quantities (coordinates, angles, and so on) by integers, favour shift operations over multiplies, and avoid division completely. This strategy will give us the speed we want, but it will pose problems in implementing the stages of the rendering pipeline. All four stages of the pipeline involve calculations on fractional quantities, which we would most naturally represent with floating-point numbers. Also, the transformation stage involves the trigonometry functions cos and sin, while the other three stages involve division operations [27]. A typical 3-D game rendering engine implementation will require programmers to master the two main techniques to solve these implementation problems: fixed-point arithmetic and pre-calculated look-up tables.

3. Opportunities

Whilst there is no doubt that developing games for mobile phones presents a great many challenges to the developer it also offers a great many opportunities for innovation over console games. Some of these opportunities are:

- anywhere, anytime entertainment;
- rapid technology advancement;
- high device penetration;
- wide ranging demographics;
- ever changing attitudes and fashion requirements;
- huge untapped market.

In this section we shall examine some of these opportunities together with providing examples of novel mobile games produced by that authors that will hopefully stimulate researchers and developers considering mobile games.

3.1 Text Games

With the billions of text messages sent every day text would seem a natural element for either a component of a game or as the basis for the game. However, the small form factor of the mobile phone and the keyboard means that it not as straight forward as it may appear and there are various methods of text input that have been tested and integrated into mainstream use.

Entering text with a standard numeric keyboard requires a 'multi-tap' functionality allowing text input with little visual confirmation. T9 predictive text offers the most commonly-used word for every key sequence you enter by default and then lets you access other choices with one or more presses of the NEXT key. T9 also learns words and expressions unique to each user. Touch screens are also becoming common on more business orientated phones; with stylus entry or in some case incorporate full QWERTY keyboard options [28]. For mobile designers, it is important to keep input when requested very short to as few characters as possible.

Text input studies at Nokia Research Center [28] show conventional keypad 'multi-tap' entry at 8 to 9 words per minute (wpm), predictive text input at 20 wpm, and QWERTY keyboard input at up to 35 wpm. This is in comparison to 50–80 wpm on a traditional PC-based keyboard.

In terms of text based games there have been commercial successes such as 'Dopewars', which is a text based trading game based upon buying and selling various drugs. One of the most interesting and innovative aspect of the text based games is demonstrated by 'Botfighters', which is a search and destroy combat game, generally regarded as the first location-based mobile phone game, launched in 2000 using Cell ID (www.botfighters.com). In the first version of Botfighters, players sent an SMS to find the location of another player and if that player was in the vicinity, they could shoot the other player's robot by sending another SMS message. While gaining some recognition, Botfighters' use of SMS and its crude location accuracy means that it suffers from delay and to limited levels of social interaction as players are physically distributed over large areas [33].

Although many designers may dismiss text as offering little innovation but in the following section we present a design that shows this is not the case and there are still opportunities to provide good gaming experiences with careful design.

3.1.1 The TxT Book

The aim of this project was to create a text based game that would appeal to a mobile game playing demographic rather than the traditional console gamer. However, as there is very little research in this area we had to look at studies of console gamers to see if there were any identifiable traits we felt could be applicable to the mobile gamer. To this end we looked at the work of International Hobo who produced a classification for both hardcore and casual gamers based on clusters from the personality types of the Myers–Briggs Dichotomies, which are used to classify individuals' personalities. They called these types [5]:

> "The Conqueror is actively interested in winning and 'beating the game'. In single-player games, completing the game generally counts as winning, while in multiplayer the goal is to beat the other players—either way, winning is the most important factor to these players."

> "The Manager is generally looking for a strategic or tactical challenge. They are interested in the mastery of the game—that is, the process oriented challenge of learning how to play well. Winning is to some extent meaningless to the Manager-type player if they have not earned it."

"The Wanderer is a player in search of a fun experience. They won't play a game they aren't enjoying, and will in fact stop playing the moment it ceases to be fun."

"The Participants are the largest group in the population. Very little is known about them except that they are often very story-oriented and will generally only play games as a social experience. They wish to participate either in the story the game is offering, or participate with other players in some emotional context."

International Hobo's research classification indicates that the majority of game players actually fall into the class of wanderer. However, more generic Myers-Briggs based research would indicate that the majority of the rest of the population are generally considered participants and International Hobo admitted this group was under represented in their study. Therefore, we have not restricted our design simply to the hardcore or casual gamer, but rather wanderers and participants that we feel are likely to be most representative of the average phone user.

Using this as the initial inspiration for the project presented, we have explored a simple to use and easily understood game that:

- can be played in short bursts;
- avoids repetitiveness and has a strong element of fun;
- is story orientated;
- has a simple user interface and can be played across a wide range of mobile phones;
- provides a social experience.

The outcome is a novel multiplayer mobile game called 'thetxtbk', based on the surrealist technique of 'Exquisite Corpses' and the old Victorian parlour game of 'Consequences'. The game builds to produce a massively multi-authored book.

The premise for the mobile game is based on Exquisite Corpse which is a word game developed by surrealists in 1920s Paris as a means of 'discovering' accidental poetry. It has its roots in the parlour game of Consequences which was designed to be played by five to nine players after dinner by Victorian ladies and gentlemen. The surrealist version is generally played by four or more people who all have a pencil and paper. To start, each player writes an adjective, then folds the paper to hide this word and then passes their paper to the next player. The process continues until the sequence of adjective, noun, adverb, verb, adjective, noun has been completed. The sentences are then read out by the players. One of the results from the first playing of the game was "Le cadavre exquis boira le vin nouveau" (The exquisite corpse will drink the new wine) which is how the technique obtained its name (www.exquisitecorpse.com).

In this project we build on the game operation previously outlined and have adapted them to produce a mobile game called 'thetxtbk', which is based around the standard SMS message length of 160 characters. Each player takes a turn writing a contribution to the book based upon only the last 160 characters entered. There are a number of reasons for basing the system on a SMS message [29]:

- Firstly, although the system is based on Wireless Application Protocol (WAP), and the length of the message is entirely a design choice, we felt WAP still has a very negative image with consumers due to the disastrous launch on GSM. SMS messaging, however, is incredibly popular and well understood, with over 85 million text messages sent everyday in the UK alone (www.textit.com).
- Secondly, allowing a short message, rather than an individual word, allows for greater creativity, increased sense of individual game play and avoids the need to control the type of word at each stage.
- Thirdly, SMS messaging is often used to support social groupings and creating a community relationship is often seen as an important aspect to the success of multiplayer games.

The system developed to support the game consists of a central server database written in MySQL, which holds both the entries in the book and names of the contributing authors [29]. To distribute the game we have utilized an SMS message which contains both an invitation to contribute to the book and the link to the WAP site as shown in Fig. 1.

The majority of new phones allow links to be opened directly from within the message itself, whilst on older models the URL link can be entered into phone's micro browser. If an SMS message is received from the known number of a friend,

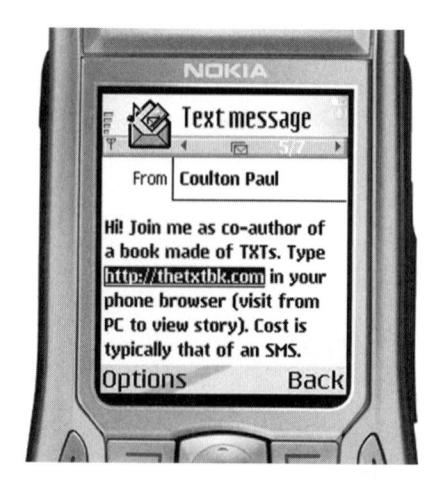

FIG. 1. SMS invitation to contribute to thetxtbk.

MOBILE GAMES: CHALLENGES AND OPPORTUNITIES

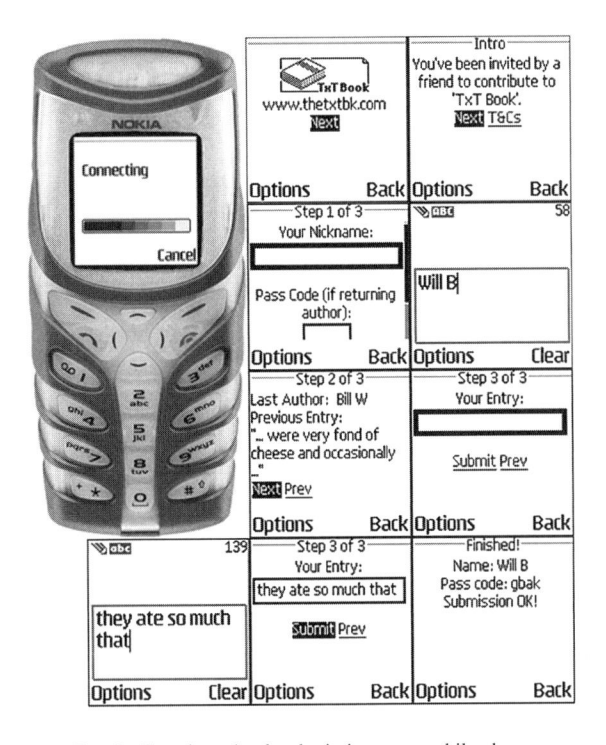

FIG. 2. Entering a book submission on a mobile phone.

it is more likely to be opened and acted upon. Therefore, we encourage contributors to personally forward the invitation SMS to their friends.

Once the player uses the link, the system detects that they are accessing the site from a phone and provides access through a series of WAP pages as shown in Fig. 2.

The process is very simple. Firstly, the prospective author enters their Nickname (and pass code if they are a returning author). They are then shown the last entry to the book from which their own entry should follow. Finally, they make their own entry. Once they have submitted their entry they are given a pass code to enable them to make future entries under the same author name. They are also encouraged to forward on the SMS invitation message to their friends. We have added other elements to the game to increase both the sense of community and a competitive element for hardcore gamers. In particular, the website offers both the ability to make an entry and also gives larger snippets of the book to allow people to identify with the larger community of authors. There is also a list of contributing players and the number of entries they have made. We felt this was analogous to the high score tables which are often used to address the competitive desire of hardcore gamers and encourage greater levels of participation in games. Although we have provided for entry via the web and have indicated in Fig. 1 that awareness of the game can also be spread via email, we have not actively encouraged this as our main focus is in mobile entry and mobile distribution [29].

Whilst the book, which is at the heart of this game, is still evolving it has already provided some interesting insights into the way the players engage within the game [30]. In this section we shall explore such interaction and the obvious place to start, as with all books, is the beginning. For our opening we decided to play homage to the Philip K Dick classic, 'Do Androids Dream of Electric Sheep?', by altering the opening line of the novel's hero *Rick Deckard* (who is listed as an author) to be in the first person past participle:

"A merry little surge of electricity piped by automatic alarm from the mood organ beside the bed awakened me.

It is interesting to note that this first line is complete, which is something that turned out to be unusual in subsequent entries. The second entry, as shown by italics and highlighting, is indicative of the main feature that dominates the style of entries in that the current author leaves the line open ended for the next author to follow. This would indicate that the players have a sense of community and that they are interested in developing the story—both attributes of the *participants*.

I felt exhausted from the strange dreams that plagued my sleep. Who was *eating my cheese, I wondered? Did I have mice? Or was there* another reason I was a few crumbs short. It occurred to me that my obsession with cheese had to stop, before I began to hallucinate with desire

The two passages that follow illustrate another common practice amongst the players, where the current author follows the story of the previous author but tries to create a humorous twist in the plot that often sends it off in a completely new direction. This would indicate it is a fun experience which is desirable from the expectation of the *wanderers*. Further, we can also see in the later of the two passages that the story is now progressing in the second person past participle. This switching generally occurs after the inclusion of a fairly random (even by the standards of this book) entry which does occur every so often.

a deep fried guinea pig that I had once enjoyed whilst trekking through deepest darkest Peru in search of a lost tribe who were reported to engage in acts so *mundane, that they remained undiscovered for centuries – I didn't find them. However, I did meet another tribe who engaged in acts so terrible that*

, the Cornish comedian, was not so lucky. The oncoming implement smashed into his chest, impaling him to a near by bench *that marked the spot of the first alien landing in Lancaster. The day they abducted Phil went down in history for* the longest ever probing session. Eventually the aliens became despondent when he pointed out their futile business model. They started data harvesting

Further, the second and third entries of the second passage were from two known members of our research group and we were able to identify an obvious in joke between them. This highlights the potential for social experience, which is another desirable trait for *participants*.

Authors tended to use a first person narrative. One of the key reasons for this is that it is almost impossible to maintain characters; omission of the character name from a single entry results in the subsequent author having no knowledge of those characters and these characters simply disappear. This is illustrated in the case of Dr Forbes in the passage below.

they'd even been shunned by their neighbours, a group of satanic cannibals with a penchant for random, drawn-out torture. So I turned to Dr Forbes and asked if *we should exterminate them using our latest "death-ray" technology. Dr Forbes suggested an easier method of neutralizing our adversaries, involving* the organs of small tree frog, ear wax, and a large roll of gaffer tape. The result was

Interestingly this was not the last appearance of Dr Forbes and regular contributors who had come across him in the past often tried to resurrect him, although this is generally a short lived experience.

the elusive Dr Forbes who seems to appear and disappear like a veritable will o' the wisp. We were particularly surprised to see him here as he had vowed to sto *ck tomato soup only on Saturdays and today was a Tuesday. 'Hey Forbsey', I whispered,*

> *'Got any of your special, out the back?'. Forbes smiled and gestured toward the trap door leading to his cellar. "They're down there mate. Careful though, one of them nearly bit my arm off the other day". I began to walk toward the hatch when*

There are a many other statistics and patterns developing for this game although they are beyond the scope of this discussion and will be presented in a future publication.

3.2 Camera Games

Cameras are now a common feature of even the most basic mobile phone and indeed a reported [1] 295.5 million were shipped in 2005 which represented nearly 40 percent of all phones shipped. There is thus a real opportunity for their use within games and there have been a number of very innovative games that have used the camera to detect movements of the phone and transferring these two movements within the game. Probably the best known are from game developer Ojum (www.ojum.com) with its games "Attack of the Killer Virus" and "Mosquitos" for Symbian S60 mobile phones. In both games, the enemy characters that the player must shoot at are superimposed on top of a live video stream from the mobile phone's camera. The player moves around this mixed reality space by moving their phone and firing using the centre key of the joy pad. Ojum has also produced a virtual version of the old children's games where you directed a marble around a maze avoiding holes called 'Action Twister'. There are others in a similar vein that utilise visual codes in conjunction with the camera to detect movement [2,3] but these appear to offer no great advantage to those that use the video input or the surroundings directly. Although, the movement detection is fun it does tend to feel 'jerky' and with the new phones emerging with in-built 3-D movement sensors, such as the Nokia 5500, they provide a much better interface with greater resolution of movement.

Using the camera as a joy pad is not the only possibility and in the following section we describe a game that utilises the camera to personalise the content.

3.2.1 Buddy Bash

Buddy Bash, shown in Fig. 3, is an addictive reaction testing boxing game with the added twist that you can add a picture of a 'friend' as the face of your opponent. The game requires a Java MIDP 2.0 enabled phone with a camera and you take a picture following the simple menu instructions. The game is played in rounds which you win by scoring a punch before your opponent. The punches are made by hitting the key number highlighted on the screen as quickly as possible. If you're fast enough

FIG. 3. Buddy Bash screen shots.

you hit your opponent but if your reactions are too slow they hit you. The game speed increases between rounds and if you reach the high-score it is preserved until your next big fight. As an added feature your friend face also appears on the ring girl who announces the new rounds. Although the game is very simple the simple one button interface and the personalisation do give it an attraction over other games in the casual mobile market.

3.3 Location Based Games

Although we often consider the requirement for providing the location of a mobile user as a new problem, in fact all mobile phone systems effectively track a user's whereabouts at the cellular level. Each cell site has a unique Cell-ID which enables the system to locate a mobile user so that it can route calls to the correct cell. To enable higher degrees of accuracy, other techniques treat location finding as a relative exercise, in other words the location of the mobile user must be estimated against some known framework. This framework could be elements such as the locations of the base stations of a mobile phone network or the satellites of the Global Positioning System (GPS). An alternative approach is to ascertain location from the user's interaction with objects of known location where their position can then be implied. The interaction could be proximity within a physical area using communication technologies such as WiFi or Bluetooth [31] or down to object level using one of the various forms of two dimensional (2D) bar codes, such as QR codes [32], or Radio Frequency Identification (RFID).

Having discussed how the position of the mobile user may be obtained we now turn our attention to one of the many possible areas in which this technology may be used which is that of location based games. A location based game is one that is aware of a user's location, can perform 'what's or who's near' queries, and then deliver information relevant to that position. There have been many innovative examples of such games reported in the literature [33,34] using all of the techniques previously described and in the following section we shall consider two specific location based games projects.

3.3.1 PAC-LAN

PAC-LAN is a version of the video game PACMAN in which human players play the game on a maze based around the Alexandra Park accommodation complex at Lancaster University [35]. Pacman was chosen for a number of reasons:

- it is widely recognized with simple but compelling game play which means that the concept behind the game can be quickly ascertained by potential players without a complex explanation;
- the virtual game maze premise transfers readily to a physical location;
- and, the Pacman character interacts with game elements (the game pills) that can be considered as physical objects at specific locations which is one of the principle advantages that RFID tags can provide.

This is not the first time Pacman has inspired the development of a location based games and the most famous is probably Pac-Manhattan (www.pacmanhattan.com). However, Pac-Manhattan differs significantly from PAC-LAN in that

- it does not incorporate actual physical objects;
- it uses mobile phones simply to provide a voice link (effectively a walkie talkie arrangement);
- the developers chose not to implement a means of location estimation as it was played around the streets of Manhattan which would have acted as urban canyons for systems such as GPS;
- and the game play of each player was controlled by a human central operator.

We have specifically chosen to avoid incorporating voice calls or SMS, and hence human game controllers, as we wanted to keep the game play as fast as possible and more akin the arcade classic. Although human controllers introduce interesting aspects of trust and acceptance, as explored by games such as Uncle Roy All Around You [36], we felt that their would be a greater possibility of the emergence of spontaneous tactics without a controller.

The other significant implementation has been Human Pacman [37] which uses an innovative combination of virtual reality goggles, GPS receivers, and portable computers with Bluetooth and WiFi access to recreate the game. In terms of differences from PAC-LAN it:

- is played over much smaller area of approximately 70 meters squared compared to 300 meters squared;
- only uses one real object used for interaction;
- is played at a much slower pace due to large amounts of equipment being carried uses human central operators to control the game play of each player.

Obviously this is highly specialized and expensive technology and in an interview with CNN in November 2004 its creator Dr. Adrian Cheok predicted that:

> 'Within two years we'll be able to see full commercial Pacman-type games on the mobile phones.'

With the commercial technology presented in this paper this has became a reality in less than a year and an equipment outlay of less than 1500 euros. This will mean that the system can be tested on large numbers of people without concern over equipment costs or the practicalities of running wearing virtual reality goggles.

In terms of playing the game one player who takes the role of the main PAC-LAN character collects game pills (using a Nokia 5140 mobile phone equipped with a Nokia Xpress-on™ RFID reader shell), which are in the form of yellow plastic discs fitted with stick-on RFID tags placed around the maze as shown in Fig. 4.

Four other players take the role of the 'Ghosts' who attempt to hunt down the PAC-LAN player. The game uses a Java 2 Platform Micro Edition (J2ME) application, running on the mobile phone which is connected to a central server using a General Packet Radio Service (GPRS) connection. The server relays to the PAC-LAN character his/her current position along with position of all ghosts based on the pills collected. The game pills are used by the Ghosts, not to gain points, but to obtain the PAC-LAN characters last known position and to reset their kill timer which must be enabled to allow them to kill PAC-LAN. In this way the ghosts must regularly interact with the server which is then able to relay their position to the PAC-LAN.

FIG. 4. PAC-LAN mixed-reality mobile phone game.

PAC-LAN sees a display with his own position highlighted by a red square around his animated icon whilst the Ghosts see both a white square highlighting their animated icon and red flashing square around PAC-LAN. These character highlights were added after pre-trials revealed players wanted a quicker way of identifying the most important information.

The Ghosts can 'kill' the PAC-LAN character by detecting him/her via an RFID tag fitted on their costume, assuming their kill timer has not run-out. Once PAC-LAN is killed the game is over and the points for the game are calculated in the form of game pills collected and time taken to do so. When PAC-LAN eats one of the red power pills, indicated by all ghost icons turning white on the screen, he/she is then able to kill the Ghosts, and thus gain extra points, using the same RFID detection process. 'Dead' ghosts must return to the central point of the game maze where they can be reactivated into the game. Figure 5 shows a number of typical screens the PAC-LAN character will experience throughout the game.

The following Fig. 6 highlights a simple game scenario for a Ghost player where he/she enters the game after a controlled delay. The Ghost player then attempts to kill PAC-LAN but his kill timer has expired and he/she then falls victim to PAC-LAN who has subsequently obtained a power pill.

The scoring in the game is simple where the PAC-LAN character gets 50 points for a normal pill, 150 points for a power pill, 1000 points for collecting all the pills,

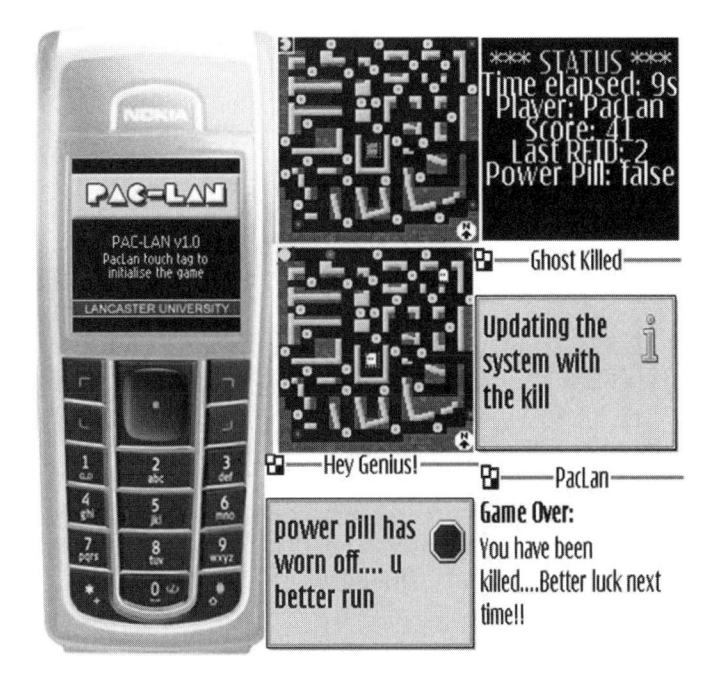

FIG. 5. PAC-LAN phone UI.

and 500 points for a Ghost kill. The Ghosts get 30 points per pill (this is linked to the length of the kill timer) and 1000 points for killing PAC-LAN. All players lose 1 point per second to ensure they keep tagging.

Overall the game has now be played by over 60 people and we concluded from feedback taken of the users' experiences that the tangible objects and the sound and vibration alerts greatly enhanced game and made the technology less intrusive in the experience. Overall, the game has been a great success being perceived as fun to play and easy to use.

3.3.2 They Howl

This is a location based game written in J2ME and utilises GPS to obtain positional information as shown in Fig. 7. The GPS unit is not an integrated part of the phone but rather we exploit the proliferation of GPS units that can be accessed via Bluetooth which is cheaper and more readily available than integrated solutions.

The premise for the game itself builds on our experiences from PAC-LAN where sounds were perceived to be the best form of feedback. The use of sounds for the

FIG. 6. Ghost phone UI.

major feedback also avoids the 'pervasive game stoop' were a player becomes completely fixated by the display on the mobile device screen they effectively ignores their surroundings, thus diluting any mixed reality experience. The sounds in this case are the howls of a wolf pack who are hunting down their prey in a mixed reality landscape. The howls from the wolves enable them to co-ordinate their efforts when following the virtual 'scent' trail left by the prey it flees from the approaching pack.

The prey leaves a trail of virtual scent markers which are automatically generated every two minutes by the application. The wolves can detect a marker by coming within 25 m of the scent marker. Once a wolf has found a scent marker they howl to the rest of the pack. The howl causes the display on the other wolves' phones to indicate the direction they must travel to reach that scent marker which appears at the centre of the screen. The wolves must then work together to find the next marker and ultimately track down the prey. The prey also gets an update to his display showing the position of the pack. The prey is caught if two or more wolves come within 25 m of the prey, having collected all the scent markers, at which point they all let out a group howl indicating their presence while the prey emits a death

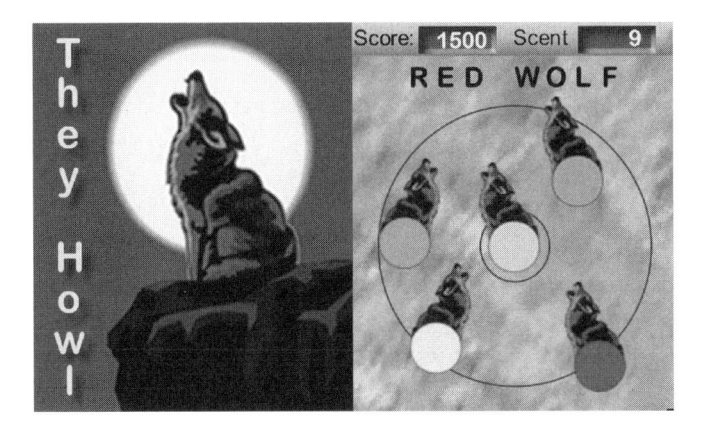

FIG. 7. They Howl splash screen and Howl screen.

gurgle. The prey gains points for the length of time within the game and the distance travelled. Wolves gain points for both finding the scent markers and capturing the prey.

Whereas PAC-LAN required a pre formed group to be present at a specific location They Howl is designed for more spontaneous game play using location although it is not tied to a specific location. To facilitate this we have create a lobby system which will allow potential players to gather in any location to play the game. The game area is normally limited to 1 Km square and potential players can be anywhere within that square. Once there are a least three players the game can begin or they can wait for others to join up to a maximum of 10. The game server randomly selects a player to act as prey whilst the remaining players become wolves.

3.4 Proximity Games

In the previous section we discussed location based games and many of them incorporate proximity, in the sense that the user's location is close to either another player or a real or virtual artefact within the game. However, none rely solely upon the proximity between either a player and another player, or the player and a game artefact, irrespective of the physical location. The most notable example of this type of game play, although not mobile phone based, is that of Pirates [38] which was an adventure game using Personal Digital Assistants and RF proximity detection. Players completed piratical missions by interacting with other players and game artefacts using RF proximity detection. One of the interesting aspects that emerged was the stimulated and spontaneous social interaction between the players. One of the research motivators for this project was to produce a game that provided the opportunity to bring this type of spontaneous social interaction to mobile phone users. We believe it is best achieved by removing the requirement for a central game server, as utilized in Pirates, and utilize a proximity detection scheme that initiates a dynamic peer to peer connection.

In terms of proximity detection the obvious choice is Bluetooth which despite previous predictions of its demise is in fact increasing its growth, with Nokia predicting a year-on-year increase of 65% in 2006. In fact there are already a small number of mobile Bluetooth proximity applications which are often described as Mobile Social Software (MoSoSo) and can be viewed as evolutions of Bluejacking. This is a phenomenon where people exploit the contacts feature on their mobile phone to send messages to other Bluetooth enabled devices in their proximity [39]. Bluejacking evolved into dedicated software applications such as Mobiluck and Nokia Sensor which provided a simpler interface, and in the case of Nokia Sensor, individual profiles that could be used to initiate a social introduction. There are various examples of using this technology for social interaction such as Nokia Sensor and Serendipity [40] but the only gaming example of this type is You-Know-Who from the University of Plymouth which provides a simple game premise to help initiate a meeting. After scanning for other users running the application and 'inviting' a person to play the game, the first player acts as a 'mystery person', who then provides clues about their appearance to the second player, who builds up a picture on their mobile phone screen. After a set number of clues have been given, the players' phones alert, revealing both players' locations and identities. Obviously, the game play is quite limited and effectively non-competitive, which is unlikely to result in repeated game play, therefore, it is closer to the other MoSoSo applications than a game [41].

3.4.1 Mobslinger

The basic game premise is simple to understand and operate, which is an essential feature in any game [5]. The relatively low cost of mobile phone games means they are often very quickly discarded by users if they cannot quickly engage with gameplay. Mobslinger runs as a background application on Symbian Series 60 smartphone which periodically scans for other users in the vicinity who are also running the mobslinger application. Once detected, a countdown timer is initiated on both phones which alerts the user by sounding an alarm and vibrating the phone. The user then has to 'draw' their mobile and enter the randomly generated number which has appeared on the screen as quickly as possible. The person with the fastest time is the winner and the loser is 'killed', which means their application is locked out from game-play for a set period of time. The game is playable in a number of different modes which we will discuss in the following paragraphs both in terms of the operation and their relative merits [41].

FIG. 8. Mobslinger Splashscreen, 'Draw' screen and 'Draw' screen in Outlaws mode.

3.4.1.1 Quick Draw. This is the basic mode of the game for two mobslingers at a time. Once the phones have detected each others presence the first mobslinger to 'draw' their mobile and enter the correct code wins, as shown in Fig. 8. This mode is intended to promote spontaneous social interaction between two players, who would generally be unknown to each other, in the form of a ludic greeting. The loser is locked out of the game for a set period of time and is unable to interact with other players. This may be a disadvantage for larger social groupings and therefore has been addressed in a subsequent mode. The optimum length of the lock out is something that can only truly be ascertained from the large scale user feedback and at present we have fixed this at 2 hours.

3.4.1.2 Blood Bath. This is the large scale battle mode (often referred to by the design team as playground mode), in which two or more mobslingers score points by beating randomly selected targets to the draw over a set time period. After the time period has expired the game is ended and the mobslingers' high score is displayed on the phone. The mobslingers can then compare scores to ascertain the winner and allows them an element of schadenfreude (amusement at the misfortune of others). Because of the intensity of the game play the constant noise may be irritating in some social situations and this mode is probably best played outside or in a regimented social setting. This mode does not have the spontaneity of Quick Draw and needs to be initiated by a preformed social grouping.

3.4.1.3 Last Man Standing. This is a less intensive battle mode than Blood Bath and is for two or more mobslingers. The aim of the game is to be the 'last man standing' after a series of quick draw encounters within a group. Duels are triggered at randomly timed intervals, so are likely to prove less irritating for the general public, and create an air of anticipation amongst the mobslingers. The random time also means that the group can be engaged in the general social proceedings of the evening and the game effectively provides a humorous side. The game play may be lengthy and players knocked out in the early stages may lose interest, although players in search of the more intense experience may opt for Blood Bath mode. As with Blood Bath mode, this requires an initiation by a preformed social grouping.

3.4.1.4 Outlaws. This is the team mode of the game where groups of mobslingers join forces to create 'Outlaw' gangs who then go out in the hope of combating other teams. The Outlaw gangs are generally comprised of two or more mobslingers and offer a mode of play where there is a greater opportunity for tactics to develop. The game play is similar to Last Man Standing, although the random interval between duels is shorter, and only mobslingers of an opposing set of Outlaws are selected to play against each other. In this mode, challenges are made and accepted between gangs. In Fig. 8 we show the 'Draw' screen for Outlaw mode where the player rating is replaced by the gang names. In essence this mode can be viewed as preformed social groups engaging in the spontaneous encounters of the Quick Draw mode.

3.4.1.5 Top Gun. We have also included a ranking system into the game to satisfy the desire of some experienced players being able to differentiate themselves from novices. The basic player starts out with a one star rating and this is increased in stages until they reach a 5 star rating as shown in Fig. 8. The levels are based upon games played, kills made and then their kill/die ratio. The levels are built up from satisfying a defined minimum for each of these criteria for each advancement level.

In this section we will not provide a highly detailed description of the implementation of mobslinger but rather than present a mere technical description. We shall provide an overview of the design and the design challenges faced. Mobslinger differs from the vast majority of mobile applications in that it is a background application that, once activated, requires no user input. The application will continue executing in the background of the users normal phone activities until another phone, with mobslinger running on it, is detected. Proximity detection is achieved using the Bluetooth discovery protocol to search for other Bluetooth devices in the near vicinity (approximately 10 meters).

Although the proximity detection process sounds fairly simple from the game description, it is actually the most difficult part to implement. Bluetooth is based upon a client server architecture which means that it must also have a state where it allows itself to be discovered in addition to being the discoverer.

To achieve this operation the Bluetooth client server architecture has to be initiated, which involves setting one phone to discovery state (server) and the other to listening (client). One technical problem faced was how best to alternatively set the phones to each of the states, and determine the length of time in which the phone should remain in that state. The optimum length of time will be dependent upon the game mode. For example, in Quick Draw mode where you are looking for spontaneous interaction you will not want to operate the Bluetooth too often otherwise it could drain the mobile battery, alternatively, in Blood Bath mode where you are creating a social event and you will wish the events to occur more quickly. There is also a need for a random element to be included in the time between switching modes as a situation could occur where two phones keep missing each other.

Once a Bluetooth device is found, a socket is opened to search for the mobslinger game. This is known as device and service detection, which plays a fundamental part in Bluetooth communication [42]. When a service has been detected, a secure socket can be set up to transfer data between the two Bluetooth devices. The useful aspect of Bluetooth service detection is that it can be made application specific; in other words, we can isolate discovery to devices running mobslinger and ignore any other Bluetooth devices that may be in the vicinity. Once a secure connection has been established between the devices, the game can generate a random number that the user has to press to 'shoot' the other person. This also activates a timer which starts recording when the random number is generated and is stopped when the user hits the correct key. The person with the fastest shoot-out time is the winner and is allowed to continue in the game, and in Blood Bath mode gain points. However, the loser dies and as a consequence is locked out of game play for a specified time, which is variable dependent upon the game mode. This operation is highlighted in the state diagram for the game shown in Fig. 9.

The different modes of game play are implemented using the client/server Bluetooth architecture which is implemented with one client and one server even for the multiplayer modes of Blood Bath and Last Man Standing. Whilst it would be feasible to use one phone as the server and all other phones as clients, this piconet structure would limit game play to 8 players. In reality, we could only ever shoot one person at once, therefore, a single connection is only ever needed between two phones, thus allowing many pairs of phones to fight simultaneously.

The server controls the game play by requesting and sending data to the client devices. The Bluetooth interaction between the two devices can be seen in Fig. 10. The game play only requires two messages (after detection) to be sent between the server and client. Firstly, the client sends a message to the server stating the time taken for the correct key to be pressed. The server then calculates the winner and returns a message with the result. The two devices are then disconnected to allow game play with other mobslingers in the area.

While the game is popular amongst our research group there are a number of different aspects related to each of the modes that can only be answered by trials involving significant numbers of users, indeed, for the game to work on a commercial

236   P. COULTON ET AL.

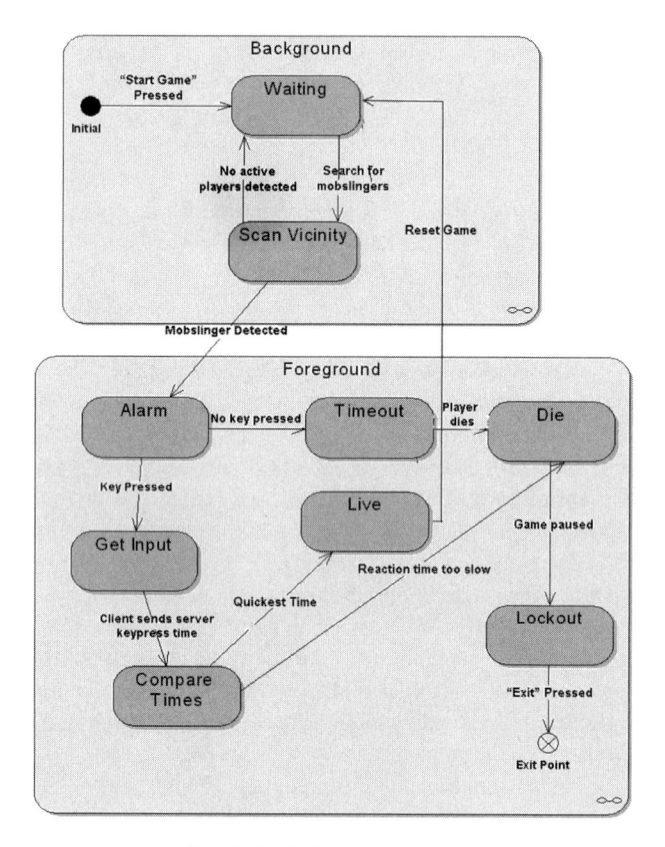

FIG. 9. Mobslinger state diagram.

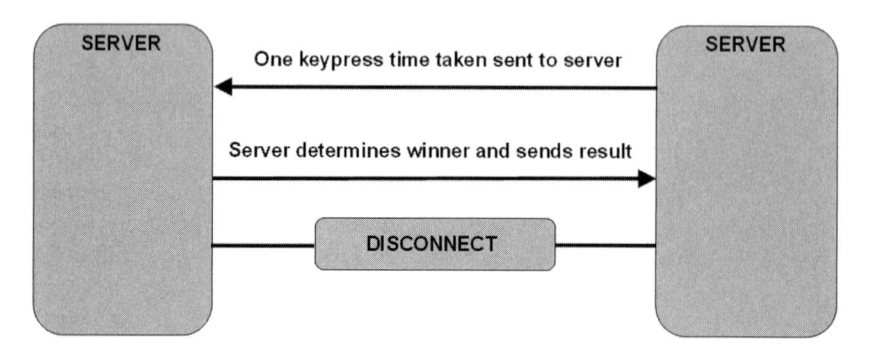

FIG. 10. Mobslinger Bluetooth client/server transfer.

level the more players running the application the better. We are therefore trying to expand the implementation of mobslingers to other mobile platforms such as UIQ, J2ME, Windows Mobile, and Python so that we can deploy the software across a wide range of users as possible.

3.5 IP Multimedia System (IMS)

IP Multimedia Subsystem (IMS) aims to merge the two successful paradigms of Internet and cellular, and provides access to all the Internet services over cellular. IMS is an architecture that in essence defines how IP networks should handle voice and data sessions by replacing circuit switched telecommunications. It is a service oriented architecture and employs a distributed component model for the applications running on top of it. This means that it aims to separate the services from the underlying networks that carry them. It originated from the Third Generation Partnership Project (3GPP) (www.3gpp.org) as means of providing 3G mobile operators migrating from the Global System for Mobile Communication (GSM) to deliver more effective data services. Since then it has been adopted by other standards organizations for both wire and wireless networks. With SIP as the backbone of the system it is widely gaining backing from services providers, vendors, application developers and infrastructure vendors.

The architecture of IMS consists of three layers, namely: the transport layer, the control layer and the application layer. Since IMS can separate the services from the underlying carrier, a GPRS enabled mobile phone can connect to IMS equally as well as a PC connected via a Digital Subscriber Line (DSL). More importantly, in a mobile environment, where a user has ability to roam, the access independence of IMS can not only allow both the physical roaming of the user but also provide the ability for his/her device to roam between various connection methods. For example, WiFi enabled mobile phones could seamlessly switch between using GPRS or WiFi and users could even switch from using their handset to PC with the same session and as the same user. Both of these features ensure that the upper layers are saved from large amounts of data traffic.

IMS is a platform for creating and delivering a wide range of services. It is only responsible for connection, state management, QoS and mobility while the actual data transfer takes place directly amongst devices. Figure 11 shows some services possible with IMS.

Some of these services can now be offered over circuit switched technology and additional J2ME APIs are available to create these services. The major drawback is not only related to the inefficiencies of the circuit switched technology but also the increased size of application and complexity. If an application was to use the IMS services API (JSR 281) these functionalities can be obtained from IMS domain

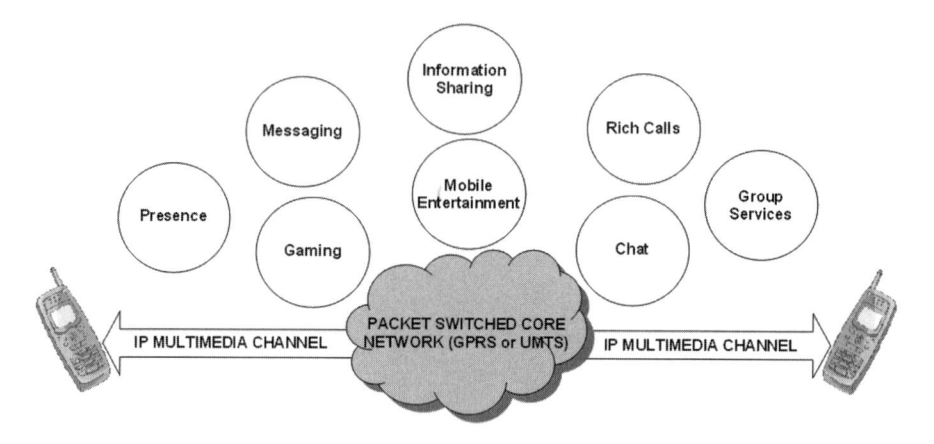

FIG. 11. P2P mobile services with IMS.

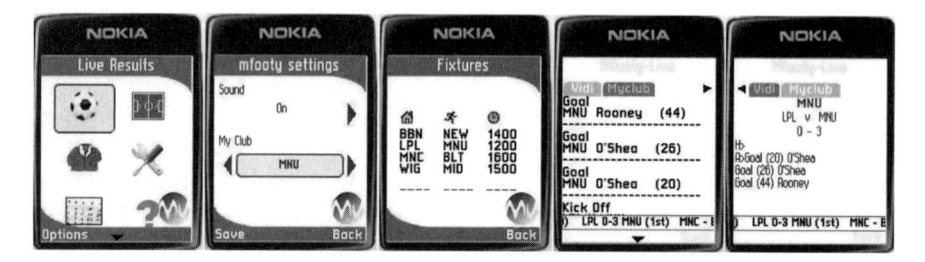

FIG. 12. Mfooty application screen shots.

itself. To illustrate this principal let us consider a networked mobile application called Mfooty (www.mfooty.com), shown in Fig. 12, as an example. Mfooty is a networked mobile application that provides live football updates from English Premier league to mobile users. The application utilizes a unique protocol to keep the cost to the end user to a bare minimum [8]. The application also provides a Fantasy Football League which provides an added community feature. The current version built using MIDP 2.0 involves the mobile device to create periodic requests over the network to check for the availability of new updates. The application is automated to suspend network connection when there are no live football games and vice versa. All these requests have to be initiated by mobile device. The following figure shows screen shots of the current application.

When IMS infrastructure is available for commercial applications in a few years time this application can be greatly enhanced and we can take advantage of the SIP, the signalling protocol for IMS, to sync with the backend game server. A SIP message initiated by the server can send out request to all mobile clients to take certain actions or wake up from running as a background process and bring updates to the display.

Since all the user information is available, Instant Messaging (IM) can be added to the application without major overheads. Presence of a user can initiate better community features encouraging users to create their own fantasy league and compete against their friends. As soon as a video highlight is available a signal from the server can prompt the user with the new information without the device or the user having to do anything. Depending upon the user settings entire application can be automated to match the schedule of the football games. So a user will probably never have to exit the application and it can keep running as a backend process, wake up on server signal and take necessary actions depending upon user settings e.g. automatically download video highlights for a certain team or all teams etc.

IMS has got some exciting new services or functionalities to offer. Mobile gaming is an ever growing industry and IMS is a step forward to utilize its potential even further.

4. Conclusions

For many years mobile games suffered from the stigma of poor functionality and usability and have been seen as the poor relations to console and PC games. That is certainly no longer true; the mobile games industry has now come of age and is creating a vibrant market in its own right built around the inherent strengths of mobile communications technology. In this chapter we have highlighted many of the inherent challenges that the mobile games developer faces but we believe theses challenges will in fact produce great innovation akin to what was seen in the early games industry. Further, if these challenges alone do not excite the possible developers, then the opportunities to create new games experiences that are uniquely mobile will more than compensate for the challenge and should provide the necessary motivation. Finally, mobile games offer enormous potential for academic research into a variety of socio-technical perspectives particularly in relation to human computer interaction and we hope that this chapter provides a useful source of reference and stimulus for innovation.

Acknowledgements

We would like to express our sincere thanks to Nokia for the general provision of software and hardware, to the Mobile Radicals research group, within the Department of Communication Systems at Lancaster University UK, to whom all the authors belong. In addition we would also like to thank Symbian for provision of UIQ phones and for their advice in developing the teaching materials used on our MSc in Mobile Games Design and M-Commerce Systems.

REFERENCES

[1] Nokia, "The mobile device market", http://www.nokia.com/nokia/0,,73210,00.htm, August, 2005.
[2] Palmer, "Mobile phone makers eye booming games market", http://news.ft.com, August, 2005.
[3] Mobile download charts available at www.elspa.com.
[4] Laramee D.F., *Secrets of the Game Business*, Charles River Media Inc., Massachusetts, USA, 2005.
[5] Bateman C., Boon R., *21st Century Game Design*, Charles River Media, Massachusetts, USA, 2005.
[6] Kashi N., "Targeting the casual gamer in the wireless world", *IGDA Online Games Quarterly, The Demographic* **1** (2) (Spring 2005).
[7] Kiely D., "Wanted: Programmers for handheld devices", *IEEE Computer* (May 2001) 12–14.
[8] Coulton P., Rashid O., Edwards R., Thompson R., "Creating entertainment applications for cellular phones", *ACM Computers in Entertainment* **3** (3) (July, 2005).
[9] Forum Nokia, "Designing single-player mobile games", Version 1.01, September 9th 2003; www.forum.nokia.com.
[10] Forum Nokia, "Multi-player mobile game performance over cellular networks", Version 1.0, January 20th 2004; www.forum.nokia.com.
[11] Forum Nokia, "Overview of multiplayer mobile game design", Version 1.1, December 17th 2003; www.forum.nokia.com.
[12] Forum Nokia, "Mobile game playing heuristics", Version 1.0, March 17th 2005; www.forum.nokia.com.
[13] Chehimi F., Coulton P., Edwards R., "C++ optimisations for mobile applications", in: *The Tenth IEEE International Symposium on Consumer Electronics, St.Petersburg, Russia*, June 29–July 1, 2006.
[14] Gooliffe P., "Register access in C++", C/C++ User, Journal, May 2005.
[15] Edwards R., Coulton P., Clemson H., Series 60 Programming: A Tutorial Guide, ISBN: 0470027657.
[16] The Linux Mobile Phones Showcase, www.linuxdevices.com/articles/AT9423084269.html, 22nd February 2006.
[17] OSDL mobile Linux initiative, www.osdl.org/lab_activities/mobile_linux/mli/.
[18] Qtopia HomePage, Trolltech, www.trolltech.com/products/qtopia.
[19] Qt, www.trolltech.com/products/qt.
[20] Heni M., Beckermann A., *Open Source Game Programming: Qt Games for KDE, PDA's, and Windows*, Charles River Media, 2005.

[21] Forum Nokia, "Flash Lite for S60—an emerging global ecosystem", white paper, May 2006; www.forum.nokia.com.
[22] Chehimi F., Coulton P., Edwards R., "Evolution of 3D games on mobile phones", in: *Proc. of IEEE Fourth International Conference on Mobile Business, Sydney Australia*, July 2005, pp. 173–179.
[23] The Khronos Group, "The OpenGL graphics system: A specification", Version 1.3, August 14th 2001; www.khronos.org.
[24] Mahmoud Q., "Getting started with the mobile 3D graphics API for J2ME", September 21, 2004; http://developers.sun.com/techtopics/mobility/apis/articles/3dgraphics/.
[25] "What's this hype about hardware acceleration?", Sony Ericsson, News & Events, December 14, 2005; http://developer.sonyericsson.com.
[26] Chehimi F., Coulton P., Edwards R., "Advances in 3D graphics for Smartphones", in: *International Conference on Information & Communication Technologies: From Theory to Applications*, Damascus, Syria, 24–28 April 2006.
[27] Forum Nokia, "Introduction to 3-D graphics on series 60 platform", Version 1.0, September 23rd 2003; www.forum.nokia.com.
[28] Silfverberg M., MacKenzie I.S., Korhonen P., "Predicting text entry speed on mobile phones", in: *Proc. of the SIGCHI Conference on Human Factors in Computing Systems, The Hague, The Netherlands*, April 01–06, 2000.
[29] Bamford W., Coulton P., Edwards R., "A massively multi-authored mobile surrealist book", in: *ACM SIGCHI International Conference On Advances In Computer Entertainment Technology, Hollywood, USA*, 14–16 June 2006.
[30] Bamford W., Coulton P., Edwards R., "A surrealist inspired mobile multiplayer game: Fact or fish?", in: *1st World Conference for Fun 'n Games, Preston, UK*, June 26–28, 2006.
[31] Rashid O., Coulton P., Edwards R., "Implementing location based information/advertising for existing mobile phone users in indoor/urban environments", in: *IEEE 4th International Conference on Mobile Business, Sydney Australia*, 2005, pp. 377–383.
[32] International Organization for Standardization, "Information Technology – Automatic Identification and Data Capture Techniques – Bar Code Symbology – QR code", ISO/IEC 18004, 2000.
[33] Rashid O., Mullins I., Coulton P., Edwards R., "Extending cyberspace: Location based games using cellular phones", *ACM Computers in Entertainment* **4** (1) (2006).
[34] Magerkurth C., Cheok A.D., Mandryk R.L., Nilsen T., "Pervasive games: Bringing computer entertainment back to the real world", *ACM Computers in Entertainment* **3** (3) (July, 2005).
[35] Rashid O., Bamford W., Coulton P., Edwards R., Scheibel J., "PAC-LAN: Mixed reality gaming with RFID enabled mobile phones", ACM CIE, in press.
[36] Benford S., Anastasi R., Flintham M., Drozd A., Crabtree A., Greenhalgh C., Tandavanitja N., Adams M., Row-Farr J., "Coping with uncertainty in a location based game", *IEEE Pervasive Computing* (July–September 2003) 34–41.
[37] Cheok A.D., Goh K.H., Liu W., Farbiz F., Fong S.W., Teo S.L., Li Y., Yang X.B., "Human pacman: A mobile, wide-area entertainment system based on physical, social, and ubiquitous computing", *Personal Ubiquitous Computing* **8** (2004) 71–81.

[38] Björk S., Falk J., Hansson R., Ljungstrand P., "Pirates!—Using the physical world as a game board", in: *Proc. of Interact 2001, IFIP TC, 13 Conference on Human–Computer Interaction, Tokyo*, 2000.

[39] Jamaluddin J., Zotou N., Edwards R., Coulton P., "Mobile phone vulnerabilities: A new generation of Malware", in: *IEEE International Symposium on Consumer Electronics*, Reading, UK, ISBN 0-7803-8527-6, 2004, pp. 1–4, ISCE_04_124.

[40] Eagle N., Pentland A., "Social serendipity: Mobilizing social software", *IEEE Pervasive Computing* **4** (2) (2005) 28–34.

[41] Clemson H., Coulton P., Edwards R., Chehimi F. "Mobslinger: The fastest mobile in the West", in: *1st World Conference for Fun 'n Games, Preston, UK*, June 26–28, 2006, in press.

[42] J. Bray, C. Sturman, *Bluetooth: Connect without cables*, Prentice Hall Inc., Upper Saddle River, NJ 07458, 2001.

Free/Open Source Software Development: Recent Research Results and Methods

WALT SCACCHI

Institute for Software Research
Donald Bren School of Information and Computer Sciences
University of California, Irvine
Irvine, CA 92697-3425
USA
wscacchi@uci.edu

Abstract

The focus of this chapter is to review what is known about free and open source software development (FOSSD) work practices, development processes, project and community dynamics, and other socio-technical relationships. It does not focus on specific properties or technical attributes of different FOSS systems, but it does seek to explore how FOSS is developed and evolved. The chapter provides a brief background on what FOSS is and how free software and open source software development efforts are similar and different. From there attention shifts to an extensive review of a set of empirical studies of FOSSD that articulate different levels of analysis. These characterize what has been analyzed in FOSSD studies across levels that examine why individuals participate; resources and capabilities supporting development activities; how cooperation, coordination, and control are realized in projects; alliance formation and inter-project social networking; FOSS as a multi-project software ecosystem, and FOSS as a social movement. Following this, the chapter reviews how different research methods are employed to examine different issues in FOSSD. These include reflective practice and industry polls, survey research, ethnographic studies, mining FOSS repositories, and multi-modal modeling and analysis of FOSSD processes and socio-technical networks. Finally, there is a discussion of limitations and constraints in the FOSSD studies so far, attention to emerging opportunities for future FOSSD studies, and then conclusions about what is known about FOSSD through the empirical studies reviewed here.

1. Introduction 244
   1.1. What Is Free/Open Source Software Development? 245
   1.2. Results from Recent Studies of FOSSD 247
2. Individual Participation in FOSSD Projects 248
3. Resources and Capabilities Supporting FOSSD 253
   3.1. Personal Software Development Tools and Networking Support 253
   3.2. Beliefs Supporting FOSS Development 254
   3.3. FOSSD Informalisms 255
   3.4. Competently Skilled, Self-organizing, and Self-managed Software Developers 256
   3.5. Discretionary Time and Effort of Developers 258
   3.6. Trust and Social Accountability Mechanisms 259
4. Cooperation, Coordination, and Control in FOSS Projects 260
5. Alliance Formation, Inter-project Social Networking and Community Development 265
   5.1. Community Development and System Development 268
6. FOSS as a Multi-project Software Ecosystem 270
   6.1. Co-evolving Socio-technical Systems for FOSS 272
7. FOSS as a Social Movement 274
8. Research Methods for Studying FOSS 277
   8.1. Reflective Practice and Industry Poll Methods 279
   8.2. Survey Research Methods 279
   8.3. Ethnographically Informed Methods 280
   8.4. Mining FOSS Artifact Repositories and Artifact Analysis Methods 281
   8.5. Multi-modal Modeling and Analysis of FOSS Socio-technical Interaction Networks 283
9. Discussion 284
   9.1. Limitations and Constraints for FOSS Research 284
10. Conclusions 286
    Acknowledgements 287
    References 287

1. Introduction

This chapter examines and compares practices, patterns, and processes that emerge in empirical studies of free/open source software development (FOSSD) projects. FOSSD is a way for building, deploying, and sustaining large software systems on a global basis, and differs in many interesting ways from the principles and practices traditionally advocated for software engineering [117]. Hundreds of FOSS systems are now in use by thousands to millions of end-users, and some of these FOSS systems entail hundreds-of-thousands to millions of lines of source code. So what is going on here, and how are FOSSD processes that are being used to build and sustain these projects different, and how might differences be employed to explain what is going on with FOSSD, and why.

One of the more significant features of FOSSD is the formation and enactment of complex software development processes and practices performed by loosely coordinated software developers and contributors. These people may volunteer their time and skill to such effort, and may only work at their personal discretion rather than as assigned and scheduled. However, increasingly, software developers are being assigned as part of the job to develop or support FOSS systems, and thus to become involved with FOSSD efforts. Further, FOSS developers are generally expected (or prefer) to provide their own computing resources (e.g., laptop computers on the go, or desktop computers at home), and bring their own software development tools with them. Similarly, FOSS developers work on software projects that do not typically have a corporate owner or management staff to organize, direct, monitor, and improve the software development processes being put into practice on such projects. But how are successful FOSSD projects and processes possible without regularly employed and scheduled software development staff, or without an explicit regime for software engineering project management? What motivates software developers participate in FOSSD projects? Why and how are large FOSSD projects sustained? How are large FOSSD projects coordinated, controlled or managed without a traditional project management team? Why and how might these answers to these questions change over time? These are the kinds of questions that will be addressed in this chapter.

The remainder of this chapter is organized as follows. The next section provides a brief background on what FOSS is and how free software and open source software development efforts are similar and different. From there attention shifts to an extensive review of a set of empirical studies of FOSSD that articulate different levels of analysis. Following this, the chapter reviews how different research methods are employed to examine different issues in FOSSD. Finally, there is a discussion of limitations and constraints in the FOSSD studies so far, attention to emerging opportunities for future FOSSD studies, and then conclusions about what is known about FOSSD through the empirical studies reviewed here.

1.1 What Is Free/Open Source Software Development?

Free (as in freedom) software and open source software are often treated as the same thing [28,29,63]. However, there are differences between them with regards to the licenses assigned to the respective software. Free software generally appears licensed with the GNU General Public License (GPL), while OSS may use either the GPL or some other license that allows for the integration of software that may not be free software. Free software is a social movement (cf. [24]), whereas OSSD is a software development methodology, according to free software advocates like Richard Stallman and the Free Software Foundation [36]. Yet some analysts also see OSS as a social movement distinct from but related to the free software movement. The hallmark of free software and most OSS is that the source code is available for remote access, open to study and modification, and available for redistribution to other with few constraints, except the right to insure these freedoms. OSS sometimes adds or removes similar freedoms or copyright privileges depending on which OSS copyright and end-user license agreement is associated with a particular OSS code base. More simply, free software is always available as OSS, but OSS is not always free software.[1] This is why it often is appropriate to refer to FOSS or FLOSS (L for *Libre*, where the alternative term "libre software" has popularity in some parts of the world) in order to accommodate two similar or often indistinguishable approaches to software development. Subsequently, for the purposes of this article, focus is directed at FOSSD practices, processes, and dynamics, rather than to software licenses though such licenses may impinge on them. However, when appropriate, particular studies examined in this review may be framed in terms specific to either free software or OSS when such differentiation is warranted.

FOSSD is mostly not about software engineering, at least not as SE is portrayed in modern SE textbooks (cf. [117]). FOSSD is not SE done poorly. It is instead a different approach to the development of software systems where much of the development activity is openly visible, and development artifacts are publicly available over the Web. Furthermore, substantial FOSSD effort is directed at enabling and facilitating social interaction among developers (and sometimes also end-users), but generally there is no traditional software engineering project management regime, budget or schedule. FOSSD is also oriented towards the joint development of an ongoing community of developers and users concomitant with the FOSS system of interest.

FOSS developers are typically also end-users of the FOSS they develop, and other end-users often participate in and contribute to FOSSD efforts (cf. [4,8]). There is also widespread recognition that FOSSD projects can produce high quality and sustainable software systems that can be used by thousands to millions of end-users [81]. Thus, it is reasonable to assume that FOSSD processes are not necessarily of the same type, kind, or form found in modern SE projects (cf. [117]). While such approaches might be used within an SE project, there is no basis found in the principles of SE laid out in textbooks that would suggest SE projects typically adopt or should practice FOSSD methods. Subsequently, what is known about SE processes, or modeling and simulating SE processes, may not be equally applicable to FOSSD

---

[1] Thus at times it may be appropriate to distinguish conditions or events that are generally associated or specific to either free software development or OSSD, but not both.

processes without some explicit rationale or empirical justification. Thus, it is appropriate to survey what is known so far about FOSSD.

1.2 Results from Recent Studies of FOSSD

There are a growing number of studies that offer some insight or findings on FOSSD practices each in turn reflects on different kinds of processes that are not well understood at this time. The focus in this chapter is directed to empirical studies of FOSSD projects using small/large research samples and analytical methods drawn from different academic disciplines. Many additional studies of FOSS can be found within a number of Web portals for research papers that empirically or theoretically examine FOSSD projects. Among them are those at MIT FOSS research community portal (opensource.mit.edu) with 200 or so papers already contributed, and also at Cork College in Ireland (opensource.ucc.ie) that features links to multiple special issue journals and proceedings from international workshops of FOSS research. Rather than attempt to survey the complete universe of studies in these collections, the choice instead is to sample a smaller set of studies that raise interesting issues or challenging problems for understanding what affects how FOSSD efforts are accomplished, as well as what kinds of socio-technical relationships emerge along the way to facilitate these efforts.

One important qualifier to recognize is that the studies below generally examined carefully identified FOSSD projects or a sample of projects, so the results presented should not be assumed to apply to all FOSSD projects, or to projects that have not been studied. Furthermore, it is important to recognize that FOSSD is no silver bullet that resolves the software crisis. Instead it is fair to recognize that most of the nearly 130,000 FOSSD projects associated with Web portals like SourceForce.org have very small teams of two or less developers [77,78], and many projects are inactive or have yet to release any operational software. However, there are now at least a few thousand FOSSD projects that are viable and ongoing. Thus, there is a sufficient universe of diverse FOSSD projects to investigate, analyze, and compare in the course of moving towards an articulate and empirically grounded theory or model of FOSSD. Consequently, consider the research findings reported or studies cited below as starting points for further investigation, rather than as defining characteristics of most or all FOSSD projects or processes.

Attention now shifts to an extensive review of a sample of empirical studies of FOSSD that are grouped according to different levels of analysis. These characterize what has been analyzed in FOSSD studies across levels that examine why individuals participate in FOSSD efforts; what resources and capabilities shared by individuals and groups developing FOSS; projects as organizational form for cooperating, coordinating, and controlling FOSS development effort; alliance formation and inter-project social networking; FOSS as a multi-project software ecosystem, and FOSS as a social movement. These levels thus span the study of FOSSD from individual participant to social world. Each level is presented in turn, though the results of some studies span more than one level. Along the way, figures from FOSSD studies or data exhibits collected from FOSSD projects will be employed to help illustrate concepts described in the studies under review.

2. Individual Participation in FOSSD Projects

One of the most common questions about FOSSD projects to date is why will software developers join and participate in such efforts, often without pay for sustained periods of time. A number of surveys of FOSS developers [33,42,65,46,45,49] have posed such questions, and the findings reveal the following.

There are complex motivations for why FOSS developers are willing to allocate their time, skill, and effort by joining a FOSS project [33,49,122]. Sometimes they may simply see their effort as something that is fun, personally rewarding, or provides a venue where they can exercise and improve their technical competence in a manner that may not be possible within their current job or line of work [12]. However, people who participate, contribute, and join FOSS projects tend to act in ways where building trust and reputation [118], achieving "geek fame" [92], being creative [32], as well as giving and being generous with one's time, expertise, and source code [3] are valued traits that accrue social capital. In the case of FOSS for software engineering design systems, participating in such a project is a viable way to maintain or improve software development skills, as indicated in Exhibit 1 drawn from the Tigris.org open source software engineering community portal.

Becoming a central actor (or node) in a social network of software developers that interconnects multiple FOSS projects is also a way to accumulate social capital and recognition from peers. Hars and Ou [46] report that based on their survey 60% or more FOSS developers participate in two or more projects, and on the order of 5% participate in 10 or more FOSS projects. Many FOSS developers therefore participate in and contribute to multiple FOSSD projects. However, a small group of developers who control the software system architecture and project direction are typically responsible for developing the majority of source code that becomes part of FOSS released by a project (cf. [81]). Subsequently, most participants typically contribute to just a single module, though a small minority of modules may be include patches or modifications contributed by hundreds of contributors [42]. In addition, participation in FOSS projects as a core developer can realize financial rewards in terms of higher salaries for conventional software development jobs [45,70]. However, it also enables the merger of independent FOSS systems into larger composite

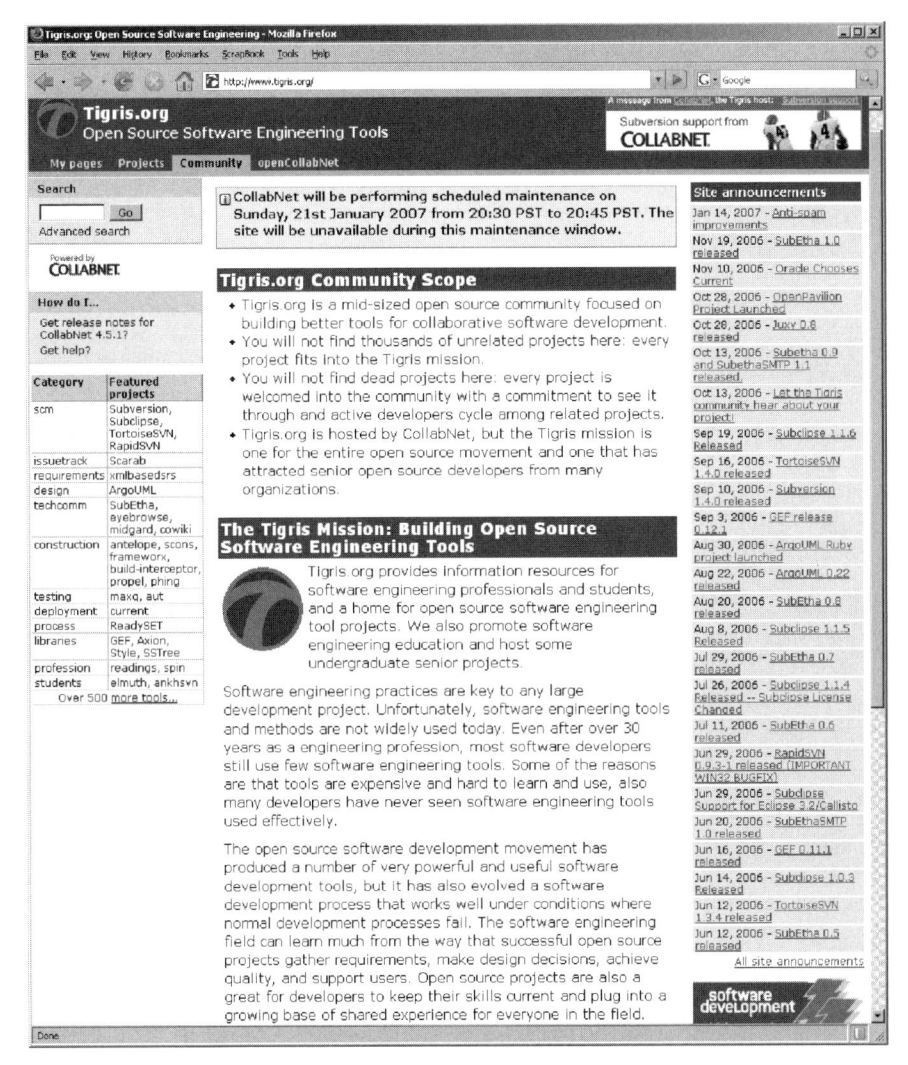

EXHIBIT 1. An example in the bottom paragraph highlighting career/skill development opportunities that encourage participation in FOSSD projects. (Source: http://www.tigris.org, June 2006.)

ones that gain the critical mass of core developers to grow more substantially and attract ever larger user-developer communities [78,103].

People who participate in FOSS projects do so within one or more roles. Classifications of the hierarchy of roles that people take and common tasks they perform

Can I join the OGRE team?
Wednesday, 01 June 2005

Probably not. The OGRE team structure reflects our emphasis on quality, design and documentation; in the OGRE project there are several distinct 'levels':

1. **Core team member**: the only people who have unlimited access to everything. This position is *by invitation only* and new appointments are very rare; to even be considered you have to have submitted several significant and high-quality patches, answered forum questions accurately, proved you have a very solid understanding of the design and principles under which OGRE is developed, and be a great team player and communicator. As such you'd have to have been an MVP for a while first.
2. **MVP**: (Most Valued Person) Experienced users and contributors who have proved their knowledge and experience time and again in the forums, on IRC, through patches, documentation and otherwise, and are an invaluable support and mentoring pool for other users. Having an MVP icon in the forum is a badge of honour and signifies that person is to be taken seriously. New MVPs are nominated by other users, and appointed by the core team. No, you can't nominate yourself - if your work speaks for itself, someone will nominate you. MVPs are moderators on the forum and get other extra permissions, but still submit patches for review.
3. **Add-on developer**: A developer on one of the **community add-on projects** - access to these is less strictly controlled and developers will be given access if the maintainer / leader of that add-on project agrees. If the project has no current maintainer you are generally free to take it over should you wish to, email a team member or ask in the forums.
4. **User**: Users have no special permissions, but are encouraged to submit patches through the patch system where they find bugs or where they would like to enhance something. Patches are reviewed by the core team before being applied.

We welcome community participation in the OGRE project within this framework, which ensures that we maximise the benefits of a distributed open source community, whilst at the same time maintaining our quality standards.

EXHIBIT 2. Joining the OGRE FOSS development team by roles/level. (Source: http://www.ogre3d.org/index.php?option=com_content&task=view&id=333&Itemid=87, June 2005.)

when participating in a FOSS project continue to appear [10,37,59,130]. Exhibit 2 from the Object-Oriented Graphics Rendering Engine (*OGRE*) project provides a textual description of the principal roles (or "levels") in that project community. Typically, it appears that people join a project and specialize in a role (or multiple roles) they find personally comfortable and intrinsically motivating [122]. In contrast to traditional software development projects, there is no explicit assignment of developers to roles, though individual FOSSD projects often post guidelines or "help wanted here" for what roles for potential contributors are in greatest need. Exhibit 3 provides an example drawn from popular FOSS mpeg-2 video player, the *VideoLAN Client* (VLC).

It is common in FOSS projects to find end-users becoming contributors or developers, and developers acting as end-users [81,83,101,121]. As most FOSS developers

VideoLAN needs your help
2006-06-19
There are many things that we would like to improve in VLC, but that we don't, because we simply don't have enough time. That's why we are currently looking for some help. We have identified several small projects that prospective developers could work on. Knowledge of C and/or C++ programming will certainly be useful, but you don't need to be an expert, nor a video expert. Existing VLC developers will be able to help you on these projects. You can find the list and some instructions on the dedicated Wiki page. Don't hesitate to join us on IRC or on the mailing-lists. We are waiting for you!

EXHIBIT 3. An example request for new FOSS developers to come forward and contribute their assistance to developing more functionality to the VLC system. (Source: http://www.videolan.org/, June 2006.)

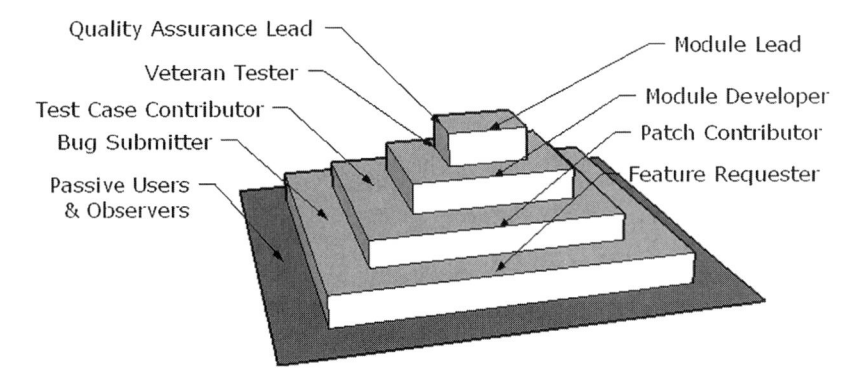

FIG. 1. A visual depiction of role hierarchies within a project community. (Source: Jensen and Scacchi [60]. Also see Kim [61].)

are themselves end-users of the software systems they build, they may have an occupational incentive and vested interest in making sure their systems are really useful. However the vast majority of participants probably simply prefer to be users of FOSS systems, unless or until their usage motivates them to act through some sort of contribution. Avid users with sufficient technical skills may actually work their way up (or "level up") through each of the roles and eventually become a core developer (or "elder"), as suggested by Fig. 1. As a consequence, participants within FOSS project often participate in different roles within both technical and social networks [75,76,96,103,113] in the course of developing, using, and evolving FOSS systems.

Making contributions is often a prerequisite for advancing technically and socially within an ongoing project, as is being recognized by other project members for having made substantive contributions [30,61]. Most commonly, FOSS project participants contribute their time, skill and effort to modify or create different types of software representations or content (source code, bug reports, design diagrams, execution scripts, code reviews, test case data, Web pages, email comments, online chat, etc.) to Web sites of the FOSS projects they join. The contribution—the authoring, hypertext linking (when needed), and posting/uploading—of different types of content helps to constitute an ecology of document genres [26,116] that is specific to a FOSS project, though individual content types are widely used across most FOSS projects. Similarly, the particular mix of online documents employed by participants on a FOSS project articulates an information infrastructure for framing and solving problems that arise in the ongoing development, deployment, use, and support of the FOSS system at the center of a project.

Administrators of FOSS project Web sites and source code repositories serve as gatekeepers in the choices they make for what information to post, when and where within the site to post it, as well as what not to post (cf. [47,51,113]). Similarly, they may choose to create a site map that constitutes a classification of site and domain content, as well as outlining community structure and boundaries.

Most frequently, participants in FOSS projects engage in online discussion forums or threaded email messages as a central way to observe, participate in, and contribute to public discussions of topics of interest to ongoing project participants [128]. However, these people also engage in private online or offline discussions that do not get posted or publicly disclosed, due to their perceived personal or sensitive content.

FOSS developers generally find the greatest benefit from participation is the opportunity to learn and share what they know about software system functionality, design, methods, tools, and practices associated with specific projects or project leaders [33,42,65]. FOSSD is a venue for learning by individuals, project groups, and organizations. Learning organizations are ones that can continuously improve or adapt their processes and practices [53,130]. However, though much of the development work in FOSSD projects is unpaid or volunteer, individual FOSS developers often benefit with higher average wages and better employment opportunities (at present), compared to their peers lacking FOSSD experience or skill [45,70].

Consequently, how and why software developers will join, participate in, and contribute to an FOSSD project seems to represent a new kind of process affecting how FOSS is developed and maintained (cf. [5,59,103,122]). Subsequently, discovering, observing, modeling, analyzing, and simulating what this process is, how it operates, and how it affects software development is an open research challenge for the software process research community.

Studies have also observed and identified the many roles that participants in an FOSSD project perform [37,60,130]. These roles are used to help explain who does what, which serves as a precursor to explanations of how FOSSD practices or processes are accomplished and hierarchically arrayed. However such a division of labor is dynamic, rather than static or fixed. This means that participants can move through different roles throughout the course of a project over time depending on their interest, commitment, and technical skill (as suggested in Fig. 1). Typically, participants start at the periphery of a project in the role of end-user by downloading and using the FOSS associated with the project. They can then move into roles like bug-reporter, code reviewer, code/patch contributor, module owner (development coordinator), and eventually to core developer or project leader. Moving through these roles requires effort, and the passage requires being recognized by other participants as a trustworthy and accomplished contributor in the progressively advancing roles.

Role-task migration can and does arise within FOSSD projects, as well as across projects [60]. Social networking, software sharing, and project internetworking enables this. But how do individual or collective processes or trajectories for role-task migration facilitate or constrain how FOSSD occurs? Role-task migration does not appear as a topic addressed in traditional SE textbooks or studies (see [112] for a notable exception), yet it seems to be a common observation in FOSSD projects. Thus, it seems that discovering, modeling, simulating or re-enacting (cf. [59]) how individual developers participate in a FOSSD effort through a role-task migration process, and how it affects or contributes to other software development or quality assurance processes, is an area requiring further investigation.

3. Resources and Capabilities Supporting FOSSD

What kinds of resources or development capabilities are needed to help make FOSS efforts more likely to succeed? Based on what has been observed and reported across many empirical studies of FOSSD projects, the following kinds of socio-technical resources enable the development of both FOSS software and ongoing project that is sustaining its evolution, application and refinement, though other kinds of resources may also be involved [101,103,105].

3.1 Personal Software Development Tools and Networking Support

FOSS developers, end-users, and other volunteers often provide their own personal computing resources in order to access or participate in a FOSS development project. They similarly provide their own access to the Internet, and may even host personal Web sites or information repositories. Furthermore, FOSS developers bring their own choice of tools and development methods to a project. Sustained commitment of personal resources helps *subsidize* the emergence and evolution of the ongoing project, its shared (public) information artifacts, and resulting open source code. It spreads the cost for creating and maintaining the information infrastructure of the virtual organization that constitute a FOSSD project [12,23,84]. These in turn help create recognizable shares of the FOSS commons (cf. [2,41,72,89]) that are linked (via hardware, software, and Web) to the project's information infrastructure.

3.2 Beliefs Supporting FOSS Development

Why do software developers and others contribute their skill, time, and effort to the development of FOSS and related information resources? Though there are probably many diverse answers to such a question, it seems that one such answer must account for the belief in the freedom to access, study, modify, redistribute and share the evolving results from a FOSS development project. Without such belief, it seems unlikely that there could be "free" and "open source" software development projects [17,16,35,36,92,127]. However, one important consideration that follows is what the consequences from such belief are, and how these consequences are realized or put into action.

In a longitudinal study of the free software project GNUenterprise.org, Elliott and Scacchi [22–24] identified many kinds of beliefs, values, and social norms that shaped actions taken and choices made in the development of the GNUe software. Primary among them were *freedom of expression* and *freedom of choice*. Neither of these freedoms is explicitly declared, assured, or protected by free software copyright or commons-based intellectual property rights, or end-user license agreements (EULAs).[2] However, they are central tenets free or open source modes of production and culture [2,41,72]. In particular, in FOSS projects like GNUenterprise.org and others, these additional freedoms are expressed in choices for what to develop or work on (e.g., choice of work subject or personal interest over work assignment), how to develop it (choice of technical method to use instead of a corporate standard), and what tools to employ (choice over which personal tools to employ versus only using what is provided by management authorities). They also are expressed in choices for when to release work products (choice of satisfaction of work quality over schedule or market imperatives), determining what to review and when (modulated by ongoing project ownership responsibility), and expressing what can be said to whom with

---

[2] EULAs associated with probably all software often seek to declare "freedom from liability" from people who want to use licensed software for intended or unintended purposes. But a belief in liability freedom is not the focus here.

or without reservation (modulated by trust and accountability mechanisms). Shared belief and practice in these freedoms of expression and choice are part of the virtual organizational culture that characterizes a FOSSD project like GNUenterprise.org [23]. Subsequently, putting these beliefs and cultural resources into action continues to build and reproduce the socio-technical interactions networks that enable sustained FOSSD projects and free software.

3.3 FOSSD Informalisms

Software informalisms [101] are the information resources and artifacts that participants use to describe, proscribe, or prescribe what's happening in a FOSSD project. They are informal narrative resources that coalesce into *online document genres* (following [64,116]) that are comparatively easy to use, and publicly accessible to those who want to join the project, or just browse around. Subsequently, Scacchi [101] demonstrates how software informalisms can take the place of formalisms like "requirement specifications" or software design notations, which traditionally are seen as necessary to develop high quality software according to the software engineering community (cf. [117]). Yet these software informalisms often capture the detailed rationale and debates for why changes were made in particular development activities, artifacts, or source code files. Nonetheless, the contents these informalisms embody require extensive review and comprehension by a developer before further contributions can be made (cf. [66]), and FOSS developers will alert one another when someone proposes a change that does not cite or acknowledge what has been previously discussed.

The most common informalisms used in FOSSD projects include (i) communications and messages within project Email, (ii) threaded message discussion forums, bulletin boards, or group blogs, (iii) news postings, (iv) project digests, and (v) instant messaging or Internet relay chat. They also include (vi) scenarios of usage as linked Web pages, (vii) how-to guides, (viii) to-do lists, (ix) FAQs, and other itemized lists, and (x) project Wikis, as well as (xi) traditional system documentation and (xii) external publications. FOSS (xiii) project copyright licenses are documents that also help to define what software or related project content are protected resources that can subsequently be shared, examined, modified, and redistributed. Finally, (xiv) open software architecture diagrams, (xv) intra-application functionality realized via scripting languages like Perl and PhP, and the ability to either (xvi) incorporate plug-in externally developer software modules, or (xvii) integrate software modules from other OSSD efforts, are all resources that are used informally, where or when needed according to the interests or actions of project participants.

All of the software informalisms are found or accessed from (xix) project related Web sites or portals. These Web environments where most FOSS software informalisms can be found, accessed, studied, modified, and redistributed [101].

A Web presence helps make visible the project's information infrastructure and the array of information resources that populate it. These include FOSSD multi-project Web sites (e.g., SourgeForge.net, Savanah.org, Freshment.org, Tigris.org, Apache.org, Mozilla.org), community software Web sites (PhP-Nuke.org), and project-specific Web sites (e.g., www.GNUenterprise.org), as well as (xx) embedded project source code Webs (directories), (xxi) project repositories (CVS [34]), and (xxii) software bug reports and (xxiii) issue tracking data base like Bugzilla (see http://www.bugzilla.org/).

Together, these two-dozen or so types of software informalisms constitute a substantial yet continually evolving web of informal, semi-structured, or processable information resources. This web results from the hyperlinking and cross-referencing that interrelate the contents of different informalisms together. Subsequently, these FOSS informalisms are produced, used, consumed, or reused within and across FOSS development projects. They also serve to act as both a distributed virtual repository of FOSS project assets, as well as the continually adapted distributed knowledge base through which project participants evolve what they know about the software systems they develop and use.

3.4 Competently Skilled, Self-organizing, and Self-managed Software Developers

Developing complex software modules for FOSS applications requires skill and expertise in a target application domain. For example, contributing to a FOSSD project like *Filezilla*[3] requires knowledge and skill in handling file transfer conditions, events, and protocols. Developing FOSS modules or applications in a way that enables an open architecture requires a base of prior experience in constructing open systems. The skilled use of project management tools for tracking and resolving open issues, and also for bug reporting and resolution, contribute to the development of a project's socio-technical system architecture. These are among the valuable professional skills that are mobilized, brought to, or drawn to FOSS development projects (cf. [12,11]). These skills are resources that FOSS developers bring to their projects.

FOSS developers organize their work as a virtual organizational form that seems to differ from what is common to in-house, centrally managed software development projects, which are commonly assumed in traditional software engineering

---

[3] See http://filezilla.sourceforge.org.

textbooks. Within in-house development projects, software application developers and end-users often are juxtaposed in opposition to one another (cf. [13,62]). Historically, Danziger [14] referred to this concentration of software development skills, and the collective ability of an in-house development organization to control or mitigate the terms and conditions of system development as a "skill bureaucracy." Such a software development skill bureaucracy would seem to be mostly concerned with rule-following and rationalized decision-making, perhaps as guided by a "software development methodology" and its corresponding computer-aided software engineering tool suite.

In the decentralized virtual organization of a large ongoing FOSSD project like the Apache.org or Mozilla.org, a "skill meritocracy" (cf. [30]) appears as an alternative to the skill bureaucracy. In such a meritocracy, there is no proprietary software development methodology or specific tool suite in use. Similarly, there are few explicit rules about what development tasks should be performed, who should perform them, when, why, or how. However, this is not to say there are no rules that serve to govern the project or collective action within it.

The rules of governance and control are informally articulated but readily recognized by project participants. These rules serve to control the rights and privileges that developers share or delegate to one another in areas such as who can commit source code to the project's shared repository for release and redistribution (cf. [34, 35]). Similarly, rules of control are expressed and incorporated into the open source code itself in terms of how, where, and when to access system-managed data via application program interfaces, end-user interfaces, or other features or depictions of overall system architecture. But these rules may and do get changed through ongoing project development.

Subsequently, FOSS project participants self-organize around the expertise, reputation, and accomplishments of core developers, secondary contributors, and tertiary reviewers and other peripheral volunteers [15,67]. This in turn serves to help create an easily assimilated basis for their collective action in developing FOSS (cf. [2, 79,86,89]). Thus, there is no assumption of a communal or egalitarian authority nor utopian spirit. Instead what can be seen is a pragmatic, continuously negotiated order that tries to minimize the time and effort expended in mitigating decision-making conflicts while encouraging cooperation through reiterated and shared beliefs, values, norms, and other mental models [23,27].

Participants nearer the core have greater control and discretionary decision-making authority, compared to those further from the core (cf. [10,15,67]). However, realizing such authority comes at the price of higher commitment of personal resources described above. Being able to make a decision stick or to convince other project participants as to the viability of a decision, advocacy position, issue or bug report, also requires time, effort, communication, and creation of project content to substantiate such an action. This authority also reflects developer experience as an interested end-user of the software modules being developed. Thus, developers possessing and exercising such skill may be intrinsically motivated to sustain the evolutionary development of their FOSS modules, so long as they are active participants in their project.

3.5 Discretionary Time and Effort of Developers

Are FOSS developers working for "free" or for advancing their career and professional development? Following the survey results of Hars and Ou [46] and others [33, 45,49,65,70], there are many personal and professional career oriented reasons for why participants will contribute their time and effort to the sometimes difficult and demanding tasks of software development. Results from case studies in free software projects like GNUenterprise.org appear consistent with these observations [22–24]. These include self-determination, peer recognition, project affiliation or identification, and self-promotion, but also belief in the inherent value of free software (cf. [17,16,35,36,92,127]).

In the practice of self-determination, no one has the administrative authority to tell a project member what to do, when, how, or why. FOSS developers can choose to work on what interests them personally. FOSS developers, in general, work on what they want, when they want. However, they remain somewhat accountable to the inquiries, reviews, and messages of others in the ongoing project, particularly with regard to software modules or functions for which they have declared their responsibility to maintain or manage as a core developer.

In the practice of peer recognition, a developer becomes recognized as an increasingly valued project contributor as a growing number of their contributions make their way into the core software modules [2,3]. In addition, nearly two-thirds of OSS developers work on 1–10 additional OSSD projects [46,78], which also reflect a growing social network of alliances across multiple FOSS development projects (cf. [82,103]). Project contributors who span multiple FOSS project communities serve as "social gateways" that increase the ongoing project's social mass ([cf. [79]), as well as affording opportunities for inter-project software composition and interoperation [58]. It also enables and empowers their recognition across multiple communities of FOSSD peers, which in turn reinforces their willingness to contribute their time and effort to FOSSD project communities.

In self-promotion, project participants communicate and share their experiences, perhaps from other application domains or work situations, about how to accomplish some task, or how to develop and advance through one's career. Being able to move from the project periphery towards the center or core of the development effort requires not only the time and effort of a contributor, but also the ability to communicate, learn from, and convince others as to the value or significance of the contributions made ([cf. [59,67]). This is necessary when a participant's contribution is being questioned in open project communications, not incorporated (or "committed") within a new build version, or rejected by vote of those already recognized as core developers (cf. [30]).

The last source of discretionary time and effort that has been reported is found in the freedoms and beliefs in FOSSD that are shared, reiterated and put into observable interactions. If a project participant fails to sustain or reiterate the freedoms and beliefs codified in the GPL, then it is likely the person's technical choice in the project may be called into question [22,23], or the person will leave the project. But understanding how these freedoms and beliefs are put into action points to another class of resources (i.e., sentimental resources) that must be mobilized and brought to bear in order to both develop FOSS systems and the global communities that surround and empower them. Social values that reinforce and sustain the ongoing project and technical norms regarding which software development tools and techniques to use (e.g., avoid the use of "non-free" software), are among the sentimental resources that are employed when participants seek to influence the choices that others in the project seek to uphold.

3.6 Trust and Social Accountability Mechanisms

Developing complex FOSS source code and applications requires trust and accountability among project participants. Though trust and accountability in a FOSSD project may be invisible resources, ongoing software and project development work occur only when these intangible resources and mechanisms for social control are present ([cf. [38,50]).

These intangible resources (or social capital) arise in many forms. They include (a) assuming ownership or responsibility of a community software module, (b) voting on the approval of individual action or contribution to ongoing project software [30], (c) shared peer reviewing [2,17,16], and (d) contributing gifts [3] that are reusable and modifiable common goods [86,41,72]. They also exist through the project's recognition of a core developer's status, reputation, and geek fame [92]. Without these attributions, developers may lack the credibility they need to bring conflicts over how best to proceed to some accommodating resolution. Finally, as a FOSSD project grows in terms of the number of contributing developers, end-users, and external sponsors, then project's socio-technical mass (i.e., web of interacting resources) becomes sufficient to insure that individual trust and accountability to the project are sustained and evolving [79].

Thus, FOSSD efforts rely on mechanisms and conditions for gentle but sufficient social control that helps constrain the overall complexity of the project. These constraints act in lieu of an explicit administrative authority or software project management regime that would schedule, budget, staff, and control the project's development trajectory with varying degrees of administrative authority and technical competence (cf. [117]).

4. Cooperation, Coordination, and Control in FOSS Projects

Getting software developers to work together, even when they desire to cooperate is not without its challenges for coordinating and controlling who does what when, and to what they do it to. Conflicts arise in both FOSSD [22,23,57] and traditional software development projects [100], and finding ways to resolve conflicts becomes part of the cost (in terms of social capital) that must be incurred by FOSS developers for development progress to occur. Minimizing the occurrence, duration, and invested effort in such conflicts quickly becomes a goal for the core developers in an FOSSD project. Similarly, finding tools and project organizational forms that minimize or mitigate recurring types of conflicts also becomes a goal for experienced core developers.

Software version control tools such as the concurrent versions system CVS—itself an FOSS system and document base [34]—have been widely adopted for use within FOSS projects (cf. [17,16,29,35,92]). Tools like CVS are being used as both (a) a centralized mechanism for coordinating and synchronizing FOSS development, as well as (b) an online venue for mediating control over what software enhancements, extensions, or architectural revisions will be checked-in and made available for check-out throughout the decentralized project as part of the publicly released version (cf. [90]).

Software version control, as part of a software configuration management activity, is a recurring situation that requires coordination but enables stabilization and synchronization of dispersed and somewhat invisible development work. This coordination is required due to the potential tension between centralized decision-making authority of a project's core developers and decentralized work activity of project contributors when two or more autonomously contributed software source code/content updates are made which overlap, conflict with one another, or generate unwanted side-effects. It is also practiced as a way to manage, track, and control both desired and undesired dependencies within the source code [15], as well as among its surrounding informalisms [101,102]. Tools like CVS thus serve to help manage or mitigate conflicts over who gets to modify what, at least as far as what changes or updates get included in the next software release from a project. However, the CVS administrator or configuration control policies provide ultimate authority and control mediated through such systems.

Each project team, or CVS repository administrator in it, must decide what can be checked in, and who will or will not be able to check-in new or modified software source code content. Sometimes these policies are made explicit through a voting scheme [30], or by reference to coding or data representation standards [55], while in others they are left informal, implicit, and subject to negotiation as needed. In either situation, version updates must be coordinated in order for a new system build and release to take place. Subsequently, those developers who want to submit updates to the project's shared repository rely extensively on online discussions that are supported using "lean media" such as threaded messages (via discussion forum, bulletin board, or similar) posted on a Web site [128], rather than through onerous system configuration control committees. Thus, software version control, system build and release are a coordination and control process mediated by the joint use of versioning, system building, and communication tools [25].

FOSSD projects teams can take the organizational form of a *layered* or *pyramid meritocracy* (cf. [30,61,102]) operating as a dynamically organized virtual enterprise [12,84]. A layered meritocracy is a hierarchical organizational form that centralizes and concentrates certain kinds of authority, trust, and respect for experience and accomplishment within the team (cf. [10]). Such an organizational form also makes administrative governance more tractable and suitable, especially when a FOSS project seeks to legally constitute a non-profit foundation to better address its legal concerns and property rights [87]. However, it does not necessarily imply the concentration of universal authority into a single individual or directorial board, since decision-making may be shared among core developers who act as peers at the top layer, and they may be arrayed into overlapping groups with other project contributors with different responsibilities and interest areas.

As seen earlier in Fig. 1, there is a layered or pyramidal form of a meritocracy common to many FOSS projects. In this form, software development work appears to be logically centralized, while being physically distributed in an autonomous and decentralized manner [84]. However, it is neither simply a "cathedral" nor a "bazaar," as these terms have been used to describe alternative ways of organizing FOSSD projects. Instead, when layered meritocracy operates as a virtual enterprise, it relies on *virtual project management* (VPM) to mobilize, coordinate, control, build, and assure the quality of FOSS development activities. It may invite or encourage system contributors to come forward and take a shared, individual responsibility that will serve to benefit the FOSS collective of user-developers. VPM requires multiple people to act in the roles of team leader, sub-system manager, or system module owner in a manner that may be short-term or long-term, based on their skill, accomplishments, availability and belief in ongoing project development. This implied

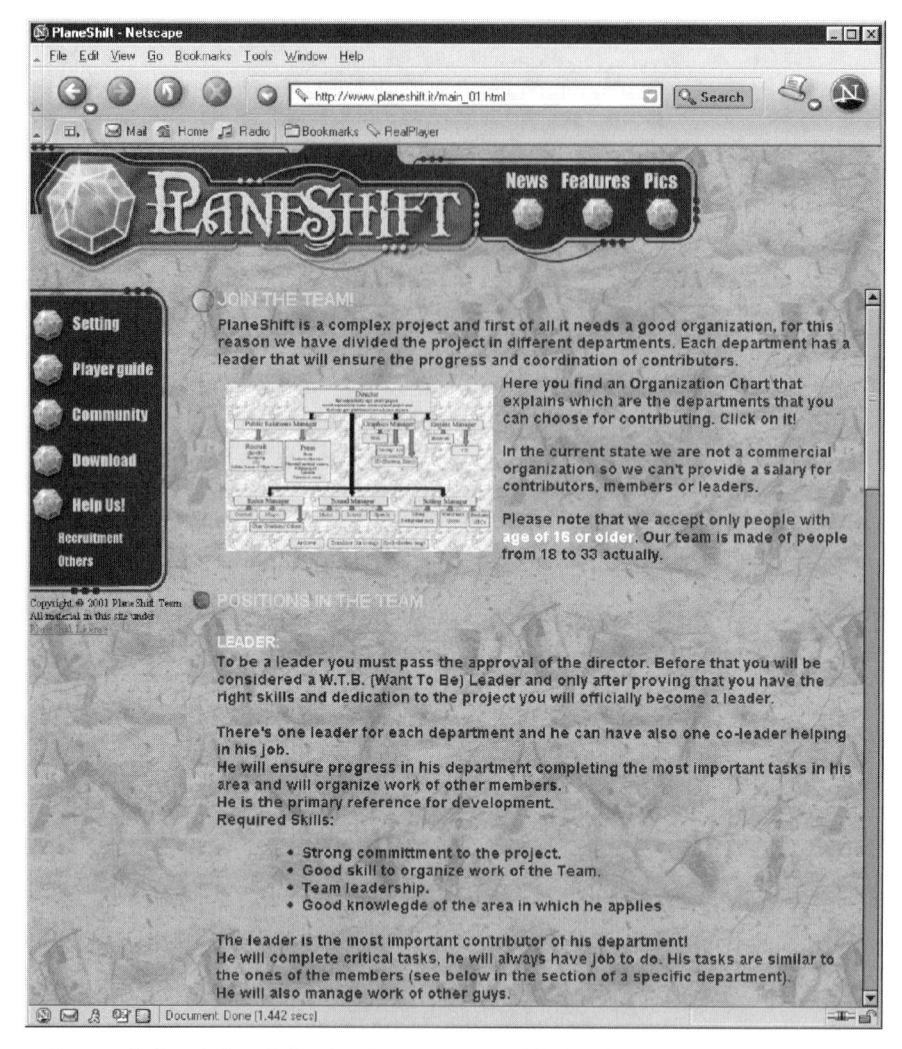

EXHIBIT 4. Description of virtual project management skills implied for a "Team Leader." (Source: http://www.planeshift.it/main_01.html, October 2003; also in Scacchi [102].)

requirement for virtual project management can be seen within Exhibit 4, from the FOSS project developing *Planeshift*, a free massively multiplayer online role-playing game.

Project participants higher up in the meritocracy have greater perceived authority than those lower down. But these relationships are only effective as long as everyone agrees to their makeup and legitimacy. Administrative or coordination conflicts that cannot be resolved may end up either by splitting or forking a new system version with the attendant need to henceforth take responsibility for maintaining that version (cf. [55]), by reducing one's stake in the ongoing project, or by simply conceding the position in conflict.

Virtual project management exists within FOSS communities to enable control via project decision-making, Web site administration, and CVS repository administration in an effective manner. Similarly, VPM exists to mobilize and sustain the use of privately owned resources (e.g., Web servers, network access, site administrator labor, skill and effort) available for shared use or collective reuse by the ongoing project. Traditional software project management stresses planning and control activities. In contrast, Lessig [71] and others [15,47,66,101] observe that source code and other online artifacts are an institutional forum for collective action [87,89] that intentionally or unintentionally realizes a mode of social control on those people who develop or use it.

In the case of FOSS development, Lessig's observation would suggest that the source code controls or constrains end-user and developer interaction, while the code in software development tools, Web sites, and project assets accessible for download controls, constrains, or facilitates developer interaction with the evolving FOSS system code. CVS is thus a tool that enables some form of social control. However, the fact that the source code to these systems is available in a free and open source manner offers the opportunity to examine, revise, and redistribute patterns of social control and interaction in ways that favor one form of project organization, system configuration control, and user-developer interaction over others.

Many FOSSD project post guidelines for appropriate and inappropriate ways of reporting and discussing bugs, unintended features, or flaws in the current FOSS system release. These guidelines are embodied in online documents/artifacts that developers choose to follow in ways that suggest these developers have elevated informalisms into community standards that act to control appropriate behavior within FOSSD projects. Exhibit 5 provides an example of such guidelines and the rules it suggests for how to best report bugs within Mozilla projects (like the Firefox Web browser or Thunderbird email client projects) when using the Bugzilla bug reporting system.

Beyond this, the ability for the eyes of many developers to review or inspect source code, system build and preliminary test results [94,95], as well as responses to bug reports, also realizes peer review and the potential for embarrassment as a form of indirect social control over the timely actions of contributing FOSS developers (cf. [92]). Thus, FOSSD allows for this dimension of VPM to be open for manipulation

Bugzilla Etiquette

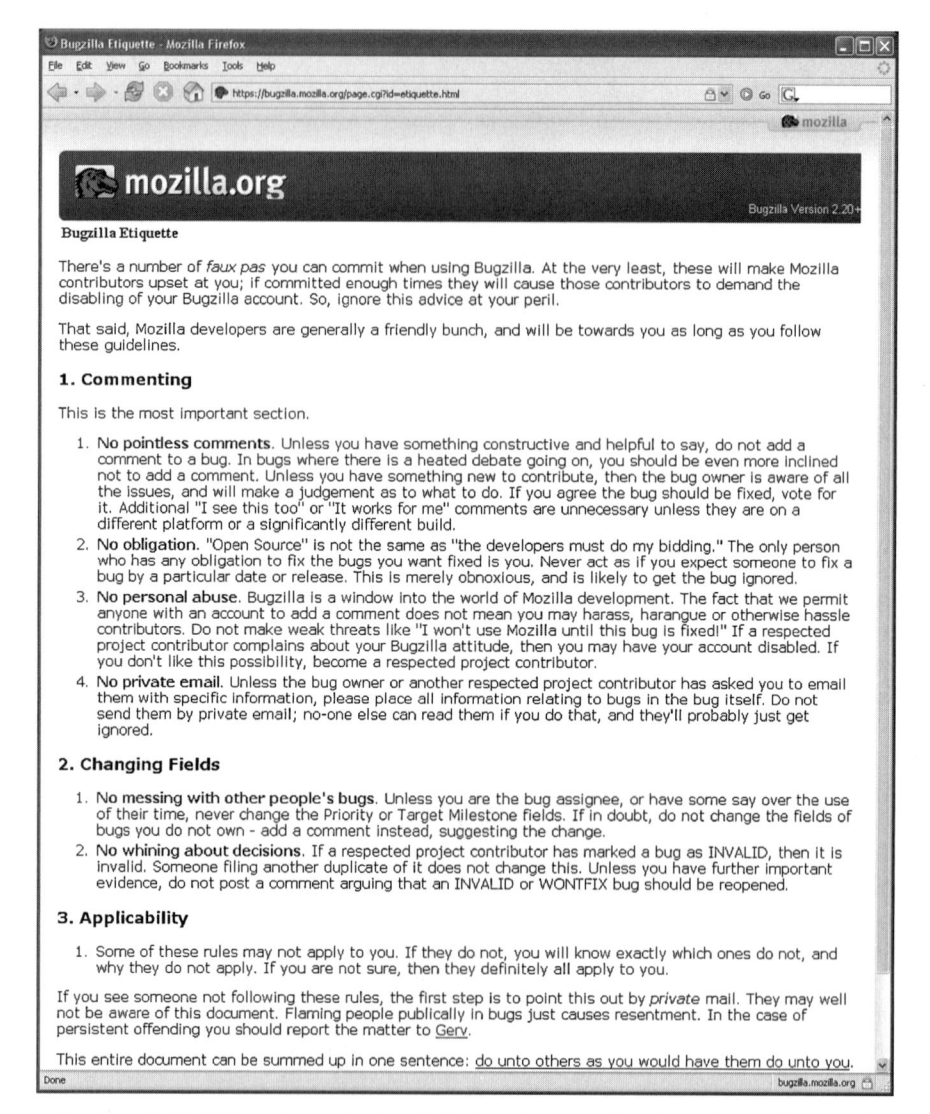

There's a number of *faux pas* you can commit when using Bugzilla. At the very least, these will make Mozilla contributors upset at you; if committed enough times they will cause those contributors to demand the disabling of your Bugzilla account. So, ignore this advice at your peril.

That said, Mozilla developers are generally a friendly bunch, and will be towards you as long as you follow these guidelines.

1. Commenting

This is the most important section.

1. **No pointless comments.** Unless you have something constructive and helpful to say, do not add a comment to a bug. In bugs where there is a heated debate going on, you should be even more inclined not to add a comment. Unless you have something new to contribute, then the bug owner is aware of all the issues, and will make a judgement as to what to do. If you agree the bug should be fixed, vote for it. Additional "I see this too" or "It works for me" comments are unnecessary unless they are on a different platform or a significantly different build.
2. **No obligation.** "Open Source" is not the same as "the developers must do my bidding." The only person who has any obligation to fix the bugs you want fixed is you. Never act as if you expect someone to fix a bug by a particular date or release. This is merely obnoxious, and is likely to get the bug ignored.
3. **No personal abuse.** Bugzilla is a window into the world of Mozilla development. The fact that we permit anyone with an account to add a comment does not mean you may harass, harangue or otherwise hassle contributors. Do not make weak threats like "I won't use Mozilla until this bug is fixed!" If a respected project contributor complains about your Bugzilla attitude, then you may have your account disabled. If you don't like this possibility, become a respected project contributor.
4. **No private email.** Unless the bug owner or another respected project contributor has asked you to email them with specific information, please place all information relating to bugs in the bug itself. Do not send them by private email; no-one else can read them if you do that, and they'll probably just get ignored.

2. Changing Fields

1. **No messing with other people's bugs.** Unless you are the bug assignee, or have some say over the use of their time, never change the Priority or Target Milestone fields. If in doubt, do not change the fields of bugs you do not own - add a comment instead, suggesting the change.
2. **No whining about decisions.** If a respected project contributor has marked a bug as INVALID, then it is invalid. Someone filing another duplicate of it does not change this. Unless you have further important evidence, do not post a comment arguing that an INVALID or WONTFIX bug should be reopened.

3. Applicability

1. Some of these rules may not apply to you. If they do not, you will know exactly which ones do not, and why they do not apply. If you are not sure, then they definitely all apply to you.

If you see someone not following these rules, the first step is to point this out by *private* mail. They may well not be aware of this document. Flaming people publically in bugs just causes resentment. In the case of persistent offending you should report the matter to Gerv.

This entire document can be summed up in one sentence: do unto others as you would have them do unto you.

EXHIBIT 5. Guidelines for appropriate behavior when reporting bugs in Mozilla.org FOSS projects when using the Bugzilla bug reporting system. (Source: https://bugzilla.mozilla.org/page.cgi?id=etiquette.html, June 2006.)

by the core developers, so as to encourage certain patterns of software development and social control, and to discourage others that may not advance the collective needs of FOSSD project participants. Subsequently, FOSSD projects are managed, coordinated and controlled, though without the roles for traditional software engineering project managers (cf. [117]).

5. Alliance Formation, Inter-project Social Networking and Community Development

How does the gathering of FOSS developers give rise to a more persistent self-sustaining organization or project community? Through choices that developers make for their participation and contribution to a FOSSD project, they find that there are like-minded individuals who also choose to participate and contribute to a project. These software developers find and connect with each other through FOSSD Web sites and online discourse (e.g., threaded discussions on bulletin boards) [82], and they find they share many technical competencies, values, and beliefs in common [12,27,23]. This manifests itself in the emergence of an alliance of FOSSD projects that share either common interests or development methods, like those for "open source software engineering" identified in the left column in Exhibit 1, in external projects that adopt a given FOSS system (e.g., OGRE) as the core system for subsequent application development as seen in Exhibit 6, or in an occupational network of FOSS developers [23].

Becoming a central node in a social network of software developers that interconnects multiple FOSS projects is also a way to accumulate social capital and recognition from peers. However, it also enables the merger of independent FOSS systems into larger composite ones that gain the critical mass of core developers to grow more substantially and attract even larger user-developer communities [77,78,103].

"Linchpin developers" [78] participate in or form gateways between multiple FOSSD projects. In so doing, they create alliances between otherwise independent FOSSD projects. Figure 2 depicts an example of a social network that clusters 24 FOSS developers within 5 FOSSD projects interconnected through two linchpin developers [78]. Multi-project clustering and interconnection enables small FOSS projects to come together as a larger social network with the critical mass [79] needed for their independent systems to be merged and experience more growth in size, functionality, and user base. It also enables shared architectural dependencies to arise (perhaps unintentionally) in the software components or sub-systems that are used/reused across projects (cf. [15,55,90]). FOSSD Web sites also serve as hubs that centralize attention for what is happening with the development of the focal FOSS

EXHIBIT 6. A partial view of an alliance of external FOSS game development projects that use the OGRE system (cf. Exhibit 2). (Source: http://www.ogre3d.org/index.php?set_albumName=album07&option=com_gallery&Itemid=55&include=view_album.php, June 2006.)

system, its status, participants and contributors, discourse on pending/future needs, etc.

Sharing beliefs, values, communications, artifacts and tools among FOSS developers enables not only cooperation, but also provides a basis for shared experience, camaraderie, and learning (cf. 27,32,65,67]). FOSS developers participate and contribute by choice, rather than by assignment, since they find that conventional software development work provides the experience of working with others who are assigned to a development effort, whether or not they find that share technical approaches, skills, competencies, beliefs or values. As a result, FOSS developers find they can choose to work with people that share their many values and beliefs in common, at least as far as software development. Further, the values and beliefs associated with free software or open source software are both signaled and institutionalized in the choice of intellectual property licenses (e.g., GPL) that FOSSD

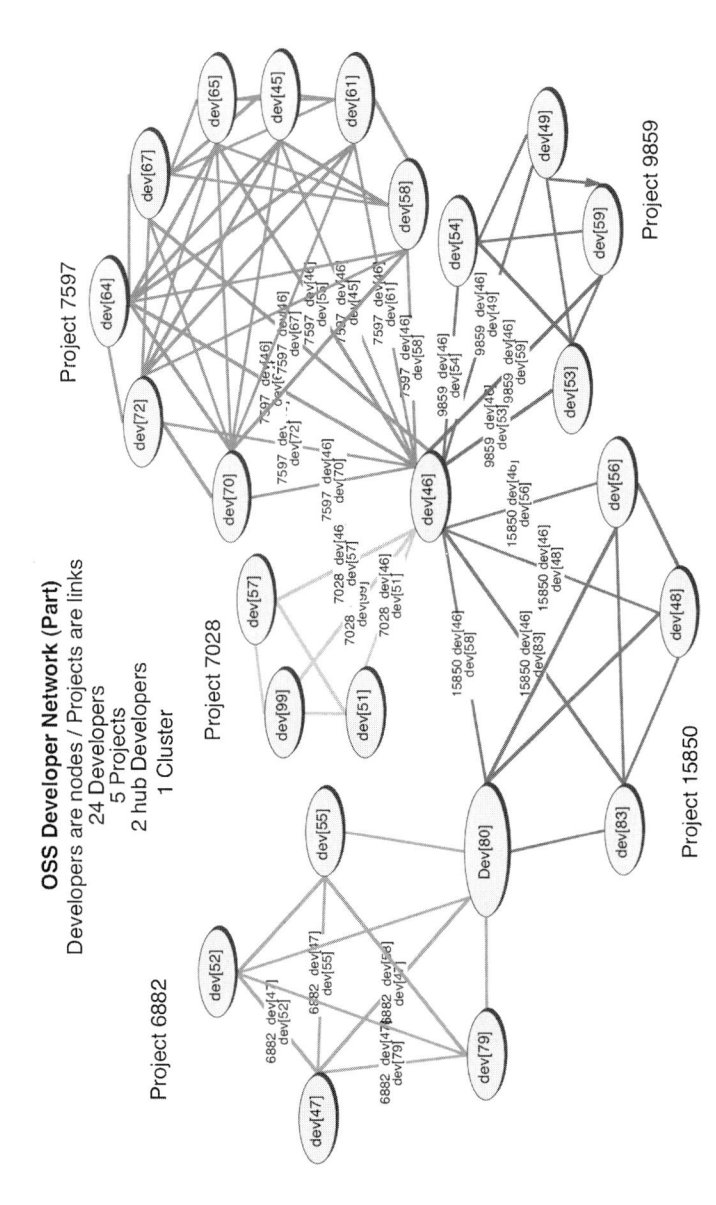

FIG. 2. A social network that clusters 24 developers in five FOSS projects through two key developers into a larger project community. (Source: Madey et al. [77]).

projects adopt and advocate. These licenses in turn help establish norms for developing free software or open source software, as well as for an alliance with other FOSSD projects that use the same licenses.

Almost half of the over 120K FOSS projects registered at SourceForce.net Web portal (as of July 2006—see Exhibit 7 later) employ the GNU General Public License (GPL) for free (as in freedom) software. The GPL seeks to preserve and reiterate the beliefs and practices of sharing, examining, modifying and redistributing FOSS systems and assets as common property rights for collective freedom [36,72,127]. A few large FOSSD projects that seek to further protect the collective free/open intellectual property rights do so through the formation of legally constituted non-profit organizations or foundations (e.g., Free Software Foundation, Apache Software Foundation, GNOME Foundation) [87]. Other OSS projects, because of the co-mingling of assets that were not created as free property, have adopted variants that relax or strengthen the rights and conditions laid out in the GPL. Dozens of these licenses now exist, with new ones continuing to appear (cf. www.opensource.org). Finally, when OSSD projects seek to engage or receive corporate sponsorship, and the possible co-mingling of corporate/proprietary intellectual property, then some variation of a non-GPL open source license is employed, as a way to signal a "business friendly" OSSD project, and thus to encourage participation by developers who want to work in such a business friendly and career enhancing project [45,111,125].

5.1 Community Development and System Development

Developing FOSS systems is a project team building process that must be institutionalized within a community [111,113,96,124,129] for its software informalisms (artifacts) and tools to flourish. Downloading, installing, and using FOSS systems acquired from other FOSS Web sites is also part of a community building process (cf. [61]), while Exhibit 6 reiterates that many external game development project use the OGRE free software. Adoption and use of FOSS project Web sites are a community wide practice for how to publicize and share FOSS project assets. These Web sites can be built using FOSS Web site content management systems (e.g., PhpNuke) to host project contents that can be served using FOSS Web servers (Apache), database systems (MySQL) or application servers (JBoss), that are increasingly accessed via FOSS Web browsers (Firefox). Furthermore, ongoing FOSS projects may employ dozens of FOSS development tools, whether as standalone systems like the software version control system CVS, as integrated development environments like NetBeans or Eclipse, or as sub-system components of their own FOSS application in development. These projects similarly employ asynchronous systems for project communications that are persistent, searchable, traceable, public and globally accessible [128].

FOSS systems, hyperlinked artifacts and tools, and project Web sites serve as online venues for socializing, building relationships and trust, sharing and learning with others. Linchpin developers [78] act as community forming hubs that enable independent small FOSS projects to come together as a larger social network with the critical mass [79] needed for their independent systems to be merged and experience more growth in size, functionality, and user base. Whether this trend is found in traditional or closed source software projects is unclear. Multi-project FOSS Web sites (e.g., Tigris.org in Exhibit 1 or SourceForge.org in Exhibit 7) also serve as hubs or "community cores" that centralize attention for what is happening with the development of focal FOSS systems, their status, participants and contributors, discourse on pending/future needs, etc. Furthermore, by their very nature, these Web sites are generally global in reach and publicly accessible. This means the potential exists for contributors to come from multiple remote sites (geographic dispersion) at different times (24/7), from multiple nations, representing the interests of participants from multiple cultures or ethnicity. Thus, multi-project FOSS Web sites help to make visible online virtual organizations, inter-project alliances, community and social networks that can share resources, artifacts, interests, and source code (cf. [32,124]).

All of these conditions point to *new kinds of requirements for software development projects*—for example, community building requirements, community software requirements, and community information sharing system (Web site and interlinked communication channels for email, forums, and chat) requirements [101,119]. These requirements may entail both functional and non-functional requirements, but they will most typically be expressed using FOSS informalisms, rather than using formal notations based on some system of mathematical logic known by few.

Community building, alliance forming, and participatory contributing are widespread and recurring activities that enable FOSSD projects to persist without central corporate authority. Thus, linking people, systems, and projects together through shared artifacts and sustained online discourse enables a sustained social network [76–78] and socio-technical community, Web-based information infrastructure [58], and network of alliances [56,82] to emerge.

Thus interesting problems arise when investigating how best to model or simulate the FOSSD processes that facilitate and constrain the co-development and co-evolution of FOSS project communities and the software systems they produce. The point is not to separate the development and evolution processes of the software system from its community, since each is co-dependent on the other, and the success of one depends on the success of the other. Thus, it appears that should best be modeled and simulated as integrating and intertwining processes.

6. FOSS as a Multi-project Software Ecosystem

As noted above, many FOSSD projects have become interdependent through the networking of software developers, development artifacts, common tools, shared Web sites, and computer-mediated communications. What can be seen to emerge from this is a kind of *multi-project software ecosystem*, whereby ongoing development and evolution of one FOSS system gives rise to propagated effects, architectural dependencies, or vulnerabilities in one or more of the projects linked to it [58]. For example, Fig. 3 depicts a software ecosystem primarily consisting of FOSS projects (each project denoted by a cloud-like shape, and the interrelationship of these project clouds denoting the ecosystem).

This particular software ecosystem highlights relationships between three large FOSS projects, the Mozilla.org Web Browser, the Apache.org Web server, and the NetBeans.org interactive development environment for Web-based Java applications. It also collectively forms a central part of the software infrastructure for the Web,[4] along with other FOSS projects that support each of these three focal projects. It further highlights examples of integration and conflict issues that have emerged as these three core Web systems as each has evolved on its own, as well as co-evolved with the others. Details on the integration and conflict issues are further described in [58].

Interdependencies are most apparent when FOSSD project share source code modules, components, or sub-systems. In such situations, the volume of source code of an individual FOSSD project may appear to grow at a super-linear or exponential rate [104,108,114] when modules, components, or sub-systems are integrated in whole into an existing FOSS system [104]. Such an outcome, which economists and political scientists refer to as a "network externality" [89], may be due to the import or integration of shared components, or the replication and tailoring of device, platform, or internationalization specific code modules. Such system growth patterns therefore seem to challenge the well-established laws of software evolution [68,69] that forecast inverse-square (i.e., sub-linear) growth for software as it evolves. Thus, software evolution in a multi-project FOSS ecosystem is a process of co-evolution of interrelated and interdependent FOSSD projects, people, artifacts, tools, code, and project-specific processes.

---

[4] Figure 3 also indicates the non-FOSS like Microsoft's Internet Explorer Web browser is a part of the software ecosystem for the Web software infrastructure. The Java Community Process (JCP) and World Wide Web Committee (W3C) respectively denote a software application coding compatibility assessment process, and a committee of diverse parties who collectively act to define Web standards for markup languages (HTML) and data communication protocols (http), which are central to the interoperation of Web browsers, Web servers, and Web applications.

FREE/OPEN SOURCE SOFTWARE DEVELOPMENT 271

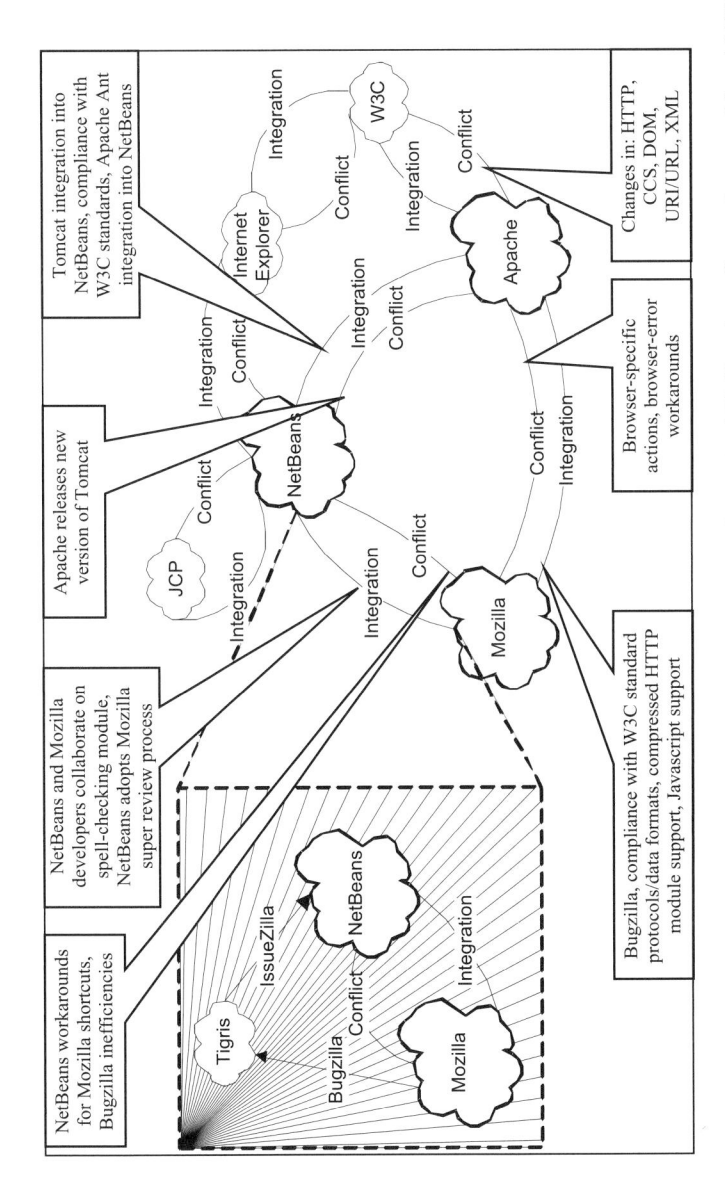

FIG. 3. Visualizing cooperative integrations and conflicts among an ecosystem of interrelated FOSS projects. (Source: Jensen and Scacchi [58].)

It seems reasonable to observe that the world FOSSD is not the only place where multi-project software ecosystems emerge, as software sharing or reuse within traditional software development enterprises is fairly common. However, the process of the co-evolution of software ecosystems found in either traditional or FOSSD projects in mostly unknown. Thus, co-evolution of interdependent software systems and standards for interoperation within an FOSS ecosystem represents an opportunity for research that investigates understanding such a software evolution process through studies supported by modeling and simulation techniques (e.g., [1,114]).

6.1 Co-evolving Socio-technical Systems for FOSS

Software maintenance, in the form of the addition/subtraction of system functionality, debugging, restructuring, tuning, conversion (e.g., internationalization), and migration across platforms, is a widespread, recurring process in FOSS development communities. Perhaps this is not surprising since maintenance is generally viewed as *the* major cost activity associated with a software system across its life cycle (cf. 117]). However, this traditional characterization of software maintenance does not do justice for what can be observed to occur within different FOSS communities. Instead, it may be better to characterize a key evolutionary dynamic of FOSS as *reinvention* (cf. [102]). Reinvention is enabled through the sharing, examination, modification, and redistribution of concepts and techniques that have appeared in closed source systems, research and textbook publications, conferences, and the interaction and discourse between developers and users across multiple FOSS projects. It is also enabled through user-contributed innovations that bring concepts or methods from other problem domains (cf. [121]). Thus, reinvention is a continually emerging source of improvement and rediscovery in FOSS functionality and quality, as well as also a collective approach to organizational learning in FOSS projects [32,65,67].

Many of the largest and most popular FOSS systems like the Linux Kernel [108], GNU/Linux distributions [55,87], GNOME user interface [39] and others are growing at an exponential rate, as is their internal architectural complexity [108]. On the other hand the vast majority of FOSS projects are small, short-lived, exhibit little/no growth, and often only involve the effort of one developer [6,78]. In this way, the overall trend derived from samples of 400–40K FOSS projects registered at the SourceForge.net Web portal reveals a power law distribution common to large self-organizing systems. This means a few large projects have a critical mass of at least 5–15 core FOSS developers [81] that act in or share project leadership roles [30] that are surrounded by dozens to hundreds of other contributors in secondary or tertiary roles, and hundreds to millions of end users in the distant periphery. The FOSS projects that attain and sustain such critical mass are those that inevitably garner the most attention, software downloads, and usage. On the other hand, the vast majority of FOSS projects are small, lacking in critical mass, and thus unlikely to thrive and grow.

The layered meritocracies that arise in FOSS projects tend to embrace incremental innovations such as evolutionary mutations to an existing software code base over radical innovations. Radical change involves the exploration or adoption of untried or sufficiently different system functionality, architecture, or development methods. A minority of code contributors who challenge the status quo of the core developers might advocate radical software system changes. However, their success in such advocacy usually implies creating and maintaining a separate version of the system, and the potential loss of a critical mass of other FOSS developers. Thus, *incremental mutation* tends to win over time (cf. [102]) and therefore represents another key dynamic mechanism affecting the evolution of FOSS.

FOSS systems seem to evolve through minor improvements or mutations that are expressed, recombined, and redistributed across many releases with short duration life cycles. End-users of FOSS systems who act as contributing developers or maintainers continually produce these mutations. These mutations appear to coalesce in daily system builds. These modifications or updates are then expressed as a tentative alpha then beta release candidates, or stable release versions that may survive redistribution and review, then subsequently be recombined and re-expressed with other new mutations in producing a new stable release version. As a result, these mutations articulate and adapt an FOSS system to what its developer-users want it to do in the course of evolving and continually reinventing the system.

Last, closed source software systems that were thought to be dead or beyond their useful product life or maintenance period may be *revitalized* through the redistribution and opening of their source code. However, this may only succeed in application domains where there is a devoted collective of enthusiastic user-developers who are willing to invest their time and skill to keep the cultural heritage of their former experience with such systems alive. Scacchi [102] provides an example for vintage arcade games now numbering in the thousands that are being revitalized, used, and evolved through FOSS systems like the Multi-Arcade Machine Emulator (*MAME*).

Overall, FOSS systems co-evolve with their development communities. This means the evolution of one depends on the evolution of the other. Said differently, a FOSS project with a small number of developers (most typically one) will not produce and sustain a viable system unless/until the team reaches a larger critical mass of 5–15 core developers. However, if and when critical mass is achieved, then it may be possible for the FOSS system to grow in size and complexity at a sustained exponential rate, defying the laws of software evolution that have held for decades [68,69,104]. Furthermore, user-developer communities co-evolve with their systems in a mutually dependent manner [23,83,87,101], and system architectures and functionality grow in discontinuous jumps as independent FOSS projects decide to join forces (e.g., [83,104]). Whether this trend is found in traditional or closed source software projects is unclear. But what these findings and trends do indicate is that it appears that the practice of FOSS development and evolution processes are different from the processes traditionally advocated for software engineering.

7. FOSS as a Social Movement

Social movements reflect sustained and recurring large-scale collective activities within a society or social world. Social movements can be characterized by (a) their recurring structural forms (e.g., boundaries around movement segments or loci of activity, multiple centers of activity, and social networks that link the segments and centers) and venues for action, (b) ideological beliefs, and (c) organizations whose purpose is to advance and mobilize broader interest in the movement [115]. The OSS movement arose in the 1990s [17,74,105,125] from the smaller, more fervent free software movement [36] started in the mid 1980s.

The OSS movement is populated with thousands of OSS development projects, each with its own Web site. Whether the OSS movement is just another computerization movement (cf. [54]), or is better recognized as a counter-movement to the proprietary or closed source world of commercial software development is unclear. For example, executives from proprietary software firms have asserted that (a) OSS is a national security threat to the US [85], or (b) that OSS (specifically that covered by the GNU Public License) is a "cancer" that attaches itself to intellectual property [44]. However, other business sources seem to clearly disagree with such characterizations and see OSS as an area for strategic investment [88], and there is growing support for recognizing that FOSS has become a matter in support of national security within the US Department of Defense [80,93]. Nonetheless, more than 120K projects are registered at OSS portals like SourceForge.org, as seen in Exhibit 7, while other OSS portals like Freshment.org, and Tigris.org contain thousands more.

The vast majority of these OSS projects at SourceForge appear to be inactive, with less than two contributing developers, as well as no software available for download, evaluation, or enhancement. However, at least a few thousand OSS projects seem to garner most of the attention and community participation, but no one project defines or leads the OSS movement. The Linux Kernel project is perhaps the most widely known FOSS project, with its celebrity leaders, like Linus Torvalds. Ironically, it is also the most studied OSS project. However, there is no basis to indicate that how things work in this project (which develops and maintains operating system kernel code for more than a dozen processor architectures or platforms, along with a

EXHIBIT 7. Home page of the SourceForge.net OSS Web portal, indicating more than 120K registered projects, and more than 1.3M registered user. (Source: http://sourceforge.net/, visited 7 June 2006.)

large base of device driver code) prescribe or predict what might be found in other successful OSS projects. Subsequently, it may be more productive to view the OSS movement as being segmented about the boundaries of each OSS project, though some of the larger project communities have emerged as a result of smaller OSS projects coming together. Finally, as already noted, a small set of studies (cf. [46,63]) indicate that upwards of 2/3 OSS developers contributes to two or more OSS projects, and perhaps as many as 5% contribute to 10 or more OSS projects. The density and interconnectedness of this social networking characterizes the membership and inbreeding of the OSS movement, but at the same time, the multiplicity of projects reflects its segmentation.

According to advocates [36], Richard M. Stallman initiated the free software movement in 1983. Participants or advocates in the free software movement identify their affiliation and commitment by openly developing and sharing their software following the digital civil liberties expressed in the GPL [21,24]. The GPL is a license agreement that promotes and protects software source code using the GPL copyright to always be available (always assuring a "copy left"), that the code is open for study, modification, and redistribution, with these rights preserved indefinitely. Furthermore, any software system that incorporates or integrates free software covered by the GPL is asserted henceforth to also be treated as free software covered by the GPL. This so-called "viral" nature of the GPL is seen by some to be an "anti-business" position, which is the most commonly cited reason for why other projects have since chose to identify them as open source software, rather than free software [31]. However, new/pre-existing software that does not integrate GPL source code is not taken over by the GPL, even if both kinds of software co-exist on the same computer or operating system, or that access one another through open or standards-based application program interfaces or some other neutral library interface.

Surveys of OSS projects reveal that about 50% or more of all OSS projects (including the Linux Kernel project) employ the GPL [33], even though there are only a few thousand of self-declared free software projects. Large OSS projects, such as the Apache Web server, KDE user interface package, Mozilla/Firefox Web browser, have chosen to not use the GPL, but to use a less restrictive, open source license. As before, free software is always open source, but open source software is not always free software. So the free software movement has emerged on its own, but increasingly it has effectively become subsumed as a segment within the larger, faster growing and faster spreading OSS movement. Subsequently, OSS licenses have become the hallmark carrier of the ideological beliefs that helps distinguish members of the free software movement, from those who share free software beliefs but prefer to be seen as open source or business-friendly developers (e.g., the Linux Kernel project). Furthermore, the use of non-GPL OSS licenses by corporate-sponsored projects (cf.

[87]) also distinguishes those who identify themselves as OSS developers, but not practitioners or affiliates of the free software movement.

A variety of organizations, enterprises, and foundations participate in encouraging the advancement and success of OSS [123]. Non-profit foundations have become one of the most prominent organizational forms founded to protect the common property rights of OSS projects. The Open Source Initiative (www.opensource.org) is one such foundation that seeks to maintain the definition of what "open source software" is, and what software licenses satisfy such a definition. OSI presents its definition of OSS in a manner that is considered business friendly [31], as opposed to "free software" which is cast by its advocates as a social movement that expresses civil liberties through software (e.g., source code as a form of free speech) [36]. The OSI's Bruce Perens who advocates that OSS is a viable economic and innovative alternative to proprietary software, often is juxtaposed or compared to Richard M. Stallman, who seeks to "put back the *free* in free enterprise" [36]. Beyond this, a sign of success of the largest OSS projects is their establishment of non-profit foundations or not-for-profit consortia that serve as the organizational locus and legal entity that can engage in contracts and intellectual property rights agreements that benefit the project and user community. A small but growing number of corporations in the IT, Financial Services, and other industries have taken on sponsorship of OSS projects, either as an external competitive strategy (e.g., IBM's Eclipse project and Sun's NetBeans project compete against Microsoft.NET products) or internal cost-reduction strategy [125].

Overall, recognizing that free software and OSS have facilitated the emergence of global-scale social (or computerization) movements, indicates that FOSS is increasingly permeating society at an industrial, governmental, and international level, and is doing so in ways that no prior software technology or development method has come close to achieving. Why this has come about, what consequences it portends for the future of FOSS, and whether corporate or public (government) policy initiatives will increasingly address the development, adoption, deployment, usage, and support of FOSS applications and projects, all require further study. But is also in clear that it is increasingly unlikely the any company, government, or nation can successfully inhibit the near-term and mid-term societal dispersion of FOSS or the FOSS movements.

8. Research Methods for Studying FOSS

Based on the survey of studies and results emerging from empirical studies of FOSSD projects, it becomes clear that there are many promising opportunities in studying, modeling, analyzing, and comparing FOSS development processes, work practices, and community dynamics, as well as project development artifacts and source code. New sources of data associated with FOSSD participants, artifacts, tools used, and development processes are available, and new systematic samples of FOSSD projects can be articulated. Empirical studies of FOSSD can therefore be examined of the research methods employed, and that is the purpose of this section.

In this chapter, different studies of FOSS development were organized and characterized according to subjects grouped into different level of analysis. Subsequently, this raises questions about what kinds of research methods have been used in these studies, or might be used in future studies of FOSS. To answer such questions, it is necessary and beneficial to review what kinds of research methods and strategies have appeared in FOSS studies, in order to identify possible categories of FOSS research methods that can be practiced by or taught to future FOSS scholars. The purpose is not to profess a treatise on how to do research or how to conduct an empirical study of FOSS, but instead to highlight which studies of FOSSD used what research methods to investigate issues at one of more levels of analysis.

As research studies of FOSS can be organized in many ways, level of analysis can be construed as a constructive element when articulating a research method. A given study may explore a single or multiple levels of analysis by research study design. Other elements in the research design include the unit of analysis, terms of analysis, and mode of analysis. The *unit of analysis* focuses on what or who is being studied, across some spatio-temporal extent within some work setting. Common foci include FOSS *developer motivations*, *project teams* or *workgroup* effort, *source code*, development or communication *artifacts*, *development processes* enacted within a project's Web Site(s) across some period of time, or a *project's trajectory* or *life history*. The choice of the unit of analysis often determines or reflects the researcher's choice for the level, terms, and mode of analysis. The *terms of analysis* refer to the choice of analytical variables and rhetorical framings that are employed to identify and describe salient features or aspects of the unit of analysis. When focusing of FOSS development processes, for example, conceptual variables like *process structure* or *process control flow* may be used to associate the partially ordered sequence of *workflow activities*, performed by participants acting in different *roles*, using *tools* to perform different activities that *access and update shared resources or artifacts*, may be used to describe observed or discovered FOSS processes. The *mode of analysis* identifies what kind of qualitative, quantitative, or triangulated schemes are employed to collect and analyze data associated with the unit of analysis.

Common FOSS research data collection and analysis modes include reflective practice and industry polls, surveys, ethnographic study, mining FOSS artifact repositories, and multi-modal modeling. As mode of analysis is core to research method, that becomes the focus here. However, as will become clear, different research methods involve trade-offs when compared to one another, so that no single research method will be best in all situations or studies.

8.1 Reflective Practice and Industry Poll Methods

FOSS research studies often focus on the interests, motivations, perceptions, and experiences of developers or end-user organizations. Typically, the unit of analysis is the *individual agent* (most commonly a person, unitary group, or firm, but sometimes a software system, tool, or artifact type) acting within a larger actor group or community. Individual behavior, personal choices, or insider views might best be analyzed, categorized, and explained in terms of volunteered or elicited statements of their interests, motivations, perceptions, or experiences. Most of the popular treatments of OSS development (e.g., [17,18,34,92]) and free software development (e.g., [35,36, 127]), provide insight and guidance for how FOSS development occurs, based on the first-hand experiences of those authors. These authors reflect on their prior experience and practice as the basis for their research findings.

Other authors informed by such practitioner studies and informal industry/government polls (like those reported in *CIO Magazine*, MITRE [80], OSBC [88], Wheeler [126], and elsewhere) seek to condense, summarize, and package the experience and wisdom of these FOSS practices into practical advice that may be offered to business executives considering when and why to adopt FOSS options [e.g., [31,43]].

As a FOSS research method, reflective practice and industry polls often tend to (a) be uncritical with respect to prior scholarship or theoretical interpretation, or (b) employ unsystematic collection of data to substantiate pithy anecdotes or proffered conclusions. Thus, by themselves such studies offer a limited basis for further research or comparative analysis. Nonetheless, they can (and often do) offer keen insights and experience reports that may help sensitize future FOSS researchers to interesting starting points or problems to further explore.

8.2 Survey Research Methods

A focus on perceptions or motivations of individual participants suggests possible attention to cognitive dimensions of FOSS development or end-user adoption. Here the quantitative survey studies of Bonaccorsi and Rossi [5], FLOSS [33,42], Hars and Ou [46], Hertel et al. [49], and Lakhani et al. [65], for example, have been key in providing broad international coverage (and descriptive statistics) of why software developers of different ages, skill bases, employment status in different countries seek to join, participate in, and help sustain FOSS development projects and their surrounding communities.

The survey research studies cited above (a) critically reflect on the data and offer alternative explanations relative to established scholarship, and (b) rely on reasonably articulated questionnaire design, survey samples, and statistical analysis to plausibly substantiate their findings and conclusions. However, these surveys typically involve hundreds of individual respondents, and thus require a significant commitment of research staff expertise, time, effort, and budget to administer the survey and process the data in the study. Furthermore, most such surveys are stand-alone studies, though Bonaccorsi and Rossi are one of the first to incorporate a comparative analysis of prior survey studies of motivations of FOSS developers and end-user firms who elect to join FOSS projects, while the Ghosh/FLOSS studies are the most international in their coverage and cross-cultural generalization of findings.

Finally, quantitative data and analyses arising from survey research of FOSS efforts are best suited for describing frequency and distribution of univariate data, as well as correlation associations among multi-variate data that characterize FOSSD. However, these data and analyses are often comparatively weak when used to characterize the structure and performance of complex socio-technical processes whose activities, participant roles, and resources are highly situated and interdependent, yet occur in relatively low frequency and evolve over time.

8.3 Ethnographically Informed Methods

While survey research methods stress collection and analysis of data that is usually easy to quantify, not all phenomena operating within or around FOSS work practices, development processes, or community dynamics are readily captured or characterized in quantitative form. Thus, qualitative research methods are needed and often better suited to such discovery-oriented or participant-observer studies of FOSS development efforts (cf. [109,120]). Central to such studies are ethnographic or ethnographically informed research methods that are intended for studies where face-to-face interviews or co-located observation are central, whereas most of the action and interactions of interest in FOSSD efforts take place online across the Internet/Web (cf. [47,51,113]) in virtual organizations represented by Web sites or portals.

Qualitative ethnographic methods are better suited to the study of the structure and performance of complex work practices, community dynamics, or socio-technical development processes whose activities and participant roles are highly situated and interdependent, yet occur in relatively low frequency and evolve over time. Here there are studies by Scacchi [101], Iannacci [56], Elliott and Scacchi [22–24], Reis and Fortes [97], Jensen and Scacchi [58–60], Lanzara and Morner [66], Longchamp [73], Duchenaunt [18], and Sack et al. [99]. A common limitation of such studies is that they tend to focus attention to a single FOSS project setting, though this is not inherent in the method. For example, Scacchi [101,102] and Jensen and Scacchi [103] examine multiple independent FOSS project settings in order to perform comparative, cross case analyses (cf. [109]). Similarly, these ethnographic studies tend to entail longitudinal data collection and devote particular attention to collection of FOSS development and communication artifacts, and thus employ methods for discourse and document genre analyses (cf. [64,116]), as well as computational or ethnographic hypermedia analyses [18,58,99,106]. As a result, (virtual) ethnographic studies are well suited to small research groups who are also equipped and competent with Web-based data mining tools for searching, crawling, indexing, coding and cross-coding textual data (cf. [109]) found in FOSSD project Web sites (e.g., development artifacts or informalisms).

8.4 Mining FOSS Artifact Repositories and Artifact Analysis Methods

Reflective practice, surveys, and ethnographic studies have been long employed in empirical studies of software development of all kinds. The world of FOSS does however provide a new opportunity for study that previously was unavailable or at least uncommon in the software research community. One such opportunity arises from the public accessibility of the source code and related development and communication artifacts associated with FOSS project Web sites or FOSS community repositories or portals like SourceForge.org and others (cf. [48]).

The accessibility of the source code and artifacts means that they can be directly subjected to various kinds of automated or semi-automated processing techniques, including text data mining, crawling and indexing, statistical analyses, and machine learning. These processing techniques give rise to not only new ways and means for analyzing large textual FOSS data sets, but also to investigate research questions or problems that heretofore could not be addressed with the established research methods for studying software development. For example, there are now studies of FOSS source code that reveal patterns of the growth and evolution of different FOSS systems over time, [6,108,91,114]. Common among the findings in these studies is growing evidence for sustained exponential growth rates for large, highly successful FOSS systems (cf. [104]), though the majority of FOSS projects fail to grow at all (cf. [77,78]). Such findings stand in contrast to the established wisdom from long-standing studies of software evolution in the world of traditional (closed-source) software, where inverse-square growth rates are more common observed (cf. [68,69]).

Other studies of FOSS repositories have focused attention to (textual) artifacts associated with different FOSS projects. For example, in a widely cited study of the development of the Apache Web server and Mozilla Web browser, Mockus, Fielding and Herbsleb [81] reported that they were able to investigate, extract, and quantify modification requests captured in change logs and bug reporting repositories associated with each of these two projects. They analyzed and compared their findings on bug frequency and severity over time identified in modification requests for the browser and server, with those found in commercial (proprietary) telecommunications systems software. Subsequently, they found these FOSS projects produce comparable or higher quality software, but without the software project management regimen used in industry.

Elsewhere, Madey et al. [77,78] and Lopez-Fernandez et al. [75,76] employ data mining techniques to extract and analyzing social network relationships between developers who communicate with each other in the course of modifying or updating FOSS project source code in stored in common transactional repositories like CVS [34]. Figure 3 from Madey and colleagues displays how a small number of FOSS developer can establish social network links through computer-mediated messaging that connect developers spanning multiple FOSS projects together. This helps create critical mass [79] that sustains their collective FOSS development efforts. However, if the linchpin developers were missing, then the multi-project cluster may dissociate or fail to link up, resulting in an insufficient collective social mass needed to go critical and enable network externalities like exponential growth of community source code. Crowston and Howison [10] similarly demonstrate how FOSS development teams often self-organize into a team hierarchy, where a small number of core developers serve as the critical center of gravity for a larger community of contributors and end-users.

Last, Ripoche and Gasser [98] demonstrate how automated mining of textual and transaction data entered into a FOSS bug tracking system (e.g., Bugzilla) can be used to extract and generate a model of the bug management process, and how it serves to help maintain and evolve the design of a FOSS system like the Mozilla Web browser.

Overall, FOSS source code and artifact repositories offer a vast array of textual and transactional data that is just beginning to be explored. For example, FOSS project meta-data is now being collected with new Web sites emerging (e.g., FLOSSmole [52] at ossmole.sourceforge.net; also see www.ohloh.net) that organize and provide access these data. This contributes to an open, shared research infrastructure for studying FOSS socio-technical characteristics, structures, and dynamics across potentially a very large sample of FOSS projects that can be analyzed quantitatively and textually. Further, as these studies employ automated tools for data collection, coding, and analysis, then these methods for mining FOSS repositories become increasingly accessible to small research groups or individual FOSS scholars. However, data in FOSS repositories like change logs [7] or modification requests associated with source code updates entered into CVS repositories [40] require careful review, cleaning, and normalization (e.g., dealing with missing or overloaded data records). Thus, mining FOSS repositories does require care and attention to both the data and their analysis, since (a) data quality problems abound which require explicit attention, (b) researchers may not have first-hand experience in using these repositories as FOSSD project participants, and (c) these repositories were not conceived or intended to be used for collecting data on FOSSD practices or processes, and thus cannot be expected to naturally meet the requirements for statistical sampling, data quality, and data analysis.

Methods for mining FOSS repositories also offer the potential for either/both in-depth (e.g., project specific) and in-breadth (scalable to large samples of projects) empirical studies of FOSS development efforts. Thus, expect to see analysis of FOSS project source or artifacts increasingly dominating large-scale quantitative studies of software development of any kind, by research groups that include experts and emerging scholars (e.g., graduate or post-doctoral students) who are motivated to develop and apply new textual data or Web mining tools/techniques to established FOSSD repositories of various kinds supporting different kinds of development activities or communities.

8.5 Multi-modal Modeling and Analysis of FOSS Socio-technical Interaction Networks

One other research method being used to study FOSS projects that is starting to gain some traction involves use of hybrid schemes involving multiple research methods. Two such efforts are those of Duchenaunt, Sack and colleagues [18,99], and Scacchi and associates [58,107]. Both of these efforts focus on collection of ethnographic data of socio-technical interaction networks or processes (cf. [18,103]) they discover in the FOSS projects identified in their studies, using virtual ethnographic techniques and computational data mining, modeling, and visualization tools. In a sense, these multi-modal research methods seek to triangulate the robust qualitative field study methods used in ethnographic studies together with techniques employing automated or semi-automated data mining and validation tools in ways that can be put into action by a small research group. However, these multi-modal methods have not yet been applied to large samples of FOSS projects, and thus it is unclear whether such methods can scale up to such challenge, or whether some other mix of research methods will be needed.

9. Discussion

One of the defining characteristics of data about the FOSSD projects is that in general is it publicly available on a global basis [48,104]. Data about FOSSD products, artifacts, and other resources is kept in repositories associated with a project's Web site. This may include the site's content management system, computer mediated communication systems (email, persistent chat facilities, and discussion forums), software versioning or configuration management systems, and networked file systems. FOSSD process data is generally either extractable or derivable from data/content in these artifact repositories. First-person data may also be available to those who participate in a project, even if just to remotely observe (e.g., through "lurking" [20]) or to electronically interview other participants about development activities, tools being used, the status of certain artifacts, and the like. The availability of such data perhaps suggest that a growing share of empirical software engineering research will be performed in the domain of FOSSD projects, rather than using traditional sources of data from in-house or proprietary software development projects. These traditional non-FOSS projects will continue to have constraints on access and disclosure via publication. FOSSD process data collection from publicly accessible artifact repositories may also be found to be more cost-effective compared to studies of traditional closed-source, proprietary, and in-house software development repositories (cf. [9]).

9.1 Limitations and Constraints for FOSS Research

FOSSD is certainly not a panacea for developing complex software systems, nor is it simply software engineering done poorly. Instead, it represents an alternative community-intensive socio-technical approach to develop software systems, artifacts, and social relationships. However, it is not without its limitations and constraints. Thus, we should be able to help see these limits as manifest within the level of analysis or research for empirical FOSSD studies examined above.

First, in terms of participating, joining, and contributing to FOSS projects, an individual developer's interest, motivation, and commitment to a project and its contributors is dynamic and not indefinite. FOSS developers are loathe to find themselves contributing to a project that is realizing commercial or financial benefits that are not available to all contributors, or that are concentrated to benefit a particular company, again without some share going to the contributors. Some form of reciprocity seems necessary to sustain participation, whereas a perception of exploitation by others can quickly dissolve a participant's commitment to further contribute, or worse to dissuade other participants to abandon an open source project that has gone astray. If linchpin developers lose interest, then unless another contributor comes forward to fill in or take over role and responsibility for the communication and coordination activities of such key developers, then the FOSS system may quickly become brittle, fragile, and difficult to maintain. Thus, participation, joining, and contributing must become sustained activities on an ongoing basis within FOSS projects for them to succeed.

Second, in terms of cooperation, coordination, and control, FOSS projects do not escape conflicts in technical decision-making, or in choices of who gets to work on what, or who gets to modify and update what. As FOSS projects generally lack traditional project managers, then they must become self-reliant in their ability to mitigate and resolve outstanding conflicts and disagreements. Beliefs and values that shape system design choices, as well as choices over which software tools to use, and which software artifacts to produce or use, are determined through negotiation rather than administrative assignment. Negotiation and conflict management then become part of the cost that FOSS developers must bear in order for them to have their beliefs and values fulfilled. It is also part of the cost they bear in convincing and negotiating with others often through electronic communications to adopt their beliefs and values. Time, effort, and attention spent in negotiation and conflict management are not spent building and improving source code, but they do represent an investment in building and sustaining a negotiated socio-technical network of dependencies.

Third, in terms of forming alliances and building community through participation, artifacts, and tools points to a growing dependence on other FOSS projects. The emergence of non-profit foundations that were established to protect the property rights of large multi-component FOSS project creates a demand to sustain and protect such foundations. If a foundation becomes too bureaucratic, then this may drive contributors away from a project. So, these foundations need to stay lean, and not become a source of bureaucratic occupational careers, in order to survive and evolve. Similarly, as FOSS projects give rise to new types of requirements for community building, community software, and community information sharing systems, these requirements need to be addressed and managed by FOSS project contributors in roles above and beyond those involved in enhancing the source code of a FOSS project. FOSS alliances and communities depend on a rich and growing web of socio-technical relations. Thus, if such a web begins to come apart, or if the new requirements cannot be embraced and satisfied, then the FOSS project community and its alliances will begin to come apart.

Fourth, in terms of the co-evolution of FOSS systems and community, as already noted, individual and shared resources of people's time, effort, attention, skill, sentiment (beliefs and values), and computing resources are part of the socio-technical web of FOSS. Reinventing existing software systems as FOSS coincides with the emergence or reinvention of a community that seeks to make such system reinvention occur. FOSS systems are common pool resources [89] that require collective action for their development, mobilization, use, and evolution. Without the collective action of the FOSS project community, the common pool will dry up, and without the common pool, the community begins to fragment and disappear, perhaps to search for another pool elsewhere.

Last, empirical studies of FOSSD are expanding the scope of what we can observer, discover, analyze, or learn about how large software systems can be or have been developed. In addition to traditional methods used to investigate FOSSD like reflective practice, industry polls, survey research, and ethnographic studies, comparatively new techniques for mining software repositories and multi-modal modeling and analysis of the socio-technical processes and networks found in sustained FOSSD projects show that the empirical study of FOSSD is growing and expanding. This in turn will contribute to and help advance the empirical science in fields like software engineering, which previously were limited by restricted access to data characterizing large, proprietary software development projects. Thus, the future of empirical studies of software development practices, processes, and projects will increasingly be cast as studies of FOSSD efforts.

10. Conclusions

Free and open source software development is emerging as an alternative approach for how to develop large software systems. FOSSD employs new types and new kinds of socio-technical work practices, development processes, and community networking when compared to those found in industrial software projects, and those portrayed in software engineering textbooks [117]. As a result, FOSSD offer new types and new kinds of practices, processes, and organizational forms to discover, observe, analyze, model, and simulate. Similarly, understanding how FOSSD practices, processes, and projects are similar to or different from traditional software engineering counterparts is an area ripe for further research and comparative study. Many new research opportunities exist in the empirical examination, modeling, and simulation of FOSSD activities, efforts, and communities.

FOSSD project source code, artifacts, and online repositories represent and offer new publicly available data sources of a size, diversity, and complexity not previously available for research, on a global basis. For example, software process modeling and simulation research and application has traditionally relied on an empirical basis in real-world processes for analysis and validation. However, such data has often been scarce, costly to acquire, and is often not available for sharing or independent re-analysis for reasons including confidentiality or non-disclosure agreements. FOSSD projects and project artifact repositories contain process data and product artifacts that can be collected, analyzed, shared, and be re-analyzed in a free and open source manner. FOSS thus poses the opportunity to favorably alter the costs and constraints of accessing, analyzing, and sharing software process and product data, metrics, and data collection instruments. FOSSD is thus poised to alter the calculus of empirical software engineering [9,48,104]. Software process discovery, modeling, and simulation research (e.g., [59]) is one area that can take advantage of such a historically new opportunity. Another would be examining the effectiveness and efficiency of traditional face-to-face-to-artifact software engineering approaches or processes for software inspections (e.g., [19,110]) compared to the online peer reviews prevalent in FOSSD efforts.

Last, through a survey of empirical studies of FOSSD projects and other analyses presented in this article, it should be clear there are an exciting variety and diversity of opportunities for new research into software development processes, work practices, project/community dynamics, and related socio-technical interaction networks. Thus, you are encouraged to consider how your efforts to research or apply FOSSD concepts, techniques, or tools can be advanced through studies that examine FOSSD activities, artifacts, and projects. Furthermore, it may also be stimulating to the larger community of software researchers to engage in free or open source research practices whereby the source data and analyses are made available for access, study, modification or extension, and redistribution (along with citation to prior results of others), together with the research publications that have helped advance collective knowledge through open science.

ACKNOWLEDGEMENTS

The research described in this chapter has been supported by grants #0083075, #0205679, #0205724, #0350754, and #0534771 from the US National Science Foundation. No endorsement implied. Mark Ackerman at University of Michigan, Ann Arbor; Les Gasser at University of Illinois, Urbana-Champaign; John Noll at Santa Clara University; Margaret Elliott, Chris Jensen, and others at the UCI Institute for Software Research are collaborators on the research described here.

REFERENCES

[1] Antoniades I.P., Samoladas I., Stamelos I., Angelis L., Bleris G.L., "Dynamic simulation models of the open source development process", in: Koch S. (Ed.), *Free/Open Source Software Development*, Idea Group Publishing, Hershey, PA, 2005, pp. 174–202.

[2] Benkler Y., *The Wealth of Networks: How Social Production Transforms Markets and Freedom*, Yale University Press, New Haven, CT, 2006.

[3] Bergquist M., Ljungberg J., "The power of gifts: Organizing social relationships in open source communities", *Info. Systems J.* **11** (2001) 305–320.
[4] Beyer H., Holtzblatt K., *Contextual Design: A Customer-Centered Approach to Systems Designs*, Morgan Kaufmann Publishers, San Francisco, CA, 1997.
[5] Bonaccorsi A., Rossi C., "Comparing motivations of individual programmers and firms to take part in the open source movement: From community to business", *Knowledge Technology & Policy* **18** (4) (Winter 2006) 40–64.
[6] Capiluppi A., Lago P., Morisio M., "Evidences in the evolution of OS projects through changelog analyses", in: *Proc. 3rd Workshop on Open Source Software Engineering*, Portland, OR, 2003.
[7] Chen K., Schach S.R., Yu L., Offutt J., Heller G., "Open source change logs", *Empirical Software Engineering* **9** (2) (2004) 197–210.
[8] Ciborra C., *The Labyrinths of Information: Challenging the Wisdom of Systems*, Oxford University Press, Oxford, UK, 2004.
[9] Cook J.E., Votta L.G., Wolf A.L., "Cost-effective analysis of in-place software processes", *IEEE Trans. Software Engineering* **24** (8) (1998) 650–663.
[10] Crowston K., Howison J., "Hierarchy and centralization in free and open source software team communications", *Knowledge Technology & Policy* **18** (4) (Winter 2006) 65–85.
[11] Crowston K., Howison J., Annabi H., "Information systems success in free and open source software development: Theory and measures", *Software Process—Improvement and Practice* **11** (2) (2006) 123–148.
[12] Crowston K., Scozzi B., "Open source software projects as virtual organizations: Competency rallying for software development", *IEE Proceedings—Software* **149** (1) (2002) 3–17.
[13] Curtis B., Krasner H., Iscoe N., "A field study of the software design process for large systems", *Communications ACM* **31** (11) (1988) 1268–1287.
[14] Danziger J., "The skill bureaucracy and intraorganizational control: The case of the data-processing unit", *Sociology of Work and Occupations* **21** (3) (1979) 206–218.
[15] De Souza C.R.B., Froehlich J., Dourish P., "Seeking the source: Software source code as a social and technical artifact", in: *Proc. ACM Intern. Conf. Supporting Group Work (GROUP 2005)*, Sanibel Island, FL, 2005, pp. 197–206.
[16] DiBona C., Cooper D., Stone M., *Open Sources 2.0*, O'Reilly Media, Sebastopol, CA, 2005.
[17] DiBona C., Ockman S., Stone M., *Open Sources: Voices from the Open Source Revolution*, O'Reilly Media, Sebastopol, CA, 1999.
[18] Ducheneaut N., "Socialization in an open source software community: A sociotechnical analysis", *Computer Supported Cooperative Work* **14** (4) (2005) 323–368.
[19] Ebenau R.G., Strauss S.H., *Software Inspection Process*, McGraw-Hill, New York, 1994.
[20] Ebner M., Holzinger A., Catarci T., "Lurking: An underestimated human–computer phenomenon", *IEEE Multimedia* **12** (4) (October–December 2005) 70–75.
[21] Elliott M.S., "Examining the success of computerization movements in the ubiquitous computing era: Free and open source software movements", in: Kraemer K.L., Elliott

M. (Eds.), *Computerization Movements and Technology Diffusion: From Mainframes to Ubiquitous Computing*, Information Today, Inc., 2007.
[22] Elliott M., Scacchi W., "Free software developers as an occupational community: Resolving conflicts and fostering collaboration", in: *Proc. ACM Intern. Conf. Supporting Group Work*, Sanibel Island, FL, November 2003, pp. 21–30.
[23] Elliott M., Scacchi W., "Free software development: Cooperation and conflict in a virtual organizational culture", in: Koch S. (Ed.), *Free/Open Source Software Development*, Idea Group Publishing, Hershey, PA, 2005, pp. 152–172.
[24] Elliott M., Scacchi W., "Mobilization of software developers: The free software movement", 2006, submitted for publication.
[25] Erenkrantz J., "Release management within open source projects", in: *Proc. 3rd Workshop on Open Source Software Engineering*, 25th Intern. Conf. Software Engineering, Portland, OR, May 2003.
[26] Erickson T., "Making sense of computer-mediated communication (CMC): CMC systems as genre ecologies", in: *Proc. 33rd Hawaii Intern. Conf. Systems Sciences*, IEEE Press, January 2000, pp. 1–10.
[27] Espinosa J.A., Kraut R.E., Slaughter S.A., Lerch J.F., Herbsleb J.D., Mockus A., "Shared mental models, familiarity, and coordination: A multi-method study of distributed software teams", in: *Intern. Conf. Information Systems*, Barcelona, Spain, December 2002, pp. 425–433.
[28] Feller J., Fitzgerald B., *Understanding Open Source Software Development*, Addison–Wesley, NY, 2002.
[29] Feller J., Fitzgerald B., Hissam S., Lakhani K. (Eds.), *Perspectives on Free and Open Source Software*, MIT Press, Cambridge, MA, 2005.
[30] Fielding R.T., "Shared leadership in the Apache project", *Communications ACM* **42** (4) (1999) 42–43.
[31] Fink M., *The Business and Economics of Linux and Open Source*, Prentice Hall PTR, Upper Saddle, NJ, 2003.
[32] Fischer G., "External and shareable artifacts as opportunities for social creativity in communities of interest", in: Gero J.S., Maher M.L. (Eds.), *Proc. Computational and Cognitive Models of Creative Design*, Heron Island, Australia, December 2001, pp. 67–89.
[33] FLOSS, "Free/libre and open source software: Survey and study", FLOSS Final Report, 2002; http://www.flossproject.org/report/ (accessed July 2006).
[34] Fogel K., *Open Source Development with CVS*, Coriolis Press, Scottsdale, AZ, 1999.
[35] Fogel K., *Producing Open Source Software: How to Run a Successful Free Software Project*, O'Reilly Press, Sebastopol, CA, 2005.
[36] Gay J. (Ed.), *Free Software Free Society: Selected Essays of Richard M. Stallman*, GNU Press, Free Software Foundation, Boston, MA, 2002.
[37] Gacek C., Arief B., "The many meanings of open source", *IEEE Software* **21** (1) (January/February 2004) 34–40.
[38] Gallivan M., "Striking a balance between trust and control in a virtual organization: A content analysis of open source software case studies", *Information Systems J.* **11** (4) (2001) 277–304.

[39] German D., "The GNOME project: A case study of open source, global software development", *Software Process—Improvement and Practice* **8** (4) (2003) 201–215.
[40] German D., "An empirical study of fine-grained software modifications", *Empirical Software Engineering* **11** (3) (2006) 369–393.
[41] Ghosh R. (Ed.), *CODE: Collaborative Ownership and the Digital Economy*, MIT Press, Cambridge, MA, 2005.
[42] Ghosh R., Prakash V.V., "The orbiten free software survey", *First Monday* **5** (7) (July 2000), http://www.firstmonday.org/issues/issue5_7/ghosh/index.html, accessed 1 June 2006.
[43] Goldman R., Gabriel R.P., *Innovation Happens Elsewhere: Open Source as Business Strategy*, Morgan Kaufmann Publishers, San Francisco, CA, 2005.
[44] Greene T.C., "Ballmer: "Linux is a cancer"", The Register, http://www.theregister.co.uk/2001/06/02/ballmer_linux_is_a_cancer/, 2 June 2001.
[45] Hann I.-H., Roberts J., Slaughter S., Fielding R., "Economic incentives for participating in open source software projects", in: *Proc. Twenty-Third Intern. Conf. Information Systems*, December 2002, pp. 365–372.
[46] Hars A., Ou S., "Working for free? Motivations for participating in open source projects", *Intern. J. Electronic Commerce* **6** (3) (2002).
[47] Hakken D., *Cyborgs@Cyberspace? An Ethnographer Looks at the Future*, Routledge, London, 1999.
[48] Harrison W., "Editorial: Open source and empirical software engineering", *Empirical Software Engineering* **6** (2) (2001) 193–194.
[49] Hertel G., Neidner S., Hermann S., "Motivation of software developers in Open Source projects: An Internet-based survey of contributors to the Linux kernel", *Research Policy* **32** (7) (July 2003) 1159–1177.
[50] Hertzum M., "The importance of trust in software engineers' assessment and choice of information sources", *Information and Organization* **12** (1) (2002) 1–18.
[51] Hine C.M., *Virtual Ethnography*, Sage Publications, Newbury Park, CA, 2000.
[52] Howison J., Conklin M., Crowston K., "FLOSSmole: A collaborative repository for FLOSS research data and analyses", *Intern. J. Info. Tech. and Web Engineering* **1** (3) (2006) 17–26.
[53] Huntley C.L., "Organizational learning in open-source software projects: An analysis of debugging data", *IEEE Trans. Engineering Management* **50** (4) (2003) 485–493.
[54] Iacono C.S., Kling R., "Computerization movements: The rise of the Internet and distant forms of work", in: Yates J.A., Van Maanen J. (Eds.), *Information Technology and Organizational Transformation: History, Rhetoric, and Practice*, Sage Publications, Newbury Park, CA, 2001.
[55] Iannacci F., "Coordination processes in open source software development: The Linux case study", *Emergence: Complexity & Organization (E:CO)* **7** (2) (2005) 21–31.
[56] Iannacci F., "Beyond markets and firms: The emergence of open source networks", *First Monday* **10** (5) (2005).
[57] Jensen C., Scacchi W., "Collaboration, leadership, and conflict negotiation in the NetBeans.org community", in: *Proc. 4th Workshop on Open Source Software Engineering*, Edinburgh, UK, May 2004.

[58] Jensen C., Scacchi W., "Process modeling across the Web information infrastructure", *Software Process—Improvement and Practice* **10** (3) (July–September 2005) 255–272.
[59] Jensen C., Scacchi W., "Discovering, modeling, and reenacting open source software development processes", in: Acuna S.T., Sanchez-Segura M.I. (Eds.), *New Trends in Software Process Modeling*, in: *Series in Software Engineering and Knowledge Engineering*, vol. 18, World Scientific Publishing, Singapore, 2006, pp. 1–20.
[60] Jensen C., Scacchi W., "Role migration and advancement processes in OSSD projects: A comparative case study", in: *Proc. 29th Intern. Conf. Software Engineering, Minneapolis, MN*, May 2007.
[61] Kim A.J., *Community-Building on the Web: Secret Strategies for Successful Online Communities*, Peachpit Press, 2000.
[62] Kling R., Scacchi W., "The Web of computing: Computer technology as social organization", in: Yovits M.C. (Ed.), *Advances in Computers* **21** (1982) 1–90.
[63] Koch S. (Ed.), *Free/Open Source Software Development*, Idea Group Publishing, Hershey, PA, 2005.
[64] Kwansik B., Crowston K., "Introduction to the special issue: Genres of digital documents", *Information, Technology and People* **18** (2) (2005).
[65] Lakhani K.R., Wolf B., Bates J., DiBona C., "The Boston consulting group hacker survey", July 2002; http://www.bcg.com/opensource/BCGHackerSurvey OSCON24July02v073.pdf.
[66] Lanzara G.F., Morner M., "Artifacts rule! How organizing happens in open source software projects", in: Czarniawska B., Hernes T. (Eds.), *Actor–Network Theory and Organizing*, Liber & Copenhagen Business School Press, Malmo, Sweden, 2005, pp. 67–90.
[67] Lave J., Wenger E., *Situated Learning: Legitimate Peripheral Participation*, Cambridge University Press, Cambridge, UK, 1991.
[68] Lehman M.M., "Programs, life cycles, and laws of software evolution", *Proc. IEEE* **68** (1980) 1060–1078.
[69] Lehman M.M., "Software evolution", in: Marciniak J. (Ed.), *Encyclopedia of Software Engineering*, second ed., John Wiley and Sons, Inc., New York, 2002, pp. 1507–1513; also see: "Software evolution and software evolution processes", *Annals of Software Engineering* **12** (2002) 275–309.
[70] Lerner J., Tirole J., "Some simple economics of open source", *J. Industrial Economics* **50** (2) (2002) 197–234.
[71] Lessig L., *Code and Other Laws of Cyberspace*, Basic Books, New York, 2000.
[72] Lessig L., *Free Culture: The Nature and Future of Creativity*, Penguin, New York, 2005.
[73] Longman J., "Open source software development process modeling", in: Acuña S.T., Juristo N. (Eds.), *Software Process Modeling*, Springer Science+Business Media Inc., New York, 2005, pp. 29–64.
[74] Ljungberg J., "Open source movements as a model for organizing", *European J. Info. Sys.* **9** (4) (2000) 208–216.
[75] Lopez-Fernandez L., Robles G., Gonzalez-Barahona J.M., "Applying social network analysis to the information in CVS repositories", in: *Proc. First Intern. Workshop on Mining Software Repositories*, Edinburgh, UK, May 2004, pp. 101–105.

[76] Lopez-Fernandez L., Robles G., Gonzalez-Barahona J.M., Herraiz I., "Applying social network analysis to community-drive libre software projects", *Intern. J. Info. Tech. and Web Engineering* **1** (3) (2006) 27–28.

[77] Madey G., Freeh V., Tynan R., "The open source development phenomenon: An analysis based on social network theory", in: *Proc. Americas Conf. Info. Systems (AMCIS2002)*, Dallas, TX, 2002, pp. 1806–1813.

[78] Madey G., Freeh V., Tynan R., "Modeling the F/OSS community: A quantitative investigation", in: Koch S. (Ed.), *Free/Open Source Software Development*, Idea Group Publishing, Hershey, PA, 2005, pp. 203–221.

[79] Marwell G., Oliver P., *The Critical Mass in Collective Action: A Micro-Social Theory*, Cambridge University Press, Cambridge, England, 1993.

[80] MITRE Corporation, "Use of free and open-source software (FOSS) in the U.S. Department of Defense", http://www.egovos.org/pdf/dodfoss.pdf, January 2003.

[81] Mockus A., Fielding R., Herbsleb J.D., "Two case studies of open source software development: Apache and Mozilla", *ACM Transactions on Software Engineering and Methodology* **11** (3) (2002) 309–346.

[82] Monge P.R., Fulk J., Kalman M.E., Flanagin A.J., Parnassa C., Rumsey S., "Production of collective action in alliance-based interorganizational communication and information systems", *Organization Science* **9** (3) (1998) 411–433.

[83] Nakakoji K., Yamamoto Y., Nishinaka Y., Kishida K., Ye Y., "Evolution patterns of open-source software systems and communities", in: *Proc. 2002 Intern. Workshop Principles of Software Evolution*, 2002, pp. 76–85.

[84] Noll J., Scacchi W., "Supporting software development in virtual enterprises", *J. Digital Information* **1** (4) (February 1999), http://jodi.tamu.edu/Articles/v01/i04/Noll/.

[85] O'Dowd D., "No defense for Linux: Inadequate security poses national security threat", *Design News* **19** (July 2004), http://www.designnews.com/article/CA435615.html.

[86] Olson M., *The Logic of Collective Action*, Harvard University Press, Cambridge, MA, 1971.

[87] O'Mahony S., "Guarding the commons: How community managed software projects protect their work", *Research Policy* **32** (7) (July 2003) 1179–1198.

[88] OSBC, "Open source business conference", http://www.osbc.com, 2006 (accessed 15 July 2006).

[89] Ostrom E., Calvert R., Eggertsson T. (Eds.), *Governing the Commons: The Evolution of Institutions for Collective Action*, Cambridge University Press, Cambridge, England, 1990.

[90] Ovaska P., Rossi M., Marttiin P., "Architecture as a coordination tool in multi-site software development", *Software Process—Improvement and Practice* **8** (3) (2003) 233–247.

[91] Paulson J.W., Succi G., Eberlein A., "An empirical study of open-source and closed-source software products", *IEEE Trans. Software Engineering* **30** (4) (April 2004) 246–256.

[92] Pavelicek R., *Embracing Insanity: Open Source Software Development*, SAMS Publishing, Indianapolis, IN, 2000.

[93] Payton S., Herz J.C., Lucas M., Scott J., "Open technology development: Roadmap plan", Final Report, Advanced Systems & Concepts, Deputy Undersecretary of Defense, http://www.acq.osd.mil/asc, April 2006 (accessed 15 August 2006).
[94] Porter A.A., Siy H.P., Toman C.A., Votta L.G., "An experiment to assess the cost-benefits of code inspections in large scale software development", *IEEE Trans. on Software Engineering* **23** (1997) 329–346.
[95] Porter A.A., Yilmaz C., Memon A.M., Krishna A.S., Schmidt D.C., Gokhale A., "Techniques and processes for improving the quality and performance of open-source software", *Software Process—Improvement and Practice* **11** (2) (2006) 163–176.
[96] Preece J., *Online Communities: Designing Usability, Supporting Sociability*, John Wiley & Sons, Chichester, UK, New York, 2000.
[97] Reis C.R., Fortes R.P.M., "An overview of the software engineering process and tools in the Mozilla project", in: *Proc. Workshop on Open Source Software Development*, Newcastle, UK, February 2002.
[98] Ripoche G., Gasser L., "Scalable automatic extraction of process models for understanding F/OSS bug repair", in: *Proc. 16th Intern. Conf. Software Engineering & Its Applications (ICSSEA-03)*, Paris, France, December 2003.
[99] Sack W., Detienne F., Ducheneaut D., Mahendran B., Barcellini F., "A methodological framework for socio-cognitive analyses of collaborative design of open source software", *Computer Supported Cooperative Work* **15** (2–3) (June 2006) 229–250.
[100] Sawyer S., "Effects of intra-group conflict on packaged software development team performance", *Information Systems J.* **11** (2001) 155–178.
[101] Scacchi W., "Understanding the requirements for developing open source software systems", *IEE Proceedings—Software* **149** (1) (February 2002) 24–39.
[102] Scacchi W., "Free/open source software development practices in the computer game community", *IEEE Software* **21** (1) (January/February 2004) 59–67.
[103] Scacchi W., "Socio-technical interaction networks in free/open source software development processes", in: Acuña S.T., Juristo N. (Eds.), *Software Process Modeling*, Springer Science+Business Media Inc., New York, 2005, pp. 1–27.
[104] Scacchi W., "Understanding free/open source software evolution", in: Madhavji N.H., Ramil J.F., Perry D. (Eds.), *Software Evolution and Feedback: Theory and Practice*, John Wiley and Sons Inc., New York, 2006, pp. 181–206.
[105] Scacchi W., "Emerging patterns of intersection and segmentation when computerization movements interact", in: Kraemer K.L., Elliott M. (Eds.), *Computerization Movements and Technology Diffusion: From Mainframes to Ubiquitous Computing*, Information Today, Inc., 2007.
[106] Scacchi W., Feller J., Fitzgerald B., Hissam S., Lakhani K., "Understanding free/open source software development processes", *Software Process—Improvement and Practice* **11** (2) (March/April 2006) 95–105.
[107] Scacchi W., Jensen C., Noll J., Elliott M., "Multimodal modeling, analysis, and validation of open source software development processes", *Intern. J. Information Technology and Web Engineering* **1** (3) (2006) 49–63.
[108] Schach S.R., Jin B., Wright D.R., Heller G.Z., Offutt A.J., "Maintainability of the Linux kernel", *IEE Proceedings—Software* **149** (1) (February 2002) 18–23.

[109] Seaman C.B., "Qualitative methods in empirical studies of software engineering", *IEEE Trans. Software Engineering* **25** (4) (July/August 1999) 557–572.
[110] Seaman C.B., Basili V., "Communication and organization: An empirical study of discussion in inspection meetings", *IEEE Trans. Software Engineering* **24** (6) (July 1998) 559–572.
[111] Sharma S., Sugumaran V., Rajagopalan B., "A framework for creating hybrid open-source software communities", *Information Systems J.* **12** (1) (2002) 7–25.
[112] Sim S.E., Holt R.C., "The ramp-up problem in software projects: A case study of how software immigrants naturalize", in: *Proc. 20th Intern. Conf. Software Engineering*, Kyoto, Japan, 19–25 April 1998, pp. 361–370.
[113] Smith M., Kollock P. (Eds.), *Communities in Cyberspace*, Routledge, London, 1999.
[114] Smith N., Capiluppi A., Ramil J.F., "Qualitative analysis and simulation of open source software evolution", in: *Proc. 5th Software Process Simulation and Modeling Workshop (ProSim'04)*, Edinburgh, Scotland, UK, May 2004.
[115] Snow D.A., Soule S.A., Kriesi H., *The Blackwell Companion to Social Movements*, Blackwell Publishers Ltd., Victoria, Australia, 2004.
[116] Spinuzzi C., *Tracing Genres through Organizations: A Sociocultural Approach to Information Design*, MIT Press, Cambridge, MA, 2003.
[117] Sommerville I., *Software Engineering*, seventh ed., Addison–Wesley, New York, 2004.
[118] Stewart K.J., Gosain S., "An exploratory study of ideology and trust in open source development groups", in: *Proc. 22nd Intern. Conf. Information Systems (ICIS-2001)*, New Orleans, LA, 2001.
[119] Truex D., Baskerville R., Klein H., "Growing systems in an emergent organization", *Communications ACM* **42** (8) (1999) 117–123.
[120] Viller S., Sommerville I., "Ethnographically informed analysis for software engineers", *Intern. J. Human–Computer Studies* **53** (2000) 169–196.
[121] von Hippel E., von Krogh G., "Open source software and the "private–collective" innovation model: Issues for organization science", *Organization Science* **14** (2) (2003) 209–223.
[122] von Krogh G., Spaeth S., Lakhani K., "Community, joining, and specialization in open source software innovation: A case study", *Research Policy* **32** (7) (July 2003) 1217–1241.
[123] Weber S., *The Success of Open Source*, Harvard University Press, Cambridge, MA, 2004.
[124] West J., O'Mahony S., "Contrasting community building in sponsored and community founded open source projects", in: *Proc. 38th Hawaii Intern. Conf. Systems Sciences*, Waikola Village, HI, 2005.
[125] West J., Dedrick J., "The effect of computerization movements upon organizational adoption of open source", in: Kraemer K.L., Elliott M. (Eds.), *Computerization Movements and Technology Diffusion: From Mainframes to Ubiquitous Computing*, Information Today, Inc., 2007.
[126] Wheeler D.A., "Why open source software/free software (OSS/FS, FLOSS or FOSS)? Look at the numbers", http://www.dwheeler.com/oss_fs_why.html, 2005, Accessed 15 November 2005.

[127] Williams S., *Free as in Freedom: Richard Stallman's Crusade for Free Software*, O'Reilly Books, Sebastopol, CA, 2002.
[128] Yamauchi Y., Yokozawa M., Shinohara T., Ishida T., "Collaboration with lean media: How open-source software succeeds", in: *Proc. Computer Supported Cooperative Work Conf. (CSCW'00)*, ACM Press, Philadelphia, PA, December 2000, pp. 329–338.
[129] Ye Y., Nakakoji K., Yamamoto Y., Kishida K., "The co-evolution of systems and communities in free and open source software development", in: Koch S. (Ed.), *Free/Open Source Software Development*, Idea Group Publishing, Hershey, PA, 2005, pp. 59–82.
[130] Ye Y., Kishida K., "Towards an understanding of the motivation of open source software developers", in: *Proc. 25th Intern. Conf. Software Engineering*, IEEE Computer Society, Portland, OR, May 2003, pp. 419–429.

Author Index

Numbers in *italics* indicate the pages on which complete references are given.

A

Abdelzaher, T., 159, 160, *189*
Adams, M., 227, *241*
Adiga, N., 90, 115, *149*
Agarwal, A., 8, *86*
Aingaran, K., 7, *85*
Allen, G., 90, *149*
Allu, B., 158, *188*
Almasi, G., 90, 115, *149*
Alon, E., 5, 12, *84*
Amarasinghe, S., 8, *86*
AMD, 126, *149*
Anastasi, R., 227, *241*
Anderson, C., 68, 69, *87*
Angelis, L., 272, *287*
Annabi, H., 256, *288*
Annavaram, M., 31, 32, *83*
Antoniades, I.P., 272, *287*
Archibald, J., 63, 64, *83*
Ardanaz, F., 32, *84*
Arief, B., 250, 253, *289*
Asanovic, K., 159, *188*
August, D.I., 96, *153*
Austin, T.M., 96, *150*
Ayguade, E., 31, *85*

B

Babb, J., 8, *86*
Baer, J.-L., 63, 64, *83*
Bailey, A.M., 91, 98, *149*
Bailey, D.H., 104, 117, 122, 123, 137, *150*
Balakrishnan, S., 31, 32, *83*
Ball, T., 167, *188*
Bamford, W., 220, 222, 226, *241*

Barcellini, F., 280, 281, 283, *293*
Barik, R., 90, 115, *149*
Barr, K., 159, *188*
Barroso, L., 7, 46, *83*
Barszcz, E., 137, *150*
Barton, J.T., 137, *150*
Barua, R., 8, *86*
Basili, V., 287, *294*
Baskerville, R., 269, *294*
Bateman, C., 193, 218, 232, *240*
Bates, J., 248, 252, 258, 272, 279, *291*
Beckermann, A., 207, *240*
Bellosa, F., 97, 98, *150*, *153*
Benford, S., 227, *241*
Benkler, Y., 254, 257–259, *287*
Bergquist, M., 248, 258, 259, *288*
Bernstein, K., 5, 12, *84*
Beyer, H., *288*
Bianchini, R., 98, 125, 148, *150*
Björk, S., 231, *242*
Blaauw, D., 156–164, 174, 182, 187, *187–189*
Bleris, G.L., 272, *287*
Blome, J.A., 96, *153*
BlueGene/LTeam, 112, 125, *150*
Bohrer, P., 125, *150*
Bonaccorsi, A., 252, 279, *288*
Boon, R., 193, 218, 232, *240*
Borkar, S., 90, 91, *150*, 159, *187*
Bose, P., 95, *150*
Bowhill, W., 22, *83*
Bray, J., 235, *242*
Briggs, F., 70, *83*
Brooks, D.M., 23, *83*, 95–98, *150*, *152*
Brown, J., 14, *86*
Bunch, J.R., 90, 117, 122, *151*

Burger, D., 9, 36, 79, *84*, *86*, 96, *150*
Burns, J., 24, 45, *83*
Butts, J.A., *188*

C

Cai, G., 96, *150*
Calder, B., 10, 12, 14, 15, 24, 38, 52, *86*
Calvert, R., 254, 257, 263, 270, 286, *292*
Cameron, K.W., 125, 126, 138, 148, *150*, *151*
Capiluppi, A., 270, 272, 281, *288*, *294*
Carrera, E.V., 98, 125, 148, *150*
Catarci, T., 284, *288*
Cekleov, M., 167, *188*
Chandra, S., 98, *150*
Chang, P.P., 167, *188*
Chaparro, P., 95, *150*
Chehimi, F., 198, 199, 214–216, 232, *240–242*
Chen, G., 159, 161, *188*
Chen, K., 282, *288*
Chen, Z., 125, *153*
Cheok, A.D., 226, 227, *241*
Chu, S., 68, *83*
Ciborra, C., *288*
Clabes, J., 68, 69, *83*, *87*
Clemson, H., 201, 202, 232, *240*, *242*
Collins, J., 45, *83*
Company, I., 114, 117, *150*
Compaq Corporation, 6, 22, *83*
Conklin, M., 282, *290*
Conte, T.M., 159, 160, 167, *187*, *188*
Cook, J.E., 284, 287, *288*
Cooper, D., 254, 258–260, *288*
Cote, W., 46, 67, *84*
Coulton, P., 195, 198, 199, 201, 202, 214–216, 218, 220, 222, 226, 232, 238, *240–242*
Crabtree, A., 227, *241*
Cronin, J., 46, 67, *84*
Crowston, K., 248, 250, 254–257, 261, 265, 281, 282, *288*, *290*, *291*
Culler, D.E., 92, *150*
Curtis, B., 257, *288*

D

Daasch, W.R., 96, *150*
Danziger, J., 257, *288*
Dawson, J., 68, *83*

Day, M.N., 8, 32, *84*
De, V., 159, 160, *187*, *189*
De Souza, C.R.B., 257, 260, 263, 265, *288*
Dedrick, J., 268, 274, 277, *294*
Degalahal, V., 157, 159, 161, 173, *188*
Despain, A.M., 158, *189*
Detienne, F., 280, 281, 283, *293*
Devgan, A., 32, *85*
Dhodapkar, A., 96, *150*
DiBona, C., 248, 252, 254, 258–260, 272, 274, 279, *288*, *291*
Diefendorff, K., 21, 22, *83*
Digital Equipment Corporation, 6, 22, *83*
DiLullo, J., 68, *83*
Dobberpuhl, D.W., 22, *83*
Dodson, S., 68, *83*
Dolbeau, R., 45, *83*
Dongarra, J.J., 90, 91, 117, 122, *151*
Dourish, P., 257, 260, 263, 265, *288*
Dramlitsch, T., 90, *149*
Drozd, A., 227, *241*
Dubois, M., 70, *83*, 167, *188*
Ducheneaut, D., 280, 281, 283, *293*
Ducheneaut, N., 279–281, 283, *288*

E

Eagle, N., 232, *242*
Ebenau, R.G., 287, *288*
Eberlein, A., 281, *292*
Ebner, M., 284, *288*
Edwards, R., 195, 198, 199, 201, 202, 214–216, 218, 220, 222, 226, 232, 238, *240–242*
Eggers, S., 8, 14, 45, *86*
Eggertsson, T., 254, 257, 263, 270, 286, *292*
Eickemeyer, R.J., 70, *84*, *85*
Elliott, M.S., 245, 254, 255, 257–260, 265, 273, 276, 280, 283, *288*, *289*, *293*
Ellis, C.S., 98, *151*
Elnozahy, E.N., 125, *150*
Elnozahy, M., 148, *151*
Emer, J., 8, 21, 22, 32, *84*, *86*
Erenkrantz, J., 261, *289*
Erickson, T., 252, *289*
Espasa, R., 32, *84*
Espinosa, J.A., 257, 265, *289*

F

Falk, J., 231, *242*
Falsafi, B., 159, 160, *187*
Fan, X., 98, *151*
Faraboschi, P., 173, *188*
Farbiz, F., 227, *241*
Farkas, K.I., 18, 23, 32, 34, 44, 69, *85*, 98, *151*
Felix, S., 32, *84*
Feller, J., 245, 260, 281, *289*, *293*
Feng, W.-C., 112, 113, 116–118, 125, 149, *151*, *152*, *153*
Feng, X., 125, 126, 138, 148, *150*, *151*
Fielding, R.T., 246, 248, 250, 252, 257–259, 261, 268, 272, 282, *289*, *290*, *292*
Finch, P., 8, *86*
Fink, M., 276, 277, 279, *289*
Fischer, G., 248, 269, 272, *289*
Fisher, J.A., 173, *188*
Fitzgerald, B., 245, 260, 281, *289*, *293*
Flanagin, A.J., 258, 265, 269, *292*
Flautner, K., 156–164, 174, 182, 187, *187*, *188*
Flinn, J., 97, *151*
Flintham, M., 227, *241*
FLOSS, 248, 252, 258, 276, 279, *289*
Floyd, M., 68, *83*
Flynn, M.J., 24, 36, *86*
Fogel, K., 254, 256–258, 260, 279, 282, *289*
Fong, S.W., 227, *241*
Fortes, R.P.M., 280, *293*
Forum Nokia, 196–198, 212, 213, 216, 217, *240*, *241*
Foster, I., 90, *149*
Frank, M., 8, *86*
Frank, S.J., 63, *84*
Freeh, V., 247, 249, 258, 265, 267, 269, 272, 281, 282, *292*
Freeh, V.W., 149, *151*
Friedrich, J., 68, *83*
Froehlich, J., 257, 260, 263, 265, *288*
Fulk, J., 258, 265, 269, *292*

G

Gabriel, R.P., 279, *290*
Gacek, C., 250, 253, *289*
Gago, J., 32, *84*
Gallivan, M., 259, *289*
Gasser, L., 282, *293*
Gaudiot, J.-L., 24, 45, *83*
Gay, J., 246, 254, 258, 268, 274, 276, 277, 279, *289*
Ge, R., 125, 126, 138, 148, *150*, *151*
German, D., 272, 283, *290*
Gharachorloo, K., 7, 46, *83*
Ghiasi, S., 31, 32, 43, *84*, *85*
Ghosh, R., 248, 252, 254, 259, 279, *290*
Gieseke, B., 22, *84*
Goh, K.H., 227, *241*
Gokhale, A., 263, *293*
Goldman, R., 279, *290*
Gonzalez, A., 95, *150*
Gonzalez, J., 95, *150*
Gonzalez-Barahona, J.M., 251, 269, 282, *291*
Gooliffe, P., 199, *240*
Gorman, G., 68, *83*
Gosain, S., 248, *294*
Goulet, M., 68, *83*
Gramunt, R., 32, *84*
Greene, T.C., 274, *290*
Greenhalgh, C., 227, *241*
Grochowski, E., 31, 32, 43, *83*, *84*
Gropp, W., 117, *151*
Grunwald, D., 31, 32, *84*, 98, *151*
Gupta, A., 92, 97, *150*, *152*
Gupta, R., 160, *189*
Gupta, S., 36, *84*
Gurumurthi, S., 97, *151*

H

Hakken, D., 252, 263, 280, *290*
Hamerly, G., 10, 12, 14, 15, 24, 38, 52, *86*
Hammond, L., 7, *84*
Hampton, M., 159, *188*
Hann, I.-H., 248, 252, 258, 268, *290*
Hansson, R., 231, *242*
Harris, T., 104, 117, 122, 123, *150*
Harrison, W., 281, 284, 287, *290*
Hars, A., 248, 258, 276, 279, *290*
HECRTF, 90, *152*
Heller, G.Z., 270, 272, 281, 282, *288*, *293*
Heni, M., 207, *240*
Hennessy, J.L., 6, *84*, 91, *152*
Heo, S., 159, *188*
Herbsleb, J.D., 246, 248, 250, 257, 265, 272, 282, *289*, *292*

Hermann, S., 248, 258, 279, *290*
Hernandez, I., 32, *84*
Herraiz, I., 251, 269, 282, *291*
Herrod, S.A., 97, *152*
Hertel, G., 248, 258, 279, *290*
Hertzum, M., 259, *290*
Herz, J.C., 274, *292*
Hine, C.M., 252, 280, *290*
Hissam, S., 245, 260, 281, *289*, *293*
Ho, R., 68, *84*
Hofstee, H.P., 8, 32, *84*
Holland, K., 46, 67, *84*
Holt, R.C., 253, *294*
Holtzblatt, K., *288*
Holzinger, A., 284, *288*
Horowitz, M., 5, 12, 68, *84*
Howison, J., 250, 256, 257, 261, 282, *288*, *290*
Hsu, C.-H., 125, 149, *152*, 158, *189*
Hu, J.S., 157, 159, 161, 171, 173, *188*
Hu, Z., 157–159, 164, 187, *187*
Huh, J., 9, *86*
Huntley, C.L., 252, *290*
Hwu, W.W., 167, *188*

I

Iacono, C.S., 274, *290*
Iannacci, F., 261, 263, 265, 269, 272, 280, *290*
IBM, 8, 63, 64, 67, 69, 70, *84*, 114, *152*
Intel, 126, 132, *152*, 168, *188*
Intel Corporation, 56, *83*
International Organization for Standardization, 226, *241*
Irwin, M.J., 96, 97, *151*, *153*, 156, 157, 159–161, 171, 173, *188*, *189*
Isci, C., 97, 98, *152*
Iscoe, N., 257, *288*
Ishida, T., 252, 261, 268, *295*

J

J, G., 117, *152*
Jamaluddin, J., 232, *242*
Jensen, C., 250, 252, 253, 258–260, 269–271, 280, 281, 283, 287, *290*, *291*, *293*
Jin, B., 270, 272, 281, *293*
John, L., 158, *189*

Johns, C.R., 8, 32, *84*
Johnson, R.E., 70, *84*
Jones, C., 125, 148, *150*
Joseph, R., 97, 98, *152*
Jouppi, N.P., 6, 18, 22, 23, 32, 34, 36, 38, 44, 45, 52, 69, 70, *84–86*, 163, *188*
Juan, T., 32, *84*

K

Kaanta, C., 46, 67, *84*
Kadayif, I., 159, 160, *189*
Kahle, J.A., 8, 32, *84*
Kale, L.V., 90, *152*
Kalla, R., 68, *83*
Kalman, M.E., 258, 265, 269, *292*
Kandemir, M., 96, 97, *151*, *153*, 156, 157, 159–161, 171, 173, *188*, *189*
Kappiah, N., 149, *151*
Karakoy, M., 159, 161, *188*
Kashi, N., *240*
Kastner, R., 156, *189*
Kaxiras, S., 157–159, 164, 187, *187*
Keaty, J., 68, 69, *87*
Keckler, S.W., 9, 36, 79, *84*, *86*
Keller, T., 32, 43, *84*, *85*, 125, *150*
Kiely, D., 195, *240*
Kim, A.J., 251, 252, 261, 268, *291*
Kim, C., 9, 79, *84*, *86*
Kim, H., 96, *153*
Kim, J., 8, *86*
Kim, N.S., 156–164, 174, 182, 187, *187*, *188*
Kircher, C., 68, 69, *87*
Kirk, C., 125, *151*
Kishida, K., 250, 252, 253, 268, 273, 274, *292*, *295*
Kistler, M., 148, *151*
Klauser, A., 21, 22, *85*
Klein, H., 269, *294*
Kling, R., 257, 274, *290*, *291*
Koch, S., 245, 276, *291*
Kollock, P., 251, 252, 268, 280, *294*
Kolodny, A., 31, *85*
Kongetira, P., 7, *85*
Korhonen, P., 218, *241*
Kotla, R., 32, *85*
Kowaleski, J., 21, *85*

AUTHOR INDEX

Krafka, B., 70, *85*
Krasner, H., 257, *288*
Kraut, R.E., 257, 265, *289*
Krauter, B., 68, 69, *87*
Kremer, U., 125, *152*, 158, *189*
Kriesi, H., 274, *294*
Krishna, A.S., 263, *293*
Krishnan, V., 45, *85*
Kumar, A., 24, *85*
Kumar, R., 5, 12, 18, 23, 32, 34, 38, 44, 45, 52, 67, 69, 70, *84*, *85*
Kumar, S., 90, *152*
Kunkel, S.R., 70, *84*, *85*
Kurita, T., 93, *152*
Kwansik, B., 255, 281, *291*

L

Lago, P., 272, 281, *288*
Lai, K., 31, 32, *83*, 156, *187*
Laird, D., 112, *152*
Lakhani, K.R., 245, 250, 252, 258, 260, 272, 279, 281, *289*, *291*, *293*, *294*
Lanzara, G.F., 255, 263, 280, *291*
Laramee, D.F., 193, *240*
Larus, J.R., 167, *188*
Laudon, J., 4, *85*
Lave, J., 257, 259, 272, *291*
LBNL, 98, *152*
Lebeck, A.R., 98, *151*
Lee, C., 173, *188*
Lee, D., 156, *189*
Lee, M., 68, *83*
Lee, P., 46, 67, *84*
Lee, V., 8, *86*
Lee, W., 8, *86*
Lehman, M.M., 270, 273, 281, *291*
Lerch, J.F., 257, 265, *289*
Lerner, J., 248, 252, 258, *291*
Lessig, L., 254, 259, 263, 268, *291*
Levis, P., 98, *151*
Levy, H., 8, 14, 45, *86*
Li, J., 43, *85*
Li, L., 159, 160, *189*
Li, Y., 227, *241*
Lim, C.H., 96, *150*
Lipasti, M., 70, *85*

Liu, D., 156, *189*
Liu, H., 9, *86*
Liu, S., 70, *84*
Liu, W., 227, *241*
Ljungberg, J., 248, 258, 259, 274, *288*, *291*
Ljungstrand, P., 231, *242*
Lo, J., 8, *86*
Longman, J., 280, *291*
Lopez-Fernandez, L., 251, 269, 282, *291*
Lorch, J.R., 98, *152*
Lovett, T., 63, *85*
Lowenthal, D.K., 149, *151*
Lowney, G., 32, *84*
Lu, S.-L., 156, *187*
Lucas, M., 274, *292*
Lusk, E., 117, *151*

M

MacKenzie, I.S., 218, *241*
Madey, G., 247, 249, 258, 265, 267, 269, 272, 281, 282, *292*
Maeurer, T.R., 8, 32, *84*
Magerkurth, C., 226, *241*
Mahendran, B., 280, 281, 283, *293*
Mahmoud, Q., 215, *241*
Mai, K., 68, *84*
Malik, S., 96, *153*, 158, *189*
Mandryk, R.L., 226, *241*
Mangione-Smith, W.H., 173, *188*
Manheim, U., 90, 122, *153*
Martin, S., 156–160, 162, 164, *187*, *188*
Martinez, J., 43, *85*
Martonosi, M., 23, *83*, 96–98, *150*, *152*, 157–159, 164, *187*, *187*
Marttiin, P., 260, 265, *292*
Marwell, G., 257–259, 265, 269, 282, *292*
Mattina, M., 32, *84*
McGill, J., 68, *83*
McMahon, F.H., 117, *152*
McNamara, R., 7, 46, *83*
McVoy, L., 117, *152*
Memon, A.M., 263, *293*
Mendelson, A., 156, *187*
Meng, Y., 156, *189*
Merritt, R., 21, 22, *85*
Michelson, K., 9, *86*
MITRE Corporation, 274, 279, *292*

Mockus, A., 246, 248, 250, 257, 265, 272, 282, *289*, *292*
Moller, C.B., 90, 117, 122, *151*
Monge, P.R., 258, 265, 269, *292*
Moore, C.R., 9, *86*
Moore, G., 5, *85*
Morad, T.Y., 31, *85*
Morisio, M., 272, 281, *288*
Morner, M., 255, 263, 280, *291*
Motorola, 168, *188*
Muchnick, S., 158, 159, 170, 185, *188*
Mudge, T., 93, 94, *152*, 156–164, 174, 182, 187, *187*, *188*
Muench, P., 68, *83*
Mulder, J.M., 24, 36, *86*
Mullins, I., 218, 226, *241*
Mullins, T., 70, *85*
Murai, H., 90, *152*

N

Nadgir, A., 171, *188*
Naffziger, S., 5, 12, *84*
Nagarajan, R., 9, *86*
Nakakoji, K., 250, 268, 273, 274, *292*, *295*
Nayfeh, B.A., 7, *84*
Neidner, S., 248, 258, 279, *290*
NERSC, 90, 122, *153*
Nilsen, T., 226, *241*
Nishinaka, Y., 250, 273, 274, *292*
Nokia, 192, 224, *240*
Noll, J., 254, 261, 283, *292*, *293*
Nowatzyk, A., 7, 46, *83*

O

Ockman, S., 254, 258–260, 274, 279, *288*
O'Dowd, D., 274, *292*
Offutt, J., 270, 272, 281, 282, *288*, *293*
Oliver, P., 257–259, 265, 269, 282, *292*
Olson, M., 257, 259, *292*
Olukotun, K., 7, *84*, *85*
O'Mahony, S., 261, 263, 268, 272, 273, 277, *292*, *294*
Onder, S., 160, *189*
OSBC, 279, *292*
Oskin, M., 9, *86*
Ostrom, E., 254, 257, 263, 270, 286, *292*
Ou, S., 248, 258, 276, 279, *290*

Ovaska, P., 260, 265, *292*

P

Palmer, 192, 224, *240*
Pan, F., 149, *151*
Pande, S., 160, *189*
Papamarcos, M., 70, *86*
Parikh, D., 159, 160, *189*
Parnassa, C., 258, 265, 269, *292*
Patel, J., 70, *86*
Patil, D., 5, 12, *84*
Patterson, D.A., 6, *84*, 91, *152*
Paulson, J.W., 281, *292*
Pavelicek, R., 248, 254, 258–260, 263, 279, *292*
Payton, S., 274, *292*
Peh, L.-S., 96, *153*
Penry, D.A., 96, *153*
Pentland, A., 232, *242*
Perelman, E., 10, 12, 14, 15, 24, 38, 52, *86*
Petrovick, J., 68, 69, *87*
Phillips, J.C., 90, *152*
Pinheiro, E., 98, 125, 148, *150*
Plass, D., 68, *83*
Pollack, F., 156, *187*
Porter, A.A., 263, *293*
Potkonjak, M., 173, *188*
Powell, L., 68, *83*
Powell, M.D., 159, 160, *187*
Prakash, V.V., 248, 252, 279, *290*
Preece, J., 251, 268, *293*

Q

Qadeer, S., 7, 46, *83*
Quach, N.T., 24, 36, *86*

R

Rabaey, J.M., 23, *86*
Rajagopalan, B., 268, *294*
Rajamony, R., 148, *151*
Rajgopal, S., 97, *153*
Rajwar, R., 31, 32, *83*
Ramil, J.F., 270, 272, 281, *294*
Ranganathan, P., 18, 23, 32, 34, 44, 69, *85*
Rashid, O., 195, 218, 226, 238, *240*, *241*

Rawson, F., 32, 43, *84*, *85*
Reis, C.R., 280, *293*
Rele, S., 160, *189*
Restle, P., 68, 69, *83*, *87*
Ripoche, G., 282, *293*
Roberts, J., 248, 252, 258, 268, *290*
Robles, G., 251, 269, 282, *291*
Ronen, R., 32, 43, *84*, 156, *187*
Rosenberg, H., 70, *85*
Rosenblum, M., 97, *152*
Rossi, C., 252, 279, *288*
Rossi, M., 260, 265, *292*
Rotenberg, E., 159, 160, *187*
Row-Farr, J., 227, *241*
Roy, K., 159, 160, *187*
Rumsey, S., 258, 265, 269, *292*
Runyon, S., 68, *83*

S

Sack, W., 280, 281, 283, *293*
Sair, S., 12, 14, 15, 52, *86*
Sakagami, H., 90, *152*
Samoladas, I., 272, *287*
Sankaralingam, K., 9, *86*
Sankaranarayanan, K., 159, 160, *189*
Sano, B., 7, 46, *83*
Saphir, W., 104, 117, 122, 123, *150*
Sarkar, V., 8, *86*
Satyanarayanan, M., 97, *151*
Sawyer, S., 260, *293*
Scacchi, W., 245, 249–263, 265, 269–274, 276, 280, 281, 283, 284, 287, *289–293*
Schach, S.R., 270, 272, 281, 282, 288, *293*
Scheibel, J., 226, *241*
Scheurich, C., 70, *83*
Schmidt, D.C., 263, *293*
Schuster, S.E., 95, *150*
Schwartz, N., 68, *83*
Schwerin, A., 9, *86*
Scott, J., 274, *292*
Scozzi, B., 248, 254, 256, 261, 265, *288*
Seaman, C.B., 280, 281, 287, *293*, *294*
Semiconductor Industry Association, 156, *187*
Seo, Y., 90, *152*
Seznec, A., 32, 45, *83*, *84*
Sharma, S., 268, *294*
Shen, J., 31, 32, 43, *83*, *84*, 156, *187*

Sherwood, T., 10, 12, 14, 15, 24, 38, 52, *86*, 156, *189*
Shinohara, T., 252, 261, 268, *295*
Shippy, D., 8, 32, *84*
Shivakumar, P., 22, 36, 70, *86*, 163, *188*
Silfverberg, M., 218, *241*
Sim, S.E., 253, *294*
Singh, D., 97, *153*
Singh, J.P., 92, *150*
Sinharoy, B., 68, *83*
Sivasubramaniam, A., 97, *151*, 159, 160, *189*
Siy, H.P., 263, *293*
Skadron, K., 159, 160, *189*
Slaughter, S.A., 248, 252, 257, 258, 265, 268, *289*, *290*
Smith, A.J., 98, *152*
Smith, J.E., 92, *153*
Smith, M., 251, 252, 268, 280, *294*
Smith, N., 270, 272, 281, *294*
Smith, S., 7, 46, *83*
Snavely, A., 14, *86*
Snow, D.A., 274, *294*
Sohi, G., *188*
Sommerville, I., 244, 246, 255, 260, 265, 280, 286, *294*
Soule, S.A., 274, *294*
Spaeth, S., 248, 250, 252, *294*
SPEC, 14, *86*
Spinuzzi, C., 252, 255, 281, *294*
Springer, R., 149, *151*
Squillante, M.S., 70, *84*
Srikrishna, D., 8, *86*
Staelin, C., 117, *152*
Stamelos, I., 272, *287*
Stamm, R., 8, *86*
Standard Performance Evaluation Corporation, 103, *152*
Stets, R., 7, 46, *83*
Stewart, G.W., 90, 117, 122, *151*
Stewart, K.J., 248, *294*
Stone, M., 254, 258–260, 274, 279, *288*
Strauss, S.H., 287, *288*
Sturman, C., 235, *242*
Su, C.L., 158, *189*
Succi, G., 281, *292*
Sugumaran, V., 268, *294*
Sun, 7, *86*
Svensson, C., 156, *189*

AUTHOR INDEX

Swanson, S., 9, *86*
Sweet, M., 68, *83*
Sylvester, D., 156, *189*

T

Takemoto, M., 93, *152*
Tan, L., 125, *153*
Tandavanitja, N., 227, *241*
Taylor, M., 8, *86*
Tennessee, U., 90, 122, *153*
Teo, S.L., 227, *241*
Thakkar, S., 63, *85*
The Khronos Group, 214, *241*
Theis, T.N., 68, *86*
Thompson, R., 195, 238, *240*
Tirole, J., 248, 252, 258, *291*
Tiwari, V., 23, *83*, 96, 97, *150*, *153*, 158, *189*
Toburen, M.C., 159, 160, *187*
Toman, C.A., 263, *293*
Top500, 90, *153*
Torrellas, J., 45, *85*
Tremblay, M., 7, *86*
Truex, D., 269, *294*
Tsai, Y., 159, 160, *189*
Tsui, C.-Y., 158, *189*
Tullsen, D.M., 8, 14, 18, 23, 24, 32, 34, 38, 44, 45, 52, 67, 69, 70, *83*, *85*, *86*
Tynan, R., 247, 249, 258, 265, 267, 269, 272, 281, 282, *292*

U

Upton, M., 31, 32, *83*

V

Vachharajani, M., 96, *153*
Vachharajani, N., 96, *153*
Valero, M., 31, *85*
Valluri, M., 158, *189*
VanderWiel, S., 70, *85*
Vargas, E., 91, *153*
Varner, D., 125, 148, *150*
Velusamy, S., 159, 160, *189*
Verghese, B., 7, 46, *83*
Vijaykrishnan, N., 96, 97, *151*, *153*, 156, 157, 159–161, 171, 173, *188*, *189*
Vijaykumar, T.N., 159, 160, *187*

Viller, S., 280, *294*
Vitale, P., 70, *85*
von Hippel, E., 250, 272, *294*
von Krogh, G., 248, 250, 252, 272, *294*
Votta, L.G., 263, 284, 287, *288*, *293*

W

Wagoner, J., 68, *83*
Waingold, E., 8, *86*
Wall, D., 15, *87*
Wang, H.-S., 32, 43, *84*, 96, *153*
Warnock, J., 68, 69, *87*
Warren, M.S., 112, 113, 116, 117, 125, *151*, *153*
Weber, S., 277, *294*
Weigle, E.H., 112, 113, 116, 117, 125, *151*, *153*
Weiser, U.C., 31, *85*
Weissel, A., 98, *153*
Wenger, E., 257, 259, 272, *291*
West, J., 268, 274, 277, *294*
Wheeler, D.A., 279, *294*
Whitley, L., 70, *85*
Willey, M., 7, *84*
Williams, S., 254, 258, 268, 279, *294*
Wilson, A., 63, *87*
Witchel, E., 97, *152*
Wolf, A.L., 284, 287, *288*
Wolf, B., 248, 252, 258, 272, 279, *291*
Wolfe, A., 158, *189*
Wright, D.R., 270, 272, 281, *293*
Wright, T., 46, 67, *84*

Y

Yamamoto, Y., 250, 268, 273, 274, *292*, *295*
Yamauchi, Y., 252, 261, 268, *295*
Yang, S., 159, 160, *187*
Yang, X.B., 227, *241*
Ye, W., 96, *153*
Ye, Y., 159, *187*, 250, 252, 253, 268, 273, 274, *292*, *295*
Yilmaz, C., 263, *293*
Yokokawa, M., 90, *152*
Yokozawa, M., 252, 261, 268, *295*
Young, C., 173, *188*
Yu, L., 282, *288*

Z

Zhang, W., 157–161, 173, *188*, *189*
Zheng, G., 90, *152*
Zhou, H., 159, 160, *187*
Zhou, Y., 125, *153*
Zhu, Q., 125, *153*
Zhu, X., 96, *153*
Zoric, B., 68, 69, *87*
Zotou, N., 232, *242*
Zyuban, V., 28, 67, 69, 70, *85*, *87*

Subject Index

A

A La Mobile, 207
ACPI, 132, 137
'Action Twister', 224
Address bus, 63, 64
Adobe, 212
Alpha cores, 6
  see also EV4 core; EV5 core; EV6 core; EV8- core
AMC cache, 160
AMD, multi-core processors, 8
Amdahl's Law, 124
ammp, 14, 27
Apache Software Foundation, 268
Apache Web server, 268, 270, 276, 282
Application memory space, 195–6
Application size, mobile game, 195
applu, 14, 24–6, 27, 28
Arbitration, 69
Arcade games, 273
Area budget, 36
Area estimation, 23–4, 36–7
Argus prototype, 112–13
  advantages, 124–5
  benchmarks, 116–17
  cost, power and space metrics, 117–19
  low power cluster metrics, 115–17
  performance limitations, 125
  performance results, 119–24
    memory system, 119–20
    message communication, 121–2
    scientific applications, 122–3
    throughput, 120–1
  system design, 113–15
ARM architecture, 215
  compilers, 204
Arrhenius' Law, 116

art, 14
Artifact analysis methods, 281–3
Artifact repository mining, 281–3
Assertive access, 61
  DCache, 60, 61
  ICache, 59
Asymmetric core clusters, 31
Athlon processor, 8
ATI, 216
'Attack of the Killer Virus', 224
Average Revenue Per User (ARPU), 192

B

Bank load queues (BLQs), 67
BenQ-Siemens, 202
Beowulf, 115
  see also DANIEL
Binary Runtime Environment for Wireless (BREW), 203–5, 215
Bitboys, 216
BlueGene/L, 90, 112, 115, 120
  limitations, 125
Bluejacking, 232
Bluetooth, 229, 232, 234–6
  latency, 197
'Botfighters', 218
Bounded-global-event, 18–19
BREW, 203–5, 215
'Buddy Bash', 224–5
Bugzilla, 256, 263, 264, 282
bzip2, 14

C

C code, performance leverage, 199
C++ code, performance leverage, 199

SUBJECT INDEX

Cache leakage energy reduction
  compiler-assisted, 160–1
    evaluation methodology, 172–3
    see also Data cache leakage reduction; Loop-based data cache leakage reduction; Static next sub-bank prediction research, 159–61
Cache sharing *see* DCache sharing; ICache sharing
CACTI, 22, 23, 36, 70, 163–4
Cai–Lim, 96
Camera games, 224–5
CASTLE, 98
Casual gamers, 194
Cell processor, 8, 32
Central address arbiter, 63
Cheok, Adrian, 227
Chip multiprocessors (CMPs), 3
CLDC, 209
Coding optimisation, for mobile games, 198–9
Coding style optimisation, 199
Coherence transactions, 63
Community development, FOSS, 268–9
Compiler-assisted leakage energy reduction
  see Cache leakage energy reduction, compiler-assisted
Component life expectancy, 116
Computational clusters, 92
  classification, 115
  power profiling, 98–112
    application characteristics, 110
    automating process, 101–2
    cluster energy-performance efficiency, 108–10
    cluster-wide measurements, 104–8
    isolating power by component, 100–1
    power measurement system, 99–102
    resource scheduling, 110–12
    single node measurements, 103–4
    see also Nodal power profiles
  see also Low power computational clusters; Power-aware computational clusters
Conditional branches with static target addresses, 167, 174–5

Configurations, J2ME, 208–9
Conflict management, 260, 285
Conjoined-core chip multiprocessors, 45–61
  baseline architecture, 46–7
  conjoined-core architectures, 47–53
    unified, 61
  floorplans
    baseline, 47
    crossbar sharing, 50
    ICache sharing, 48–9
  intelligent sharing, 58–61
  modeling, 52–3
  multiple sharings, 61
  simple sharing, 53–8
Connected Device Configuration (CDC), 209
Connected Limited Device Configuration (CLDC), 209
conquerors, 218
'Consequences', 219
Copy left, 276
Core Duo processor, 8
Core switching, 19, 21
  oracle heuristics, 26–9
  realistic heuristics, 29–31
CPUSPEED daemon, 128–9, 137–8
crafty, 14, 27
Crossbar interconnection systems, 62, 66
Crossbars, 66
  design issues, 77–8
  overheads, area and power, 76–8
  performance, 78–9
  sharing, 50–1
  simple, 57–8
  width reduction, 51, 58
Crusoe TM5600, 115
Current versions system (CVS), 259–60, 263, 268, 283

D

DANIEL, 117
  cost, power and space metrics, 117–19
  performance results, 119–24
    memory system, 119–20
    scientific applications, 122–3
    throughput, 120–1
Data bus, 64

Data cache leakage reduction
  hardware-based, 157–8, 161
    experimental results, 182–4
  see also Loop-based data cache leakage reduction
Data cache sharing see DCache sharing
Data queues, 64
DCache sharing, 49–50
  intelligent sharing, 60
  overheads, area and power, 76–8
  performance, 78–9
  simple sharing, 55–6
Delay, 92
  normalized, 108
Device and service detection, 235
DirectX Mobile Edition, 206
Distribution, mobile games, 200–1
Doja, 209
'Dope-wars', 218
DRI-cache, 160
Drowsy instruction caches, 158
  see also Static next sub-bank prediction
DVFS, 94, 125
  in cache cells, 160
DVS
  external control, 126
    system-driven, 128–9, 137–8
    user-driven, 129–30, 139–43
  in high-performance clusters, 126–8
  internal control, 126, 130–1, 143–8
    energy savings, 144–5, 146–8
    performance, 143–4, 145–6
  scheduling strategies, 126, 128–31
  transition automation, 149
Dynamic power, 156
Dynamic voltage and frequency scaling see DVFS
Dynamic voltage scaling see DVS

E

E-ten, 205
Earth Simulator, 91
Eclipse, 268, 277
ED2P metric, 111–12, 137, 141–3
ED3P metric, 137, 140

EDP metric, 111–12, 134, 141
Elcor, 173
End-user license agreements (EULAs), 254
Energy, 94–5
Energy conservation, 98
Energy-delay product (EDP), 27–8, 181–2, 184
Energy-delay^2 metric, 28–9
Energy efficiency optimization, 26–7
Energy per instruction (EPI), reduction, 32
Energy-performance efficiency metrics, 111–12, 137
Energy-performance microbenchmarks, 133–6
  communication-bound, 135–6
  CPU-bound, 134–5
  memory-bound, 134
Energy-performance tradeoffs, 111–12
Energy profiling, 132
eon, 14
EPIC architecture, 173
EPOC software, 201
Ericsson, 201
Ethnographically informed methods, 280–1
EULAs, 254
EV4 core, 20–1
  configuration, 22
  power and area statistics, 23
EV5 core, 6, 20–1, 35, 46
  configuration, 13, 22
  FPU area, 51
  power and area statistics, 23
EV6 core, 6, 20–1
  configurations, 13, 22
  power and area statistics, 23
EV8- core, 20–1
  configuration, 22
  power and area statistics, 23
Event-based power estimation, 98
Execution time
  parallel, 92–3
  sequential, 92
'Exquisite Corpses', 219

F

Fetch width, ICache, 49, 53–5
Filezilla, 256
Firefox, 263, 276

310 SUBJECT INDEX

Flash, 212, 213
Flash Lite, 212–13
Floating-point units *see* FPUs
Floorplan, co-design with architecture, 45
FLOSS, 246
FLOSSmole, 282
fma3d, 27
FOSS
  co-evolving socio-technical systems for, 272–4, 283, 285–6
  development *see* FOSSD
  as multi-project software ecosystem, 270–4
  research limitations and constraints, 284–6
  research methods for studying, 277–83
    artifact analysis methods, 281–3
    artifact repository mining, 281–3
    ethnographically informed methods, 280–1
    industry poll methods, 279
    mode of analysis, 278–9
    multi-modal analysis of socio-technical interaction networks, 283
    reflective practice, 279
    survey research methods, 279–80
    terms of analysis, 278
    unit of analysis, 278, 279
  as social movement, 274–7
  systems growth, 281
FOSS commons, 254
FOSS project property licenses, 255, 266, 268
FOSSD
  alliance formation, 265, 269, 285
  beliefs supporting, 254–5, 259
  bug reporting guidelines, 263, 264
  business-friendly OSSD projects, 268
  community development, 268–9
  coordination and control, 260–5, 285
  data availability, 284
  description, 245–7
  individual participation in projects, 248–53, 284–5
  informalisms, 255–6, 269
  inter-project social networking, 265–8
  networking support, 253–4

personal software development tools, 253–4
project numbers, 247
project roles, 250, 253, 272
research opportunities, 286–7
results from recent studies, 147–8
rules of control, 257
sentimental resources, 259
software development project requirements, 269
software version control, 260–1
trust and accountability mechanisms, 259–60
Web sites/portals, 256, 265–6, 268, 269, 274
see also FOSS; Software developers, FOSS
Foundation Profile, 209
Foundations, 268, 276–7, 285
FPUs
  sharing, 45, 51–2
    simple sharing, 55–6
Free/open source software development *see* FOSSD
Free software
  licensing, 245, 266, 268
  OSS vs, 245–6, 276–7
Free Software Foundation, 246, 268
Free software movement, 274, 276
Freedom
  of choice, 254–5
  of expression, 254–5
Freshment.org, 274
Frontier instructions, 165–6, 174–5
Functional simulator, 70

G

Game API, 210
Game pills, 226, 227
Gated V_{dd}, 160–1
gcc, 10
General Packet Radio Service (GPRS), 227
Gigabyte, 205
Gizmondo, 205
GNOME, 272
GNOME Foundation, 268

GNU General Public License (GPL), 245, 266, 268, 274, 276
GNU/Linux distributions, 272
GNUenterprise.org, 254, 255, 256
GPL, 245, 266, 268, 274, 276
GPML50, 133
GPS, 226, 229
Green Destiny, 112, 113, 115
   cost, power and space metrics, 117–19
   limitations, 125
   performance results, 119
GSM, 237

H

Handango, 200
Hardcore gamers, 193–4
   types, 194
Hardware sharing, benefits, 45
Heap, 196
High-end computing systems, 115
   peak performance, 90
   power consumption, 91
Hint bits, 169
HPC, 205
HPL, 122
HTML, 270
http, 270
'Human Pacman', 227
Hydra, 7

I

IBM
   BlueGene/L *see* BlueGene/L
   Eclipse project, 268, 277
   see also Cell; Power4; Power5; Power6
ICache sharing, 48–9
   intelligent sharing, 59
   overheads, area and power, 76–8
   performance, 78–9
   simple sharing, 53–5
ILP, 6, 185
Imagination Technologies, 215
Immersion, 214
IMPACT, 173
IMS, 237–9

In-order cores, 35
Incremental mutation, 273
Industry poll methods, 279
Information Module Profile (IMP), 209
Instant Messaging (IM), 239
Instruction cache addressing schemes, 167–8
Instruction cache sharing *see* ICache sharing
Instruction-level parallelism (ILP), 6, 185
Instructions per cycle *see* IPC
Integrated Development Environments, 211
Intel
   4004 microprocessor, 5
   dual-core processors, 8
International Hobo, 218–19
Internet
   access over cellular, 237–9
   see also Web
Internet Explorer, 270
IP Multimedia System (IMS), 237–9
IPC, 14, 120
IX Bus, 113, 114
IXIA chassis, 113–14

J

J2EE, 208
J2ME, 208–10, 215
   extension to BREW, 203
   in IMS services, 237
   in mobile games, 226, 229
J2SE, 208
Java Community Process (JCP), 208, 270
Java Specification Request (JSR), 208
Java Virtual Machine (JVM)
   for BREW, 203
   for Windows Mobile, 206
JCP, 208, 270
Joule Watcher, 98
JVM *see* Java Virtual Machine

K

KDE user interface package, 276
Kernel application benchmarks, 117
Keypads, mobile phone, 197
Khronos Group, 214
Kilobyte virtual machine (KVM), 208

L

Latency, 196
LBoss, 268
Leakage modes, 171
Leakage power, 156
　see also Cache leakage energy reduction
Learning organizations, 252
LINPACK, 117, 119, 122
Linux, on mobile phones, 206–7, 215
Linux Kernel, 272, 274–6
Linux Phone Standards Forum (LiPS), 207
Livermore Loops, 117, 121–2
LMBENCH, 117, 119
Location, mobile user, 225–6
Location based games, 225–31
Loop-based data cache leakage reduction, 170–2
　experimental results, 182–6
　　energy-delay product, 184
　　impact of compiler optimizations, 185–6
　　leakage energy savings, 183–4
　　with varying data cache sizes, 184–5
Loop distribution, 185–6
Loop tiling, 185–6
Loops, time spent in, 170
Low power computational clusters, 112–25
　design analysis, 117–24
　design lessons, 124–5
　metrics, 115–17
　see also Argus prototype
LSE, 96

M

M3G, 214–15
MAJC 5200 processor, 7
Managers, 218
Matsushita, 201
MC68030, 168
mcf, 10, 14, 27
Mean time between failures (MTBF), 91
Media API, 210
Mediabench, 173
　des, 177
Memory optimisation, 198
Memory sub-bank prediction buffer, 162–5
　dynamic energy overhead, 163–5
Meritocracy
　layered/pyramid, 261–3
　skill, 257
Mfooty, 238
mgrid, 135
Micro benchmarks, 117
Microprocessor power simulators, 96
Microsoft, 205–6, 215, 277
　Internet Explorer, 270
MIDP, 209–10, 215
MMAPI, 210
Mobile, ITU definition, 192
Mobile games, 192–239
　3-D, 214–17
　　hardware support, 215–16
　　optimisation, 216–17
　　software support, 214–15
　development see Software development environments
　opportunities, 217
　physical constraints, 195–201
　　coding optimisation, 198–9
　　distribution, 200–1
　　memory, 195
　　networking, 196–7
　　phone user interface, 197
　　processors, 196
　　screen aspect ratio, 197
　　screen size, 197
　　sound support, 197–8
　　testing, 199–200
　players, 193–4
　　personalities, 218–19
　see also Camera games; IP Multimedia System; Location based games; Proximity games; Text games
Mobile Information Device Profile (MIDP), 209–10, 215
Mobile Linux Initiative (MLI), 207
Mobile Multimedia API (MMAPI), 210
Mobile Social Software (MoSoSo), 232
Mobiluck, 232
'Mobslinger', 232–7
　Blood Bath, 233, 234, 235
　Last Man Standing, 233–4, 235
　Outlaws, 234
　Quick Draw, 233, 235

Top Gun, 234
Monotonicity, 34
  performance cost, 44
MonsterMob, 200
Moore's law, 5
'Mosquitos', 224
Motorola, 201, 205, 206–7
Mozilla, 268, 270, 276
  bug management, 263, 264, 282
MPCore processor, 9
MPICH, 133, 143, 145
MPPTEST, 117, 121
Multi-Arcade Machine Emulator (MAME), 273
Multi-core interconnect
  example holistic approach, 79–81
  interconnection mechanisms, 62
  modeling cores, 69–70
    workload, 70
  modeling interconnect overhead, 67–9
    latency, 69
    logic area, 68
    power, 68–9
    wiring area, 67–8
  see also Crossbar interconnection system; P2P links; Shared bus fabric
"Multi-core-oblivious" designs, 4
  overprovisioning, 45
Multi-core processors, 3
  design from ground up, 32–44
    customizing cores to workloads, 34–5
    design space navigation approach, 32–4
    fixed area budget results, 38–43
    modeling CPU cores, 35–6
    modeling performance, 37–8
    modeling power and area, 36–7
    non-monotonic design impact, 43–4
  early efforts, 7–9
  holistic design approach, 4–5
  move to, 5–9
  naive design methodology, 3–4
  see also Conjoined-core chip multiprocessor; Multi-core interconnect; Multi-ISA heterogeneous multi-core

architectures; Single-ISA heterogeneous multi-core architectures
Multi-ISA heterogeneous multi-core architectures, 32
Multi-speed disks, 125
Mutation, incremental, 273
Myers–Briggs Dichotomies, 218–19
MySQL, 220, 268

N

NAS Parallel Benchmarks *see* NPB
NEC, 206–7
Negotiation, 285
NEMO, 131–2
  power, energy and performance profiling, 132–3
.NET Compact Framework, 206
NetBeans, 268, 270, 277
Network externalities, 270
Network power simulators, 96–7
Networking, issues in mobile game design, 196–7
Next sub-bank prediction
  compiler-assisted *see* Static next sub-bank prediction
  hardware-based, 161–2
    experimental results, 174, 176–7, 179–82
  hybrid approach, 179–82
Niagara, 7
Nodal power profiles
  for heterogeneous workloads, 106
  over time, 104–5
  for varying node counts, 106–7
  for varying problem sizes, 106
Nokia
  a05140 phone, 227
  a05500 phone, 224
  a06630 phone, 215
  Bluetooth prediction, 232
  Flash Lite and, 212
  N-Gage games, 195
  OpenGL ES and, 215
  Python script testing emulator, 211
  S60/Series, 202, 210

Sensor, 232
Series 80, 210
SNAP, 210
Symbian and, 201, 202
Non-monotonicity, 34–5, 43
  impact, 43–4
Normalized delay, 108
Normalized system energy, 108–9
NPB
  BT, 109, 123–4, 137–42
  CG, 109, 123–4, 137–43, 145–8
  EP, 108–10, 123–4, 137–42
  FT, 104–9, 110–11, 137–45, 148
  IS, 109, 123–4, 137–43
  LU, 109, 123–4, 137–42
  MG, 108–9, 110–11, 123–4, 137–42
  SP, 106, 109, 123–4, 137–43
NSIEVE, 117, 121
NTT DoCoMo, 207, 209, 212
NVIDIA GoForce 3D 4800, 216

O

Object-Oriented Graphics Rendering Engine (OGRE), 250, 265, 266, 268
Object-oriented optimisation, 198
Octeon processor, 9
OGRE, 250, 265, 266, 268
Ojom, 224
On-chip diversity, 38–9
On-chip interconnect *see* Multi-core interconnect
Online document genres, 255
Open Source Initiative (OSI), 276–7
Open source software *see* OSS
OpenGL ES, 202, 205, 214–15
OpenSL ES, 214
Opteron processor, 8
Orange, 202
Orion, 96–7
OSI, 276–7
OSS
  free software vs, 245–6, 276–7
  licenses, 276
OSS movement, 274, 276
Out-of-order cores, 35
Over-The-Air (OTA) installation, 195

Overprovisioning, 45

P

P2P links, 62, 65–6
PA-RISC, 9
'PAC-LAN', 226–9, 231
'Pac-Manhattan', 226
'Pacman', 226
Panasonic, 202, 206–7
Parallel Benchmarks *see* NPB
Parallel speedup, 93
  see also Normalized delay
Participants, 219, 222, 223
PD2P metric, 95
PD3P metric, 95
PDAs
  entry level, 209
  stylus input, 206
    Windows Mobile compatibility with, 205
PDP metric, 95
Peak activity power, 36
Peer review, 263, 287
Pentium D processor, 8
Pentium Extreme processor, 4, 8
Pentium M processor, with reduced power consumption, 132
Perens, Bruce, 277
Performance/cost ratio, 115
Performance efficiency, 108, 110
Performance profiling, 133
Personal Digital Assistants (PDAs), 192
Personal Profile, 209
Philips, 206
PhP-Nuke, 268
Piranha processor, 7, 46, 57
'Pirates', 231–2
Planeshift, 262
Planet three, 200
Point-to-point links *see* P2P links
Port processors, 113–14
Portals, 200–1
Power-aware computational clusters, 125–45
  design analysis, 137–45
    CPUSPEED daemon scheduling, 137–8
    internal scheduling, 143–8
    user-driven external scheduling,

SUBJECT INDEX 315

139–43
  design lessons, 148–9
  experimental framework, 131–7
  see also DVS
Power-aware computing, 93–4
Power budget, 36
Power consumption
  CMOS circuit, 93
  see also Dynamic power; Leakage power
Power measurement, 132–3
Power modeling, 23, 36–7
Power reduction, 98
Power4 processor, 8, 67
Power5 processor, 8, 64, 67
  cache density, 69
Power6 processor, 8
PowerAnalyzer, 102
PowerMeter control thread, 101
PowerNow, 94
PowerPack, 101, 130, 131, 144
  software enhancements, 133
PowerPC 750CXe, 115
Power-performance modes, 93–4
Power-performance tradeoffs, 95
Powerscope, 97
PowerVR MBX, 215
Processor power, direct measurement, 97–8
Processors, mobile device, 196
Profiles, Java, 209
Proximity games, 231–7
Psion, 201
PyS60, 210–11, 212
Python, 210–12

Q

Qt, 207
Qtopia, 207
Qualcomm, 203–5, 216
Quality of Service (QoS), 196
Queuing simulator, 70

R

R10000 core, 35
RAW processor, 8–9
Rebel, 173

Reflective practice, 279
Register-bit equivalents (rbe), 52
Reinvention, 272, 285
Reliability, multi-core system, 4
Rendering pipeline, 216–17
Request queues, 64
Response bus, 64, 65
Revitalization, software system, 273
RFID tags, 226, 227, 228
Role-task migration, 253

S

S60 (Series 60), 202, 210, 224, 232
Sampling phase, 18
Samsung, 202, 205, 206–7
Savaje, 215
SBF *see* Shared bus fabric
Scalable Network Application Package
  (SNAP), 210
Serendipity, 232
Series 60 *see* S60
Series 80, 202, 210
Series 90, 202
Shared bus fabric (SBF), 62–5, 79–81
  design issues, 75–6
  elements, 64–5
  overheads, 71–6
    area, 72–3
    latency, 74–6
    power, 73–4
  performance, 74–6
  split, 80–1
  typical transaction, 63–4
Shared memory multiprocessors, 63, 64
SiByte processor, 9
Siemens, 201, 206
SimOS, 97
SimplePower, 96
SimpleScalar, 96
Simpoint tool, 14, 24, 52
Simulator-based power estimation, 96–7
Single-ISA heterogeneous multi-core
    architectures, 10–19
  advantages, 11–12, 31–2
  dynamic scheduling for intra-thread
    diversity, 17–19
  energy benefits, 31–2

evaluation methodology, 12–15
  hardware assumptions, 13–14
  metrics, 14–15
  multi-programming support, 12–13
  simulation approach, 14
  workload construction, 14
power advantages, 19–31
  chip area estimation, 23–4
  choice of cores, 20
  core switching, 19, 21
  modeling CPU cores, 22
  modeling performance, 24
  modeling power, 23
  oracle dynamic switching heuristics, 26–9
  practical heterogeneous architectures, 31
  realistic dynamic switching heuristics, 29–31
  variation in core performance and power, 24–6
static scheduling for inter-thread diversity, 15–17
Single processor system profiling, 96–8
SIP, 237, 238–9
Skill bureaucracy, 257
Skill meritocracy, 257
SMP, 91
SMS messaging, 220
SMTSIM, 14, 24, 38, 52
SNAP, 210
Snoop bus, 64, 65
Snoop queue, 64
Social capital, 259, 265
Social gamers, 194
Social gateways, 258
Social networking, inter-project, 265–8
Socio-technical interaction, in FOSS projects, 272–4, 283, 285–6
Software developers, FOSS
  community development, 268–9
  discretionary time and effort, 258–9
  linchpin, 265, 269, 284–5
  organization, 256–8
  peer recognition, 258, 265
  self-determination, 258
  self-promotion, 258–9
  skills, 256–8

Software development environments, 201–14
  BREW, 203–5, 215
  choosing, 213–14
  Flash Lite, 212–13
  J2ME, 208–10, 215
  Linux, 206–7, 215
  Python, 210–12
  Symbian, 201–3, 215
  Windows Mobile, 205–6, 215
Software development studies, 283, 286, 287
Software ecosystems, multi-project, 270–4
Software engineering, 246, 256–7
Software informalisms, 255–6
Software inspections, 287
Software maintenance, 272
Software version control, 260–1
Softwatt, 97
Sony Ericsson, 201, 202, 212
  W900 Walkman phone, 215, 216
Sounds, in games, 197–8
SourceForge, 269, 274–5, 281
  FOSS projects registered at, 247, 268, 272, 274
SPEC 95 benchmarks, 173
SPEC FPT 92 benchmarks, btrix, 186
SPEC2000 benchmarks, 10, 14, 24, 37, 173
  164.gzip, 177
  CFP2000, 56
  CINT2000, 56
  CPU2000, 52
  SPECfp, 24
  SPECint, 24
SpeedStep, 94, 132
Speedup
  parallel, 93
  see also Normalized delay
  weighted, 14–15
SPLASH benchmark, 52, 57
Stallman, Richard M., 246, 276, 277
Static next sub-bank prediction, 161–9
  compiler support, 169
  experimental results, 174–82
    energy-delay product, 181–2
    leakage energy savings, 176–7
    overall energy savings, 181
    performance overhead reduction, 174–6, 179–80

sensitivity analysis, 177–9
instruction cache addressing schemes, 167–8
ISA support, 169
memory sub-bank prediction buffer, 162–5
  dynamic energy overhead, 163–5
transitional instructions, 165–7, 174
  identification, 168
Static port assignment, 60–1
Static power *see* Leakage power
Steady phase, 18
STREAM benchmark, 52, 58
StrongARM, 168
Sun, 208–10
  NetBeans, 268, 270, 277
Survey research methods, 279–80
swim, 127
Symbian, 201–3, 215
  see also S60; Series 80; Series 90
Symbian Signed, 202
SymbianGear, 200
System development, community development and, 269
System energy, normalized, 108–9
System performance, 92–3
System power simulators, 97

T

T-Mobile, 202
T9 predictive text, 217
Tarantula processor, 32
TCO metrics, 116, 118–19
TEM^{2}P^{2}EST, 96
Text games, 217–24
'thetxtbk', 218–24
'They Howl', 229–31
Third Generation Partnership Project (3GPP), 237
Thread bias, 80
Thread-level parallelism (TLP), 12
Throughput, 120
  peak, 120
Thunderbird, 263
Tigris.org, 248, 249, 274
Torvalds, Linus, 274

Total cost of ownership *see* TCO
Touch screens, 217–18
Transitional instructions, 165–7, 174–5
  identification, 168
Tree Code, 119
Trimaran, 173
TRIPS processor, 8, 9
Trolltech, 207
TRUE BREW testing, 204
Turion processor, 8
twolf, 27

U

UIQ, 202, 210
UltraSparc-IV, 7
'Uncle Roy All Around You', 227
Unconditional branches with static target addresses, 166, 174–5

V

Vector processing unit (VMX), 45
VideoLAN Client (VLC), 250–1
Virtual project management (VPM), 261–3
VLIW architectures, 172–3
Vodafone live, 200
VPM, 261–3

W

W900 Walkman phone, 215, 216
wanderers, 219, 222
water, 52, 57
Wattch, 23, 96
WaveScalar processor, 8, 9
Web, software infrastructure for, 270
Web sites, FOSSD, 256, 265–6, 268, 269, 274
Weighted speedup, 14–15
Windows CE, 205
Windows Mobile, 205–6, 215
Wireless Application Protocol (WAP), 220
Wireless Toolkit, 210
Workloads
  all different, 37, 39–44
  all same, 37, 39–44
  commercial, 70
  diversity, 10

World Wide Web Committee (W3C), 270
wupwise, 14

X

Xbox, 9
Xeon processor, 8

XLR processor, 9
XScale, 168

Y

'You-Know-Who', 232

Contents of Volumes in This Series

Volume 42

Nonfunctional Requirements of Real-Time Systems
    TEREZA G. KIRNER AND ALAN M. DAVIS
A Review of Software Inspections
    ADAM PORTER, HARVEY SIY, AND LAWRENCE VOTTA
Advances in Software Reliability Engineering
    JOHN D. MUSA AND WILLA EHRLICH
Network Interconnection and Protocol Conversion
    MING T. LIU
A Universal Model of Legged Locomotion Gaits
    S. T. VENKATARAMAN

Volume 43

Program Slicing
    DAVID W. BINKLEY AND KEITH BRIAN GALLAGHER
Language Features for the Interconnection of Software Components
    RENATE MOTSCHNIG-PITRIK AND ROLAND T. MITTERMEIR
Using Model Checking to Analyze Requirements and Designs
    JOANNE ATLEE, MARSHA CHECHIK, AND JOHN GANNON
Information Technology and Productivity: A Review of the Literature
    ERIK BRYNJOLFSSON AND SHINKYU YANG
The Complexity of Problems
    WILLIAM GASARCH
3-D Computer Vision Using Structured Light: Design, Calibration, and Implementation Issues
    FRED W. DEPIERO AND MOHAN M. TRIVEDI

Volume 44

Managing the Risks in Information Systems and Technology (IT)
    ROBERT N. CHARETTE
Software Cost Estimation: A Review of Models, Process and Practice
    FIONA WALKERDEN AND ROSS JEFFERY
Experimentation in Software Engineering
    SHARI LAWRENCE PFLEEGER
Parallel Computer Construction Outside the United States
    RALPH DUNCAN
Control of Information Distribution and Access
    RALF HAUSER
Asynchronous Transfer Mode: An Engineering Network Standard for High Speed Communications
    RONALD J. VETTER

Communication Complexity
  EYAL KUSHILEVITZ

Volume 45

Control in Multi-threaded Information Systems
  PABLO A. STRAUB AND CARLOS A. HURTADO
Parallelization of DOALL and DOACROSS Loops—a Survey
  A. R. HURSON, JOFORD T. LIM, KRISHNA M. KAVI, AND BEN LEE
Programming Irregular Applications: Runtime Support, Compilation and Tools
  JOEL SALTZ, GAGAN AGRAWAL, CHIALIN CHANG, RAJA DAS, GUY EDJLALI, PAUL HAVLAK, YUAN-SHIN HWANG, BONGKI MOON, RAVI PONNUSAMY, SHAMIK SHARMA, ALAN SUSSMAN, AND MUSTAFA UYSAL
Optimization Via Evolutionary Processes
  SRILATA RAMAN AND L. M. PATNAIK
Software Reliability and Readiness Assessment Based on the Non-homogeneous Poisson Process
  AMRIT L. GOEL AND KUNE-ZANG YANG
Computer-Supported Cooperative Work and Groupware
  JONATHAN GRUDIN AND STEVEN E. POLTROCK
Technology and Schools
  GLEN L. BULL

Volume 46

Software Process Appraisal and Improvement: Models and Standards
  MARK C. PAULK
A Software Process Engineering Framework
  JYRKI KONTIO
Gaining Business Value from IT Investments
  PAMELA SIMMONS
Reliability Measurement, Analysis, and Improvement for Large Software Systems
  JEFF TIAN
Role-Based Access Control
  RAVI SANDHU
Multithreaded Systems
  KRISHNA M. KAVI, BEN LEE, AND ALLI R. HURSON
Coordination Models and Language
  GEORGE A. PAPADOPOULOS AND FARHAD ARBAB
Multidisciplinary Problem Solving Environments for Computational Science
  ELIAS N. HOUSTIS, JOHN R. RICE, AND NAREN RAMAKRISHNAN

Volume 47

Natural Language Processing: A Human–Computer Interaction Perspective
  BILL MANARIS
Cognitive Adaptive Computer Help (COACH): A Case Study
  EDWIN J. SELKER
Cellular Automata Models of Self-replicating Systems
  JAMES A. REGGIA, HUI-HSIEN CHOU, AND JASON D. LOHN
Ultrasound Visualization
  THOMAS R. NELSON

Patterns and System Development
   BRANDON GOLDFEDDER
High Performance Digital Video Servers: Storage and Retrieval of Compressed Scalable Video
   SEUNGYUP PAEK AND SHIH-FU CHANG
Software Acquisition: The Custom/Package and Insource/Outsource Dimensions
   PAUL NELSON, ABRAHAM SEIDMANN, AND WILLIAM RICHMOND

Volume 48

Architectures and Patterns for Developing High-Performance, Real-Time ORB Endsystems
   DOUGLAS C. SCHMIDT, DAVID L. LEVINE, AND CHRIS CLEELAND
Heterogeneous Data Access in a Mobile Environment – Issues and Solutions
   J. B. LIM AND A. R. HURSON
The World Wide Web
   HAL BERGHEL AND DOUGLAS BLANK
Progress in Internet Security
   RANDALL J. ATKINSON AND J. ERIC KLINKER
Digital Libraries: Social Issues and Technological Advances
   HSINCHUN CHEN AND ANDREA L. HOUSTON
Architectures for Mobile Robot Control
   JULIO K. ROSENBLATT AND JAMES A. HENDLER

Volume 49

A Survey of Current Paradigms in Machine Translation
   BONNIE J. DORR, PAMELA W. JORDAN, AND JOHN W. BENOIT
Formality in Specification and Modeling: Developments in Software Engineering Practice
   J. S. FITZGERALD
3-D Visualization of Software Structure
   MATHEW L. STAPLES AND JAMES M. BIEMAN
Using Domain Models for System Testing
   A. VON MAYRHAUSER AND R. MRAZ
Exception-Handling Design Patterns
   WILLIAM G. BAIL
Managing Control Asynchrony on SIMD Machines—a Survey
   NAEL B. ABU-GHAZALEH AND PHILIP A. WILSEY
A Taxonomy of Distributed Real-time Control Systems
   J. R. ACRE, L. P. CLARE, AND S. SASTRY

Volume 50

Index Part I
Subject Index, Volumes 1–49

Volume 51

Index Part II
Author Index
Cumulative list of Titles
Table of Contents, Volumes 1–49

Volume 52

Eras of Business Computing
 ALAN R. HEVNER AND DONALD J. BERNDT
Numerical Weather Prediction
 FERDINAND BAER
Machine Translation
 SERGEI NIRENBURG AND YORICK WILKS
The Games Computers (and People) Play
 JONATHAN SCHAEFFER
From Single Word to Natural Dialogue
 NEILS OLE BENSON AND LAILA DYBKJAER
Embedded Microprocessors: Evolution, Trends and Challenges
 MANFRED SCHLETT

Volume 53

Shared-Memory Multiprocessing: Current State and Future Directions
 PER STEUSTRÖM, ERIK HAGERSTEU, DAVID I. LITA, MARGARET MARTONOSI, AND MADAN VERNGOPAL
Shared Memory and Distributed Shared Memory Systems: A Survey
 KRISHNA KAUI, HYONG-SHIK KIM, BEU LEE, AND A. R. HURSON
Resource-Aware Meta Computing
 JEFFREY K. HOLLINGSWORTH, PETER J. KELCHER, AND KYUNG D. RYU
Knowledge Management
 WILLIAM W. AGRESTI
A Methodology for Evaluating Predictive Metrics
 JASRETT ROSENBERG
An Empirical Review of Software Process Assessments
 KHALED EL EMAM AND DENNIS R. GOLDENSON
State of the Art in Electronic Payment Systems
 N. ASOKAN, P. JANSON, M. STEIVES, AND M. WAIDNES
Defective Software: An Overview of Legal Remedies and Technical Measures Available to Consumers
 COLLEEN KOTYK VOSSLER AND JEFFREY VOAS

Volume 54

An Overview of Components and Component-Based Development
 ALAN W. BROWN
Working with UML: A Software Design Process Based on Inspections for the Unified Modeling Language
 GUILHERME H. TRAVASSOS, FORREST SHULL, AND JEFFREY CARVER
Enterprise JavaBeans and Microsoft Transaction Server: Frameworks for Distributed Enterprise Components
 AVRAHAM LEFF, JOHN PROKOPEK, JAMES T. RAYFIELD, AND IGNACIO SILVA-LEPE
Maintenance Process and Product Evaluation Using Reliability, Risk, and Test Metrics
 NORMAN F. SCHNEIDEWIND
Computer Technology Changes and Purchasing Strategies
 GERALD V. POST
Secure Outsourcing of Scientific Computations
 MIKHAIL J. ATALLAH, K. N. PANTAZOPOULOS, JOHN R. RICE, AND EUGENE SPAFFORD

Volume 55

The Virtual University: A State of the Art
  LINDA HARASIM
The Net, the Web and the Children
  W. NEVILLE HOLMES
Source Selection and Ranking in the WebSemantics Architecture Using Quality of Data Metadata
  GEORGE A. MIHAILA, LOUIQA RASCHID, AND MARIA-ESTER VIDAL
Mining Scientific Data
  NAREN RAMAKRISHNAN AND ANANTH Y. GRAMA
History and Contributions of Theoretical Computer Science
  JOHN E. SAVAGE, ALAN L. SALEM, AND CARL SMITH
Security Policies
  ROSS ANDERSON, FRANK STAJANO, AND JONG-HYEON LEE
Transistors and 1C Design
  YUAN TAUR

Volume 56

Software Evolution and the Staged Model of the Software Lifecycle
  KEITH H. BENNETT, VACLAV T. RAJLICH, AND NORMAN WILDE
Embedded Software
  EDWARD A. LEE
Empirical Studies of Quality Models in Object-Oriented Systems
  LIONEL C. BRIAND AND JÜRGEN WÜST
Software Fault Prevention by Language Choice: Why C Is Not My Favorite Language
  RICHARD J. FATEMAN
Quantum Computing and Communication
  PAUL E. BLACK, D. RICHARD KUHN, AND CARL J. WILLIAMS
Exception Handling
  PETER A. BUHR, ASHIF HARJI, AND W. Y. RUSSELL MOK
Breaking the Robustness Barrier: Recent Progress on the Design of the Robust Multimodal System
  SHARON OVIATT
Using Data Mining to Discover the Preferences of Computer Criminals
  DONALD E. BROWN AND LOUISE F. GUNDERSON

Volume 57

On the Nature and Importance of Archiving in the Digital Age
  HELEN R. TIBBO
Preserving Digital Records and the Life Cycle of Information
  SU-SHING CHEN
Managing Historical XML Data
  SUDARSHAN S. CHAWATHE
Adding Compression to Next-Generation Text Retrieval Systems
  NIVIO ZIVIANI AND EDLENO SILVA DE MOURA
Are Scripting Languages Any Good? A Validation of Perl, Python, Rexx, and Tcl against C, C++, and Java
  LUTZ PRECHELT

Issues and Approaches for Developing Learner-Centered Technology
    CHRIS QUINTANA, JOSEPH KRAJCIK, AND ELLIOT SOLOWAY
Personalizing Interactions with Information Systems
    SAVERIO PERUGINI AND NAREN RAMAKRISHNAN

Volume 58

Software Development Productivity
    KATRINA D. MAXWELL
Transformation-Oriented Programming: A Development Methodology for High Assurance Software
    VICTOR L. WINTER, STEVE ROACH, AND GREG WICKSTROM
Bounded Model Checking
    ARMIN BIERE, ALESSANDRO CIMATTI, EDMUND M. CLARKE, OFER STRICHMAN, AND YUNSHAN ZHU
Advances in GUI Testing
    ATIF M. MEMON
Software Inspections
    MARC ROPER, ALASTAIR DUNSMORE, AND MURRAY WOOD
Software Fault Tolerance Forestalls Crashes: To Err Is Human; To Forgive Is Fault Tolerant
    LAWRENCE BERNSTEIN
Advances in the Provisions of System and Software Security—Thirty Years of Progress
    RAYFORD B. VAUGHN

Volume 59

Collaborative Development Environments
    GRADY BOOCH AND ALAN W. BROWN
Tool Support for Experience-Based Software Development Methodologies
    SCOTT HENNINGER
Why New Software Processes Are Not Adopted
    STAN RIFKIN
Impact Analysis in Software Evolution
    MIKAEL LINDVALL
Coherence Protocols for Bus-Based and Scalable Multiprocessors, Internet, and Wireless Distributed Computing Environments: A Survey
    JOHN SUSTERSIC AND ALI HURSON

Volume 60

Licensing and Certification of Software Professionals
    DONALD J. BAGERT
Cognitive Hacking
    GEORGE CYBENKO, ANNARITA GIANI, AND PAUL THOMPSON
The Digital Detective: An Introduction to Digital Forensics
    WARREN HARRISON
Survivability: Synergizing Security and Reliability
    CRISPIN COWAN
Smart Cards
    KATHERINE M. SHELFER, CHRIS CORUM, J. DREW PROCACCINO, AND JOSEPH DIDIER

Shotgun Sequence Assembly
   MIHAI POP
Advances in Large Vocabulary Continuous Speech Recognition
   GEOFFREY ZWEIG AND MICHAEL PICHENY

Volume 61

Evaluating Software Architectures
   ROSEANNE TESORIERO TVEDT, PATRICIA COSTA, AND MIKAEL LINDVALL
Efficient Architectural Design of High Performance Microprocessors
   LIEVEN EECKHOUT AND KOEN DE BOSSCHERE
Security Issues and Solutions in Distributed Heterogeneous Mobile Database Systems
   A. R. HURSON, J. PLOSKONKA, Y. JIAO, AND H. HARIDAS
Disruptive Technologies and Their Affect on Global Telecommunications
   STAN MCCLELLAN, STEPHEN LOW, AND WAI-TIAN TAN
Ions, Atoms, and Bits: An Architectural Approach to Quantum Computing
   DEAN COPSEY, MARK OSKIN, AND FREDERIC T. CHONG

Volume 62

An Introduction to Agile Methods
   DAVID COHEN, MIKAEL LINDVALL, AND PATRICIA COSTA
The Timeboxing Process Model for Iterative Software Development
   PANKAJ JALOTE, AVEEJEET PALIT, AND PRIYA KURIEN
A Survey of Empirical Results on Program Slicing
   DAVID BINKLEY AND MARK HARMAN
Challenges in Design and Software Infrastructure for Ubiquitous Computing Applications
   GURUDUTH BANAVAR AND ABRAHAM BERNSTEIN
Introduction to MBASE (Model-Based (System) Architecting and Software Engineering)
   DAVID KLAPPHOLZ AND DANIEL PORT
Software Quality Estimation with Case-Based Reasoning
   TAGHI M. KHOSHGOFTAAR AND NAEEM SELIYA
Data Management Technology for Decision Support Systems
   SURAJIT CHAUDHURI, UMESHWAR DAYAL, AND VENKATESH GANTI

Volume 63

Techniques to Improve Performance Beyond Pipelining: Superpipelining, Superscalar, and VLIW
   JEAN-LUC GAUDIOT, JUNG-YUP KANG, AND WON WOO RO
Networks on Chip (NoC): Interconnects of Next Generation Systems on Chip
   THEOCHARIS THEOCHARIDES, GREGORY M. LINK, NARAYANAN VIJAYKRISHNAN, AND MARY JANE IRWIN
Characterizing Resource Allocation Heuristics for Heterogeneous Computing Systems
   SHOUKAT ALI, TRACY D. BRAUN, HOWARD JAY SIEGEL, ANTHONY A. MACIEJEWSKI, NOAH BECK, LADISLAU BÖLÖNI, MUTHUCUMARU MAHESWARAN, ALBERT I. REUTHER, JAMES P. ROBERTSON, MITCHELL D. THEYS, AND BIN YAO
Power Analysis and Optimization Techniques for Energy Efficient Computer Systems
   WISSAM CHEDID, CHANSU YU, AND BEN LEE
Flexible and Adaptive Services in Pervasive Computing
   BYUNG Y. SUNG, MOHAN KUMAR, AND BEHROOZ SHIRAZI

Search and Retrieval of Compressed Text
    AMAR MUKHERJEE, NAN ZHANG, TAO TAO, RAVI VIJAYA SATYA, AND WEIFENG SUN

Volume 64

Automatic Evaluation of Web Search Services
    ABDUR CHOWDHURY
Web Services
    SANG SHIN
A Protocol Layer Survey of Network Security
    JOHN V. HARRISON AND HAL BERGHEL
E-Service: The Revenue Expansion Path to E-Commerce Profitability
    ROLAND T. RUST, P.K. KANNAN, AND ANUPAMA D. RAMACHANDRAN
Pervasive Computing: A Vision to Realize
    DEBASHIS SAHA
Open Source Software Development: *Structural Tension in the American Experiment*
    COSKUN BAYRAK AND CHAD DAVIS
Disability and Technology: Building Barriers or Creating Opportunities?
    PETER GREGOR, DAVID SLOAN, AND ALAN F. NEWELL

Volume 65

The State of Artificial Intelligence
    ADRIAN A. HOPGOOD
Software Model Checking with SPIN
    GERARD J. HOLZMANN
Early Cognitive Computer Vision
    JAN-MARK GEUSEBROEK
Verification and Validation and Artificial Intelligence
    TIM MENZIES AND CHARLES PECHEUR
Indexing, Learning and Content-Based Retrieval for Special Purpose Image Databases
    MARK J. HUISKES AND ERIC J. PAUWELS
Defect Analysis: Basic Techniques for Management and Learning
    DAVID N. CARD
Function Points
    CHRISTOPHER J. LOKAN
The Role of Mathematics in Computer Science and Software Engineering Education
    PETER B. HENDERSON

Volume 66

Calculating Software Process Improvement's Return on Investment
    RINI VAN SOLINGEN AND DAVID F. RICO
Quality Problem in Software Measurement Data
    PIERRE REBOURS AND TAGHI M. KHOSHGOFTAAR
Requirements Management for Dependable Software Systems
    WILLIAM G. BAIL
Mechanics of Managing Software Risk
    WILLIAM G. BAIL

The PERFECT Approach to Experience-Based Process Evolution
    Brian A. Nejmeh and William E. Riddle
The Opportunities, Challenges, and Risks of High Performance Computing in Computational Science and Engineering
    Douglass E. Post, Richard P. Kendall, and Robert F. Lucas

Volume 67

Broadcasting a Means to Disseminate Public Data in a Wireless Environment—Issues and Solutions
    A.R. Hurson, Y. Jiao, and B.A. Shirazi
Programming Models and Synchronization Techniques for Disconnected Business Applications
    Avraham Leff and James T. Rayfield
Academic Electronic Journals: Past, Present, and Future
    Anat Hovav and Paul Gray
Web Testing for Reliability Improvement
    Jeff Tian and Li Ma
Wireless Insecurities
    Michael Sthultz, Jacob Uecker, and Hal Berghel
The State of the Art in Digital Forensics
    Dario Forte

Volume 68

Exposing Phylogenetic Relationships by Genome Rearrangement
    Ying Chih Lin and Chuan Yi Tang
Models and Methods in Comparative Genomics
    Guillaume Bourque and Louxin Zhang
Translocation Distance: Algorithms and Complexity
    Lusheng Wang
Computational Grand Challenges in Assembling the Tree of Life: Problems and Solutions
    David A. Bader, Usman Roshan, and Alexandros Stamatakis
Local Structure Comparison of Proteins
    Jun Huan, Jan Prins, and Wei Wang
Peptide Identification via Tandem Mass Spectrometry
    Xue Wu, Nathan Edwards, and Chau-Wen Tseng